NANOTECHNOLOGY IN DRUG DELIVERY

Fundamentals, Design, and Applications

NANOTECHNOLOGY IN DRUG DELIVERY

Fundamentals, Design, and Applications

Saurabh Bhatia, PhD

APPLE ACADEMIC PRESS

Apple Academic Press Inc. | Apple Academic Press Inc.
3333 Mistwell Crescent | 9 Spinnaker Way
Oakville, ON L6L 0A2 | Waretown, NJ 08758
Canada | USA

First issued in paperback 2021

Exclusive worldwide distribution by CRC Press, a member of Taylor & Francis Group

No claim to original U.S. Government works

ISBN 13: 978-1-77463-615-2 (pbk)
ISBN 13: 978-1-77188-360-3 (hbk)

Library and Archives Canada Cataloguing in Publication

Bhatia, Saurabh, author
Nanotechnology in drug delivery : fundamentals, design, and applications / Saurabh Bhatia, PhD.

Includes bibliographical references and index.
Issued in print and electronic formats.
ISBN 978-1-77188-360-3 (hardcover).--ISBN 978-1-77188-361-0 (pdf)
1. Drug delivery systems. 2. Nanoparticles. 3. Protein drugs.
4. Peptide drugs. 5. Pharmaceutical biotechnology. I. Title.

RS201.N35B54 2016   615'.6   C2016-903211-6   C2016-903212-4

Library of Congress Cataloging-in-Publication Data

Names: Bhatia, Saurabh, author.
Title: Nanotechnology in drug delivery : fundamentals, design, and applications / author, Saurabh Bhatia.
Description: Toronto ; New Jersey : Apple Academic Press, 2016. |
Includes bibliographical references and index.
Identifiers: LCCN 2016020820 (print) | LCCN 2016021534 (ebook) | ISBN 9781771883603 (hardcover : alk. paper) | ISBN 9781771883610 (eBook) | ISBN 9781771883610 ()
Subjects: | MESH: Drug Delivery Systems | Nanoparticles
Classification: LCC RS199.5 (print) | LCC RS199.5 (ebook) | NLM QV 785 | DDC 615.1/9--dc23
LC record available at https://lccn.loc.gov/2016020820

Apple Academic Press also publishes its books in a variety of electronic formats. Some content that appears in print may not be available in electronic format. For information about Apple Academic Press products, visit our website at **www.appleacademicpress.com** and the CRC Press website at **www.crcpress.com**

CONTENTS

LIST OF ABBREVIATIONS

ACE	angiotensin converting enzyme
AChR	acetylcholine receptor-related peptide
ACPP	activatable cell penetrating peptide
ACTH	adrenal corticotropic hormone
ADH	alcohol dehydrogenases
ALI	acute lung injury
AM	adrenomedullin
AMPs	antimicrobial peptides
ANEP	anti-neuroexcitation peptide
APTS	aminopropyltriethoxysilane
BBB	blood brain barrier
BBN	bombesin
BCEC	brain capillary endothelial cells
BDNF	brain derived neurotrophic factor
BLM	bleomycin
BSA	bovine serum albumin
CAM	cell adhesion molecule
CART	cocaine- and amphetamine-regulated transcript
CCK	cholecystokinin
CCP	cytochrome c peroxidase
CD	cyclodextrin
CDE	clathrin-dependent endocytosis
CDI	clathrin-independent
CGA	chromogranin A
CGRPa	calcitonin gene-related peptide
CNS	central nervous system
CSII	continuous, subcutaneous insulin infusion
CTL	cytotoxic T-lymphocyte
CTPs	cell-targeting peptides
DDAVP	1-deamino-8-D-arginine vasopressin
DDS	drug delivery systems

DIDS	disothiocyano-stilbene-2
DNP	2,4-dinitrophenol
DOPA	L-3,4-dihydroxyphenylalanine
DOPE	dioleoyl phosphatidyl ethanolamine
DSPC	distearoyl phosphatidylcholine
DVDAVP	deamino 4-valine, 8-D-arginine-vasopressin
EC	ethyl cellulose.
EDC	1-ethyl-33-dimethylaminopropyl carbodiimides hydrochloride
EDTA	ethyl/enediamine tetra acetic acid
EGFR	epidermal growth factor receptor
EPA	Environmental Protection Agency
ES	embryonic stem
FDA	Food and Drug Administration
FHV	flock house virus
FPE	fluid-phase endocytosis
FSHR	follicle-stimulating hormone receptor
GAGs	glycosaminoglycans
GALP	galanin-like peptide
GALT	gut-associated lymph tissues
GAS	guaiazulene-3-sulfonate
GFRs	growth factor receptors
GI-MAPS	gastrointestinal mucoadhesive patch system
GIP	gastric inhibitory
GIT	gastrointestinal tract
GLP-1	glucagon-like peptide
GMAP	galanin-message-associated peptide
GPCRs	G-protein coupled receptors
GRP	gastrin releasing peptide
HEK293	human embryonic kidney cells
HGC	hydrophobic glycol chitosan
HPMC	hydroxypropyl methylcellulose
HSP20	heat shock protein
HSV-1	herpes simplex virus type 1
HUVEC	human umbilical vein endothelial cells

IAASF	interfacial activity assisted surface functionalization
ICAM-1	intercellular cell-adhesion molecule-1
IF	intrinsic factor
IGF-I	insulin-like growth factor-I
LATS	long-acting thyroid stimulator
LCST	low critical solution temperature
LEAP	liver-expressed antimicrobial peptides
LHRH	luteinizing hormone releasing hormone
LPS	lipopolysaccharide
LRP1	lipoprotein receptor-related protein-1
MMP	matrix metalloprotease
MSC	mesenchymal stem cells
MT1-MMP	membrane type-1 matrix metalloproteinase
NEM	N-ethylamaleimide
NHS	N-hydroxysulfo-succinimide
NLS	nuclear localization signal
NMIIA	nonmuscle myosin II A
NP	nanoparticles
NPY	neuropeptide Y
NTA	nitrilotriacetic acid
OSHA	occupational safety and health administration
PACA	polyalkylcyanoacrylate
PACAP	pituitary adenylate cyclase activating peptide
PDAC	pancreatic ductal adenocarcinoma
PE	pseudomonas exotoxin A
PEG	polyethylene glycol
PEGs	poly(ethylene glycol)
PEI	poly-ethylene imine
PEO-PPO-PEO	poly (ethylene oxide)-poly (propylene oxide)-poly(ethylene oxide)
PIBCA	poly(isobutylcyanoacrylate)
PLA	poly(lactide)
PLGA	poly(lactic glycolic acid)
PMAA-gEG	poly(methacrylic acid g-ethylene glycol)
PNA	peptide nucleic acid

PNIPAAM	poly(n-isopropylacrylamide)
PTD	protein transduction domain
PTHrP	parathyroid hormone-related peptide
PTP	plectin-1 targeting peptide
PVP	polyvinyl pyrrolidone
PYY	peptide YY
QD	quantum dot
QSFR	quantitative structure function relationship
QSSR	quantitative structure stability relationship
RES	reticuloendothelial cells
RME	receptor-mediated endocytosis
RVG	rabies virus glycoprotein
SCK	shell cross-linked knedel-like
SCL	shell-cross-linked
SELEX	systematic evolution of ligands by exponential enrichment
SPIO	superparamagnetic iron oxide
SPPS	solid phase peptide synthesis
SSCHE	Slovak Society of Chemical Engineering
TAF-I	template activating factor I
TAMs	tumor-associated macrophages
TAT	transcriptional activator of transcription
TATp	transactivating transcriptional activator peptide
TDDS	transdermal drug delivery systems
TEER	transepithelial electrical resistance
TF	transferrin
TFA	trifluoroacetic acid
TJs	tight junctions
TLR	toll-like receptor
TMC	N-trimethyl chitosan chloride
TMT	tumor metastasis targeting
TNF	tumor necrosis factor
TPIS	transdermal periodic iontotherapeutic system
TPP	tripolyphosphate
TRAIL	tumor related apoptosis-inducing ligand
TRH	thyrotropin releasing hormone

TT	tetanus toxoid
USDA	U.S. Department of Agriculture
VDA	vascular disrupting agent
VIP	vasoactive intestinal peptide
VTA	vascular antiangiogenic agent
WGA	wheat germ agglutinin

PREFACE

Drug delivery is a term that refers to the delivery of pharmaceutical compound to humans or animals. The most common methods of delivery include the preferred noninvasive oral (through the mouth), nasal, inhalation, and rectal routes. Many medications, however, cannot be delivered using these routes because they might be susceptible to degradation or not incorporated efficiently. For this reason any protein and peptide drugs have to be delivered by injection.

Human cellular and tissue membrane always resist the entry of foreign material (medicine). Selection of efficient drug delivery system allows efficient delivery for medicine into the body. Due to high stability, high carrier capacity, feasibility of incorporation of both hydrophilic and hydrophobic substances, and feasibility of variable routes of administration, including oral application and inhalation, nanoparticles are always preferred. Small peptides can easily transport drug, protein to a specific region in the cell, including the nucleus, mitochondria, endoplasmic reticulum (ER), chloroplast, apoplast, peroxisome and plasma membrane. Some target peptides are cleaved from the protein by signal peptidases after the proteins are transported. For the specified targeting of the nanoparticle at particular site, its bioconjugation with peptide is essential. This type of bioconjugation is widely used to deliver drugs, proteins, DNA/RNA, viruses, and vaccines to the human body to improve the quality of human health.

This book provides the fundamental understanding of drug delivery systems of peptide and proteins with a special focus on their nanotechnology applications. Since this book was first conceived, there has been increasing research efforts dedicated to peptide and protein drug delivery systems in both academics and industrial laboratories worldwide. This book was written to fulfill a need for a comprehensive review and assessment of conventional and nonconventional routes of administration.

To this end I have organized Chapter 1 so that it gives fundamental and comprehensive knowledge regarding the synthesis, stability, degradation, and physicochemical aspects of proteins and peptides. In addition this part also provides the comprehensive information related with barriers to peptide and protein delivery and their respective formulation and delivery strategies.

Chapter 2 discusses the pharmacokinetic consideration, methods of preparation, design and synthesis of peptides, structure function relationship, peptide polymer conjugate, peptide-encapsulating nanoparticles, biomedical applications of peptide polymer conjugate, and nonpeptide targeting.

Chapter 3 discusses the CPP and CTP in drug delivery and cell targeting.

This book fulfill my expectations if it creates interest in the area of peptide and protein drug delivery and provides a framework for the pharmaceutical or biotechnological scientist to formulate a strategy to deliver peptide and protein drugs to their site of action.

I am highly thankful to my caring and lovable parents, **Mr. R. L. Bhatia** (Father), **Vinod Bhatia** (Mother), my dearest brother, **Sanjay Bhatia**, and my sweet sister-in-law, **Aastha Popli Bhatia**, for their caring, valuable suggestions and timely input. This book would not have seen the light of the day without the moral support and patience of my family. In addition **Ashish Kumar, Sandy Jones Sickels**, and **Rakesh Kumar**, all of Apple Academic Press, must be praised for their active work and support for our effort. We are sure that readers of this book—students, researchers, and industrialists—will find it interesting and useful. The publication of this book would not have been possible without the valuable work of earlier researchers.

—*Saurabh Bhatia, PhD*

ABOUT THE AUTHOR

 **Saurabh Bhatia, PhD,** is currently an Assistant Professor at the PDM College of Pharmacy in Bahadurgarh, Haryana, India. He has several years of academic experience, teaching such specialized subjects as pharmacognosy, traditional concepts of medicinal plants, plant tissue culture, modern extraction and isolation methodologies, natural polymers, parasitology (Leishmania), medicinal and pharmaceutical values of marine and fresh water algae, and nanoparticles and peptide-mediated drug delivery systems. He has investigated several marine algae and their derived polymers throughout India. In addition, he was the first person from India who has explored the medicinal values of various natural polymers isolated from marine red algae. He has written more than 30 international publications in these areas and has been an active participant of more than 35 national and international conferences. His published books include *Modern Applications of Plant Biotechnology in Pharmaceutical Science* and *Practical Applications of Plant Biotechnology.*

CHAPTER 1

PROTEIN AND PEPTIDE-BASED DRUG DELIVERY SYSTEMS

CONTENTS

ABSTRACT

Protein and peptides are the main elements of biopharmaceuticals that are recently explored in current research in various drug delivery systems. Structural conformation and other physico-chemical properties of these biomolecules determine the biological activity, which may ultimately decides its utility in various drug delivery system. Peptides synthesis is now possible by using such as solid phase peptide synthesis methods and various ligation methods. Prodigious advances in peptide synthesis cause improvement in stability. Within the context of biomedicine and pharmaceutical sciences, the stability issue of protein assumes particular relevance. Stabilization of protein and protein-like molecules is essential for their proper function at desirable site. This is attained through establishment of a thermodynamic equilibrium with the (micro)environment. Thus study related degradation pathways indicating instability of proteins & peptides is essentially required for designing suitable delivery system. This chapter also covers intracellular targets and intracellular drug delivery and delivery of proteins and nucleic acids using a non-covalent peptide-based strategy. In addition transduction technology, methods for various nanoconjugate preparation, recent progress and major challenges related with several delivery considerations of peptide and protein are also discussed in this chapter.

1.1   INTRODUCTION

In the recent years therapeutic peptides and proteins have risen to prominence as potential drugs of the future. Management of illness through this class of pharmaceuticals has entered an era of rapid growth. Biotechnology has played a key role in the development of peptide and proteins drugs. A new series of peptide based and protein based pharmaceuticals have made their presence felt with the advent of recombinant DNA and hybridoma techniques and with the advances in large scale fermentation and purification process. These entities are now available in a much purer form in significant quantities, at a reasonable cost. The problem of immunogenicity and antigenicity has also been considerably reduced. Simultaneously, great advance made in the understanding of the physiologic and pharmaco-logic behavior of these biologic response modifiers and regulatory agents. Ailments that might be treated more effectively with this class of therapeutics include autonomous diseases, cancer, mental disorders, hypertension and certain cardiovascular and metabolic disease. The peptides and protein drugs produced by recombinant DNA technology are the replicas of that obtained from natural sources. In spite of their potency and specificity in physiological functions, a majority of these therapeutics are difficult to administer clinically. The chemical and structural complexities involved demand an effective delivery system so that physicochemical and biological properties are duly considered. Peptides and proteins possess complex architecture structure. The twenty different naturally occurring amino acids join with each other by peptide bonds and build polymers referred to peptides and proteins. Although the distinction between peptides and proteins is arbitrary, a peptide contains less than 20 amino acids, having a molecular weight less than 5000, while a protein possess 50 or more amino acids and its molecular weight lies above this value. Different fermentation, purification processes and recombination technology produced potential protein drugs at acceptable cost which can be useful in various diseases through various routes like oral, transdermal, nasal, pulmonary, ocular, buccal, rectal. As these therapeutic proteins and peptides are made available, it will be essential to formulate these drugs into safe and effective delivery systems. Due to its wider applications in pharmaceutical industries, they will replace many existing organic based pharmaceuticals. Now days, many drugs are in the world market, while several hundred are in clinical trials.

In majority of the cases chronic therapy of these peptides and proteins is warranted. Generally they have extremely short biological half-life. This trait precludes the parenteral delivery, as daily multiple injections would be required to maintain the therapeutic levels of drug, which has its inherent drawbacks. Oral administration is limited due to enzymatic degradation. Only after the development of viable novel delivery system to improve their systematic bioavailability, these peptides and protein drugs can be of therapeutic importance. There is an urgent need to develop alternative non-parental routes of administration like buccal, nasal, pulmonary, ophthalmic, rectal, vaginal and transdermal routes. Alternatively other approach such as implants, self-regulatory delivery systems can be exploited. However the transmucosal route mentioned in the preceding lines may impose additional biological barriers to this class of difficult drugs in terms of tissue permeability, protease activity, etc.

In the last three decades therapeutic peptides and proteins have risen in prominence as potential drug of future. The recent advance in large-scale fermentation and purification processes and analytical characterization has widened the horizons. Ailments that might be treated with this type of therapeutics include auto-immune diseases, cancer, mental disorder, hypertension and certain cardiovascular and metabolic diseases. Protein drug must be highly purified and concentrated and have extremely short half-life and should have a shelf life of at least two years. Recombinant technology has allowed the production of many potential protein drugs at an acceptable cost, allowing the treatment of severe, chronic and life-threatening diseases, such as diabetes, rheumatoid arthritis, hepatitis, etc. Currently, over 160 protein drugs are available on the world market, and several hundreds more are in clinical trials. The total protein drug market already exceeds 30 billion, and is expected to rise by at least 10% a year. One of the biggest opportunity areas in the Protein Therapeutics Market will be in the field of biogenerics, which is expected to create a multi-billion dollar market in future. Current research and advancement lead to the discovery of various biologically active protein molecules. These proteins are classified in Figure 1.1.

1.2 STRUCTURE OF PROTEINS

The proteins are large molecules complex architecture. The peptide and protein are seldom linear and adapt a variety of specific folded three-dimensional patterns and conformation. Structure of a protein is directly related to its function, so that anything that severely disrupts the shape will also disrupt the function. There are four types of protein structure (Figure 1.2) and they are as follows:

- primary structure;
- secondary structure;
- tertiary structure; and
- quaternary structure.

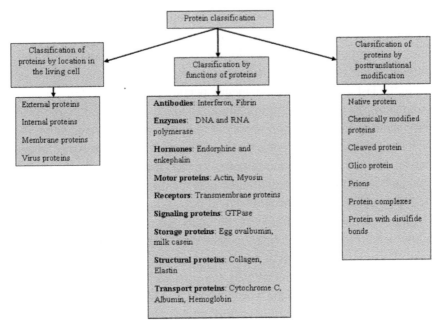

FIGURE 1.1   General classification of proteins.

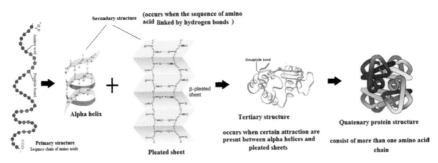

FIGURE 1.2   Structure of proteins.

The peptide chains in peptides and proteins are seldom linear and adopt a variety of specific folded 3-D patterns or conformations. Conformation of the peptide chain is determined by the covalently bonded amino acid sequence, by disulfide bridges between cysteine residues and by total conformation energy (the sum of electrostatic energy, hydrogen-bonded energy, non bonded energy and torsional energy). The properties that are affected by conformation include:

- Physical properties: such as solubility and spectral properties such as circular dichroism.
- Chemical properties: since folding may stabilize reactive groups by hydrogen bonding or sterically shield them from reagents
- Biological properties, as the 3-D structure places catalytic groups into proper orientation for enzymatic activity or places backbone and side chain groups into proper orientation for hormone-receptor interaction
- Stability to enzymatic cleavages since some of the amide groups susceptible to proteolysis may be sterically in a folded peptide chain orientation.

1.2.1 LEVELS OF PROTEIN STRUCTURE

All peptides and proteins are polymers of amino acids connected via amide linkages referred to as peptide bonds. The primary structure of a peptide or protein is the sequence of amino acids in the same. The structure is determined genetically by the sequence of nucleotides in DNA. The other form or conformation is the secondary structure. This is the respective arrangement of individual amino acids along the polypeptide backbone. At times, results in a very specific and well ordered structures as helices, loops, β-strands and β-turns. Tertiary structure is the 3-D arrangement of a single protein molecule. It refers to the way that secondary structural elements interacts with each other as well as with random portions of the molecule to form stable domains. And finally, quaternary structure is the form of protein as it exists in the solid state or in solution. It refers to the noncovalent interaction or spatial arrangement of individual protein monomers with each other to form oligomers and also the interaction of monomers and oligomers with the solvent and other solute components that affect the molecular weight distribution of proteins. Proteins can be divided in to two major classes depending on the conformation: fibrous and globular. Fibrous proteins are composed of polypeptide chains that are arranged in parallel position along a single axis and thereby yielding long fibers or sheets. These are insoluble in aqueous systems. Keratin, collagen, and elastin are the representative examples of this class. On the other hand, the polypeptide chains are tightly folded into compact globular or spherical shapes. A majority of these proteins are soluble in water. Most of the peptide/protein therapeutic agents are example of this class.

1.3 CHEMICAL SYNTHESIS OF PROTEIN AND PEPTIDES

Proteins have become accessible targets for chemical synthesis. The basic strategy is to use native chemical ligation, Staudinger ligation, or other orthogonal chemical reactions to couple synthetic peptides. The ligation reactions are compatible with a variety of solvents and proceed in solution or on a solid support. Chemical synthesis enables a level of control on protein composition that greatly exceeds that attainable with ribosome-mediated biosynthesis. Accordingly, the chemical synthesis of proteins is providing previously unattainable insight into the structure and function of proteins. The human genome contains approximately 30,000 genes [1]. Scientists from a broad range of disciplines are now working to reveal the structure and function of the proteins encoded by these genes. Their findings could lead to the solution of a multitude of problems in biology and medicine. In addition to structure-function analyzes of extant proteins, protein chemists are working to create new proteins with desirable properties, either by de novo design or by altering natural frameworks. The study of natural proteins and the creation of nonnatural ones require the ability to access and manipulate proteins. The isolation of proteins from their natural source is often tedious, idiosyncratic, and impractical. In contrast, the production of proteins with recombinant DNA (rDNA) technology, either in a heterologous host or in vitro, can provide access to large quantities of protein or allow for the exchange of 1 of 20 common amino acid residues for another. However, aggregation often limits the yield of properly folded proteins produced with rDNA. Moreover, the restrictions of the genetic code severely limit the possible modifications. The chemical synthesis and semisynthesis of proteins harbor the potential to overcome many of the disadvantages of current protein production methods [2]. In particular, chemical synthesis using established solid-phase techniques are rapid to effect, easily automated, and facilitate purification. Accordingly, the application of existing and emerging synthetic methods could facilitate research in all aspects of protein science. Chemical synthesis enables the facile incorporation of nonnatural functionality into proteins. The genetic code limits the components of natural proteins to approximately 20 α-amino acids. Methods that overcome this limitation but still rely on the ribosome are similarly limited to a subset of α-amino acids and α-hydroxy acids [3]. In marked contrast, the nonnatural functionality made available by chemical synthesis is limited only by the constraints of the periodic table and the imagination of protein chemists. The desire to synthesize proteins is not new. On December 12, 1902, Emil Fischer [4] delivered his Nobel Prize lecture in Stockholm, Sweden, saying in part: Of the chemical aids in the living organism the ferments—mostly referred to nowadays as enzymes—are so pre-eminent that they may justifiably be claimed to be involved in most of the chemical transformations in the living cell. The examination of the synthetic glucosides has shown that the action of the enzymes depends to a large extent on the geometrical structure of the molecule to be attacked, that the two must match like lock and key. Consequently, with their aid, the organism is capable of performing highly specific chemical transformations, which can never be accomplished with the customary agents. A century later, Fischer's vision is becoming reality. Enzymes and other proteins not only are accessible targets for synthetic chemistry, but are poised to become dominant targets of the twenty-first century. Herein, we discuss current efforts toward preparing proteins synthetically, focusing on the development of powerful new methodologies for splicing peptide fragments in a convergent strategy for the total chemical synthesis of proteins.

1.3.1 PEPTIDE SYNTHESIS

The chemical synthesis of proteins is now possible because of the prodigious advances in peptide synthesis that have occurred over the last century. Fischer's 1901 synthesis of glycylglycine is the first reported synthesis of a dipeptide and is also the first instance of the term "peptide" used to refer to a polymer of amino acids [5]. His 1907 synthesis of pentadecapeptide consisting of 15 glycine and 3 leucine residues was a remarkable achievement, despite his inability to control its amino acid sequence. An important advance in peptide synthesis was Bergmann and Zervas' (1932) introduction of reversible protection for the α-amino group [6]. With the emergence of protecting group strategies, it became possible to synthesize small peptide hormones. For example, in 1953 Du Vigneaud and coworkers [7] reported a solution-phase synthesis of the octapeptide hormone oxytocin. Even though fifty years had passed since Fischer's first synthesis of a peptide, these types of syntheses were still only accomplished with considerable effort [8]. The advent of solid-phase methods heralded a revolution for peptide synthesis. In 1963, Merrifield [9] described the first solid-phase synthesis of a peptide, a tetrapeptide. He attached an amino acid to an insoluble support via its carboxyl group and then coupled the next aminoacid, which had a protected amino group and an activated carboxyl group. The aminoprotecting group was removed, and the next amino acid was coupled in a similar manner. Within a few years, Merrifield [9] reported the development of an instrument for the automated synthesis of peptides. In short order, Gutte and Merrifield [10] used this new strategy to achieve the first synthesis of an enzyme, ribonuclease A (*RNase A*), albeit in low overall yield. Concurrently, a team led by Hirschman [11] reported the chemical synthesis of RNase S by solution-phase segment condensation reactions. Automated solid-phase peptide synthesis is commonplace today. In the most common strategy, an amino acid with both α-amino group and side chain protection is immobilized to a resin. The α-amino-protecting group is typically an acid-sensitive tert-butoxycarbonyl (Boc) group or a base-sensitive 9-fluorenylmethyloxycarbonyl (Fmoc) group. These α-amino-protecting groups can be removed quickly and completely, and a protected amino acid with an activated carboxyl group can then be coupled to the unprotected resin-bound amine. The coupling reactions are forced to completion by using an excess of activated soluble aminoacid. The cycle of deprotection and coupling is repeated to complete the sequence. With side chain deprotection and cleavage, the resin yields the desired peptide. The efficiency of solid-phase peptide synthesis continues to improve. New solid supports have increased the length of accessible peptides. New linkers between the support and the peptide have diversified the conditions that can be used to liberate a synthetic peptide. New side chain protection strategies have minimized deleterious side reactions. Finally, new carboxyl-activating groups have increased the s peed and efficiency of amino acid couplings while reducing the risk of epimerization. Solid-phase peptide synthesis alone has enabled the total chemical synthesis of some proteins. Since the pioneering work of Merrifield, proteins that contain as many as 166 amino acid residues have been synthesized in this fashion. These syntheses have, in some cases, been critical to the structure-function analyzes of the target proteins. Notably, the chemical synthesis of HIV-1 protease enabled the structural characterization of protease inhibitor complexes [12]. In addition, an enantiomer of this protease was synthesized with D-amino acids to demonstrate its chiral specificity for a peptide substrate derived from D-amino acids. Another notable example is a recent synthesis of a modified β-1 domain of streptococcal G protein. This synthesis incorporated a completely nonnatural dibenzofuran-based β-turn mimic as a conformational probe. These studies highlight the true power and potential of the total chemical synthesis of proteins in enhancing our understanding of protein structure and function. Despite the significant gains made in advancing the technology of solid-phase peptide synthesis, there remain limitations. Modern peptide synthesis is typically limited to peptides consisting of no more than 40 residues. Peptides and proteins of greater length can be prepared, although not routinely. Hence, most proteins cannot by synthesized by the stepwise assembly of amino acid monomers. The convergent assembly of protected or partially protected peptide segments, both in solution and on solid phase, is one approach used to access proteins that contain more than 40 residues [13]. The 238-residue precursor of the green fluorescent protein was synthesized with this segment-condensation approach. Twenty-six peptide fragments corresponding to parts of the green fluorescent protein were synthesized, assembled in solution, and deprotected. The resulting protein exhibited a fluorescence spectrum indistinguishable from that of the biosynthetic protein upon standing in solution. In addition to producing a large, complex protein in its native form by chemical synthesis, this study also demonstrated unequivocally that the formation of the green fluorescent protein fluorophore is not dependent on any external cofactors.

A peptide is a chain of special acids called amino acids linked together by bonds known as amide bonds. A protein consists of one peptide folded in a particular way, or several peptides folded together. Such peptides are synthesized very rapidly within living cells, but until recently could only be artificially synthesized in very long, slow processes that had poor yields and gave impure products. Recently a new technique known as solid phase peptide synthesis (SPPS) has been developed. Solid phase peptide synthesis results in high yields of pure products and works more quickly than classical synthesis, although still much more slowly in than living cells. This technique is discussed below. Peptides are the long molecular chains that make up proteins. Synthetic peptides are used either as drugs or in the diagnosis of disease. Peptides are difficult to make as the synthetic chemist must ensure that the amino acids that make up the chain are added in the correct order and that they don't undergo any other reactions. This involves adding one amino acid, washing away any unreacted acid then adding the next and so on. As can be imagined, this is very time consuming and only gives very low yields. A technique that has been relatively recently developed involves attaching one end of the peptide to a solid polymer, meaning that the peptide cannot get washed away along with the excess acid. This is much quicker than classical synthesis, and leads to dramatically improved yields. The process consists of five steps carried out in a cyclic fashion.

Step 1 – Attaching an amino acid to the polymer
The amino acid is reacted with a molecule known as a "linkage agent" that enables it to attach to a solid polymer, and the other end of the linkage agent is reacted with the polymer support.

Step 2 – Protection
An amino acid is an acid with a basic group at one end and an acid group at the other. To prevent an amino acid from reacting with itself, one of these groups is reacted with something else to make it unreactive.

Step 3 – Coupling
The protected amino acid is then reacted with the amino acid attached to the polymer to begin building the peptide chain.

Step 4 – Deprotection
The protection group is now removed from the acid at the end of the chain so it can react with the next acid to be added on. The new acid is then protected (Step 2) and the cycle continues until a chain of the required length has been synthesized.

Step 5 – Polymer removal
Once the desired peptide has been made the bond between the first amino acid and the linkage agent is broken to give the free peptide. Peptides synthesized at Massey University are widely used in medical research, as they can be synthesized quickly and accurately.

1.3.1.1 Solid Phase Peptide Synthesis

Peptide synthesis is much more complicated than simply forming amide bonds by mixing the desired amino acids together in a test tube. With twenty natural amino acids and a number of unnatural ones as well the possible combinations formed with this technique are numerous. This complexity makes the synthesis of peptides both fascinating and challenging. If solutions containing two amino acids are mixed together, four different dipeptides will be formed. To ensure that only the desired dipeptide is formed the basic group of one amino acid and the acidic group of the other must both be made unable to react. This 'deactivation' is known as the protection of reactive groups, and a group that is unable to react is spoken of as a protected group. In classical organic synthesis the acids are protected, allowed to react and deprotected, then one end of the dipeptide is protected and reacted with a new protected acid and so on. In SPPS the amino acid that will be at one end of the peptide is attached to a water-insoluble polymer and remains protected throughout the formation of the peptide, meaning both that fewer protection/deprotection steps are necessary and that the reagents can easily be rinsed away without losing any of the peptide.

Step 1 – Attaching an amino acid to the polymer
Peptide chains have two ends, known respectively as the N-terminus and the C-terminus 3, and which end is attached to the polymer depends on the polymer used. This article assumes that polyamide beads are used and hence that the C-terminus of the peptide is attached to the polymer. The attachment is done by reacting the amino acid with a linkage agent and then reacting the other end of the linkage agent with the polymer. This means that a peptide polyamide link can be formed that will not be hydrolyzed during the subsequent peptide forming reactions.

Step 2 – Protection

The next amino acid also needs to have its amino group protected to prevent the acids reacting with each other. This is done by protecting it with FMOC (9-fluorenylmethoxycarbonyl). In addition, any amino acid side chains that are aromatic, acid, basic or highly polar are likely to be reactive. These must also be protected to prevent unwanted branched chains from forming.

Step 3 – Coupling

The FMOC protected amino acid is then reacted with the last amino acid attached to the polyamide. The reaction is catalyzed by DCC (1,3 dicyclohexylcarbodiimide), which is itself reduced to DCU (1,3-dicyclohexylurea).

Step 4 – Deprotection

Excess DCC is washed off the insoluble polymer with water, then the FMOC group removed with piperidine (a cyclic secondary amine). This is a trans-amidification reaction. Steps 2 to 4 are repeated as each new amino acid is added onto the chain until the desired peptide has been formed.

Step 5 – Polymer removal

Once the peptide is complete it must be removed from the polyamide. This is done by cleaving the polyamide – peptide bond with a 95% solution of trifluoroacetic acid (TFA). The side-chain protecting groups are also removed at this stage.

1.3.2   *PROTEIN SYNTHESIS BY PEPTIDE LIGATION*

The formation of an amide bond between protected peptide fragments by the attack of an amino group on an activated carboxyl group can be problematic, especially with large peptides. In other words, the reactivity between an amino group and an activated carboxyl group is not inherently high enough for acyl transfer to overcome the relatively low solubility of peptides. In nature, peptide bond formation occurs by sequential transfer of the C-terminal acyl group of a nascent peptide chain to the α-amino group of an aminoacyl tRNA. The C-terminal acyl group is activated only as an ester with a tRNA, and the transfer is aided, in large part, simply by the reactants being held in close proximity by the ribosome [14].

Proteases have been engineered to perform in a similar manner and used for the convergent synthesis of proteins as large as *RNase* A. New chemical strategies have emerged that take advantage of enforced proximity to couple peptides. These peptide ligation methods provide a practical and powerful means to assemble synthetic peptides into proteins. The peptides can be protected or unprotected, and coupling can occur in an aqueous or organic solvent, in solution, or on a solid support. Peptide ligation strategies utilize three steps. An initial capture step links the peptides by a chemical reaction that is more rapid than intermolecular acyl transfer to an amine and that uses functional groups with reactivity that is orthogonal to that in proteins. The pivotal acyl transfer step is thus rendered intramolecular and hence can occur with maximal efficiency [15]. In the last step, the capture moiety is released, either in a discrete chemical process or spontaneously. Below, we discuss current strategies for peptide ligation, along with their relative advantages and disadvantages. We also highlight some notable applications of these strategies. Our focus is on ligation strategies that yield native amide bonds. We do not discuss approaches that yield nonnatural bonds between peptide fragments [16].

1.3.3 PEPTIDE LIGATION WITH SULFUR

1.3.3.1 Prior Thiol Capture

Modern methods of peptide ligation trace their origin to Kemp's prior thiol capture strategy. In this approach, 4-hydroxy-6-mercaptodibenzofuran is used as the association element to bring the peptide coupling partners together [17]. This strategy uses a highly efficient thiol-disulfide exchange reaction in a capture step prior to the acyl transfer reaction. Acyl transfer occurs with a half-life ranging from 0.1 to 50 h, depending on the bulk of the side chain of the C-terminal acyloxy-derived residue, and without racemization of the coupled amino acids [18]. Upon completion of the acyl transfer reaction, phosphine reduction of the mixed disulfide yields the native peptide. The utility of this approach was demonstrated by the ligation of a variety of peptide fragments. In one study, the C-terminal 29-residue fragment of bovine pancreatic trypsin inhibitor was synthesized from four segments, each possessing an N-terminal cysteine residue and a C-terminal dibenzofuran [19]. The synthesis commenced with the two fragments at the extreme C terminus and proceeded sequentially toward the N terminus. The N-terminal cysteine residues were protected orthogonally until they were needed for coupling, when they were deprotected and derivatized as mixed disulfides. In a separate study, a 39-residue peptide and a 25-residue peptide were synthesized in high yield, utilizing side chain protection only at cysteine residues. The ligation of fully protected peptides was 50-fold slower. Kemp's prior thiol capture strategy is a seminal contribution to the development of peptide ligation concepts and methodology. It represents the first demonstration of the chemoselective ligation of unprotected peptide fragments. In addition, it represents the first systematic application of the use of proximity effects to evoke acyl transfer via an intramolecular reaction.

1.3.4 PEPTIDE LIGATION WITH SELENIUM

Selenocysteine is a natural amino acid residue with a low natural abundance [20, 21]. During the biosynthesis of selenoproteins, selenocysteine (Sec or U) is incorporated by the ribosomal translation of mRNA and has its own tRNA Sec and codon, UGA (which is also the opal stop codon). Decoding a UGA codon for selenocysteine requires a unique structure in the 3′-untranslated region of the mRNA called a selenocysteine insertion sequence element. The production of eukaryotic selenocysteine-containing proteins in prokaryotes is problematic because eukaryotic and prokaryotic cells use a different selenocysteine insertion sequence element [22]. The feasibility of using selenocysteine in native chemical ligation has been demonstrated with model systems [23–25]. pH rate profiles have demonstrated that a ligation with a selenocysteine residue occurs much more readily than does one with cysteine, as expected from the lower pK a of selenols and the somewhat higher reactivity of selenates. For example, at pH 5.0 the reaction with selenocysteine is 10^3-fold faster than that with cysteine. Selenocysteine has been used to mediate expressed protein ligation [26, 27]. rDNA technology was used to prepare a fragment corresponding to residues 1–109 of RNase A with a C-terminal thioester. Standard solid-phase methods were used to synthesize a peptide corresponding to residues 110–124, but with selenocysteine rather than cysteine as residue 110. The thioester fragment and the peptide fragment were then ligated, and the product was folded and purified. The integrity of the desired C110U variant was verified by mass spectrometry and its wild-type enzymatic activity. The data indicate that C110U RNase A is not only an intact protein, but also a correctly folded enzyme with a selenosulfide (Se-S) bond between Sec-110 and Cys-56. A semisynthesis of a selenocysteine-containing variant of azurin has also been achieved with expressed protein ligation [28]. Cys-112 was replaced with selenocysteine, and fragment 112–128 was ligated to fragment 1–111. The variant azurin protein displayed unique spectral properties that were used to reveal the nature of the enzyme-copper interaction in the native protein.

The isomorphous replacement of sulfur with selenium can stabilize a protein. Specifically, proteins containing selenosulfide bonds should have greater conformational stability in a reducing environment than proteins with disulfide bonds, as the reduction potential of a selenosulfide bond is much less than that of the corresponding disulfide bond [29]. This use of selenocysteine to stabilize an enzyme represents another form of protein prosthesis. Selenomethionine has been incorporated into bovine pancreatic polypeptide with a strategy similar to that used for methionine [30]. Here, native chemical ligation of a peptide fragment having an N-terminal selenohomocysteine residue was used to effect native chemical ligation. Subsequent Se-methylation yielded a selenomethionine

1.3.5 GENERAL STRATEGIES FOR PEPTIDE LIGATION

A limitation of native chemical ligation is its intrinsic reliance on having a cysteine residue at the ligation juncture. Cysteine is the second least common amino acid, comprising only 1.7% of all residues in proteins. Hence, most proteins cannot be prepared in their native form by any method that allows peptides to be coupled only at cysteine residues. However, non-native cysteine residues have been added to enable protein synthesis by native chemical ligation or semisynthesis by expressed protein ligation [31]. The addition of nonnative cysteine residues can incur risk. Of the functional groups in the 20 common amino acids, the thiolate of cysteine is by far the most reactive toward disulfide bonds, O_2 (g), and other electrophiles [32]. In addition, cysteine can suffer base-catalyzed β-elimination to form dehydroalanine, which can undergo further deleterious reactions [33]. The side chains of nonnative cysteine residues have been alkylated [e.g., with bromoacetate to form 4-thiahomoglutamate] to minimize these risks. The removal of the cysteine limitation by applying a more general ligation technology would extend greatly the utility of protein synthesis. An ideal technique would enable complete and rapid ligation between any two amino acid residues without detectable epimerization. Below, we describe three such strategies

1.4   INSTABILITY PROBLEMS ASSOCIATED WITH PROTEINS AND PEPTIDES

Within the context of biomedicine and pharmaceutical sciences, the issue of (therapeutic) protein stabilization assumes particular relevance. Stabilization of protein and protein-like molecules translates into preservation of both structure and functionality during storage and/or targeting, and such stabilization is mostly attained through establishment of a thermodynamic equilibrium with the (micro)environment. The basic thermodynamic principles that govern protein structural transitions and the interactions of the protein molecule with its (micro)environment are, therefore, tackled in a systematic fashion. Highlights are given to the major classes of (bio)therapeutic molecules, viz. enzymes, recombinant proteins, (macro)peptides, (monoclonal) antibodies and bacteriophages. Modification of the microenvironment of the biomolecule via multipoint covalent attachment onto a solid surface followed by hydrophilic polymer co-immobilization, or physical containment within nanocarriers, are some of the (latest) strategies discussed aiming at full structural and functional stabilization of said biomolecules

In its essence, all life forms are polymeric, since their most important components are all biopolymers. And nature uses polymers both as building bioblocks and as part of the highly complex cell machinery. While nowadays proteins and peptides are well known to be the not-so-secret secret of life, the real secrets and mystery of life are hidden in an extremely complex structure of natural biomacromolecules, viz DNA and RNA. Besides those naturally occurring (bio)polymers, humans now possess the technology to produce synthetic macromolecules via polymerization processes to form polymer-based artificial materials, usually aiming at replacing natural ones. But both natural and synthetic polymers display a strong structure–property relationship and, therefore, a detailed knowledge of the structure of macromolecules is of utmost importance in modern (bio)polymer chemistry and biotechnology. It is the fact that such biopolymers are ubiquitous in the metabolic machinery of all living beings that makes them so much attractive for biotechnological and industrial applications. In fact, the lifetime of proteins inside the cells must be limited, since constitutive proteolysis is the main source of amino acids for denovo protein synthesis [34, 35]. But we are interested in proteins in nonnative conditions for biotechnological applications. Protein stabilization has a tremendous importance due to the increasing number of protein applications in almost all areas [36], but especially in the biopharmaceutical and biomedicine areas [37–40]. However, the only moderate stability of proteins, and specially enzymes, is the major drawback hindering the generalized application of these bioactive molecules at the industrial scale [41, 42]. The causes of poor biocatalyst stability are closely associated with the process conditions prevailing, and may include extreme temperatures or pH values [43], or even the presence of organic solvents, that are outside the operating stability window of the biomolecule, but that are often necessary to solubilize poorly-water-soluble substrates in high concentration values [44]. In particular, within the context of biomedicine and pharmaceutical sciences, the issue of (therapeutic, recombinant) protein stabilization assumes particular relevance [45–48]. Protein instability is one of the major drawbacks that hinder the (more appealing) oral administration of protein pharmaceuticals. When preparing and applying (biopharmaceutical) protein preparations for use in biomedicine, one usually faces three aspects of protein instability, viz. operational stability and in vivo stability [49–51]. Stabilization of protein and protein-like molecules translates into preservation of both structure and functionality during storage and/or specific targeting, and such stabilization is mostly attained through establishment of a thermodynamic equilibrium between said molecules and their (micro)environment [52]. Therefore, to satisfy the increasing demand of biomolecules with biopharmaceutical applications, a basic understanding of the interactions between such biomolecules and their (micro)environment is in order. There is a delicate balance between stability and flexibility needed for enzyme function [53–58], added to the increasing awareness of the importance of the protein surface for stability, since it is through this interface that a protein entity senses the "external world." It is generally accepted that functionally important amino acid residues are mainly solvent accessible residues on the protein surface, while structurally important residues are likely part of the protein core [59]. According to Jaenicke [60] and Vieille and Zeikus [61], the core packing in native (folded) protein molecules is so well arranged that virtually all solvent molecules are essentially excluded, making the protein core more like a crystalline solid than a non-polar fluid. Damborsky [62] investigated the effect of the changes in structure of a protein on its function and stability. By using quantitative structure function relationship (QSFR) or quantitative structure stability relationship (QSSR) analyzes, Damborsky [62] aimed at investigating and mathematically describing the effect of the changes in structure of macromolecules such as proteins on their activity, specificity or stability. Foit and colleagues, Becktel and Schellman [63], Somero [64] and Lee and Vasmatzis [65] extensively discussed the intrinsic stability of a protein molecule, for example, the stability that can be obtained by mutating its amino acid residues, and concluded that by mutating some amino acid residues of the hydrophobic-core packing of a protein molecule, it is possible to stabilize a protein molecule to a remarkable degree. Protein thermal stability usually increases if amino acid substitution and/or modification result in increased internal and decreased external hydrophobicity [66–71]. Similarly, Campos and colleagues [72] used one of the most promisingly general methods of protein stabilization, the optimization of surface charge–charge interactions, a method which relies in identifying by theoretical methodologies, mutations of a protein that are expected to lower the electrostatic energy of the native state. In this way, charge mutations involving the replacement by lysine of negatively charged residues not involved in either salt bridges or hydrogen bonds, and located at the protein surface, were introduced in the apo flavodoxin from Anabaena PCC 7119 [72], and the resulting increase in stability of the native, folded state, duly verified relative to its unfolded counterpart. More than 95% of all charged moieties are located on the surface of the protein, consisting mostly of hydrophilic moieties and since the static or dynamic conditions of the physicochemical microenvironment of the protein are sensed intrinsically through those moieties, the protein surface thus constitutes a very challenging and attractive target for protein engineering aiming at enhancing its stability [75]. In the same line of thought, Palmer and colleagues [76] mutated a protein-G aiming at improving stability towards caustic alkali, demonstrating that strategies for stabilization of proteins at extreme alkaline pH should consider thermodynamic stabilization that will retain the tertiary structure of the protein and modification of surface electrostatics, as well as mutation of alkali-susceptible moieties. In a similar fashion, Villegas and colleagues demonstrated that by using a helix/coil algorithm to design helix-stabilizing mutations on the solvent-exposed moiety patches of protein helices, it was possible to rationally increase the stability of proteins. Bioactive entities such as insulin, enzymes, fibrinogen, monoclonal antibodies, interferons, and bacteriophages, among others, are mainly produced via synthetic methods using biofactories such as cloned transgenic animals, cloned transplastomic plants, microbial fermentation and mammalian cell culture. Of these, monoclonal antibodies constitute the second largest biopharmaceutical product category currently in clinical trials. Highly complex production processes and, concomitantly, high production costs associated with such complex protein molecules, all require that they are fully stabilized and preserved at high efficiency levels during long periods of time. The structural stability of protein entities is extensively controlled by the interactions between the protein molecules and the surrounding solvent molecules [77]. Protein stabilization is based upon dampening the molecular motions and, therefore, eliminating conformational transitions while the molecule is still in the native state. A further, and especially complex problem, lies in the stabilization of multimeric proteins, with dissociation of the subunits producing enzyme inactivation with concomitant product contamination [78]. The stability of multimeric proteins is very dependent on the concentration, due to their multimeric nature, which translates into a low thermal stability under diluted conditions due to a subunit dissociation mechanism [78, 79]. Due to the high glutathione concentration inside the cell, there is no disulfide bond in intracellular proteins, hence many of the cross-links stabilizing the three-dimensional structure of proteins are clusters of non-covalent bonds. The three-dimensional structure of a protein molecule depends mostly on two types of interactions: intramolecular interactions between amino acid moieties (and intermolecular interactions with solute and/or solvent molecules present in its environment. Makhatadze and Privalov [80], Liaoetal [81] and later Miyawaki [82], pointed out in their studies, the importance of intramolecular hydrogen bonding in the stabilization of proteins. When looking at the (non-specific, indirect) effects of solutes on the molecular motions of proteins, one can envisage modification of the solvent promoted by the former at four levels:

- molecular motions;
- chemical properties;
- physical properties; and
- thermodynamic properties.

Water is the universal and natural solvent for proteins, dictating the molecular motions, the structure and function of these molecules. Globular proteins are only marginally stable, and such metastability makes proteins difficult to handle experimentally [83–86]. According to Bizzarri and Cannistraro [87], a threshold level of hydration (0.4 g water/g protein) is required to fully activate the functionality of globular proteins, an amount less than would be necessary for a complete coverage of the surface of the protein. Later on, the issue of the essential hydration shell of all proteins, a role played beautifully by such universal solvent is addressed. Proteins are strongly hydrated in aqueous solutions, and the hydration state affects their stability, function and three-dimensional conformation [88]. Since water is the environment in which protein molecules do exist and operate, the structure and dynamics of the water hydration shell are directly linked to both protein flexibility and stability [89]. Protein stability is directly correlated with the ability of water molecules in the hydration layer to fluctuate among different equilibrium structures and, additionally, internal water molecules contribute also to protein stability by providing the necessary flexibility for biological activity, by acting as lubricant and by rendering to the protein a certain level of plasticity [90]. Additionally, water can take part in the reaction indirectly by providing solvation to polar residues of the enzyme (or other protein) and other intervening molecules in order to facilitate protein conformational changes during the biocatalytic process.

A protein molecule can only move if the nearest molecular neighbors also move, a collective phenomenon resembling a continuous search for escape out of a cage rather than a discontinuous jump across an energetic barrier [90]. The (liquid) cage becomes a trap when the density reaches a critical value, a moment when the liquid is arrested on a macroscopic scale. Water, therefore, acts as a plasticizer to protein motions, expanding the accessible protein conformational space by decreasing friction [90], via changing allegiances of hydrogen bonds between donors and acceptors [91]. Structural plasticity is thus dominated by polar interactions [90]. Stabilizing a biomolecule involves dampening its molecular motions, and this can be achieved by reducing the chemical activity of the water present in its microenvironment at the expense of either:

- changing the thermodynamic phase of water;
- exposing the water to specific solutes (such as the disaccharide trehalose); or
- by completely removing such water (viz. by freeze drying).

Dehydration thus leads to a virtually infinite viscosity, or glassy state [90], implying that embedding a protein molecule in a rigid solvent or removing the solvent entirely reduces small-scale liquid-like motions. Proteins are characterized by a small thermodynamic stability. The thermodynamic stability of a protein entity conformation is the result of several non-covalent interactions, which may occur intramolecularly or with the solvent. Thermodynamic stability of a protein is, in effect, the work required to disrupt the tertiary structure of the protein [92], where the transition from the native (folded) state to the denatured (unfolded) conformation is a highly cooperative process involving disruption of intramolecular hydrogen bonds, hydrophobic interactions and other types of non-covalent interactions. In freeze-drying, as the water content decreases, hydroxyl groups from excipient molecules might be expected to gradually approach and hydrogen bond with protein entities [93], thus accounting for the preservation of the native structure of proteins. Stabilization mechanisms for preservation of viable cells also involve modification of the water within their microenvironment, via, for example, freeze drying with replacement of water molecules close to the membrane lipid head groups with trehalose [93]. Ganjalikhany and colleagues [94] studied the effects of trehalose and magnesium sulfate on the structural stability and function of luciferase from firefly and noticed that the stability of this enzyme increased in the presence of the additives. They concluded that magnesium sulfate and trehalose can be used, respectively, for short- and long-term stabilization of the enzyme. Kohda and colleagues [95] proposed an intriguing method for the stabilization of immobilized enzymes, viz. the coimmobilization of hyperthermophilic chaperonin from Thermococcus strain KS-1 [96]. Hyperthermophilic chaperonins are expected to stabilize proteins because they are thermostable and suppress the thermal inactivation of enzymes [97]. Irrespective of the method being utilized, stabilization procedures aim at preserving the function of a protein by stabilizing both its structure and functionality during storage. Protein structures are composed of ordered regions, alpha-helices and beta-sheets, which are connected by disordered turns, and involve four different domains: primary, secondary, tertiary and quaternary. The primary structure encompasses a specific linear sequence of amino acids that form the protein, and remains totally unaltered during preservation processes since the peptide (covalent) bonds between the amino acids are quite strong and are not broken by changes in hydration status or temperature. The secondary structure is the three-dimensional construct assumed by certain parts of the primary structure. These local constructs are mainly determined by hydrogen bonds that are established within the primary domain [98]. The two most common constructs encompassing the secondary domain are the α-helix and the β-pleated sheet. This secondary construct is also sometimes called a classic Pauling–Corey–Branson α-helix [99]. Among types of local structure in proteins, the α-helix is the most regular and the most predictable from sequence, as well as the most prevalent. The β-pleated sheet is the second form of regular secondary construct in proteins, only somewhat less common than the alpha helix. β-Pleated sheets consist of β-strands connected laterally by at least two or three backbone hydrogen bonds, forming a generally twisted, pleated sheet. A β-strand is a stretch of polypeptide chain typically 3–10 amino acid long with backbone in an almost fully extended conformation. The higher-level association of βsheets has been implicated in formation of the protein aggregates and fibrils observed in many human diseases, notably the amyloidosis such as Alzheimer's disease. Generally, protein aggregation involves mostly β-sheets while α-helix constructs seem to be less likely to form aggregates. The denaturation and aggregation of proteins, observed during either preservation processes or storage, are often associated with changes in the populations of α-helixes, β-sheets and random coil structures within the protein [100]. From the industrial pharmaceutical biotechnology point of view, protein aggregation is undoubtedly the most common and troubling manifestation of protein instability encountered in almost all phases of protein drug development, which hinders the rapid commercialization of potential protein-based drug candidates. The tertiary domain of a protein encompasses its three-dimensional folded-shape, with the hydrophobic side-groups of the amino acid moieties hidden within the core and the hydrophilic groups being exposed to the surrounding microenvironment. In monomeric proteins, it is the tertiary domain of the protein that determines its function. The primary and secondary domains of a (monomeric) protein both contribute to its tertiary domain. Any changes occurring in the physicochemical and thermodynamic properties of the protein's microenvironment will affect the secondary and tertiary domains of the protein, therefore causing a change in its three-dimensional conformation, followed by unfolding with concomitant loss of function [100, 101]. The secondary construct of a protein entity is generally more resistant to environmental aggressions compared to the associated tertiary domain [100]. The contribution of water activity was proved to be always positive for stabilization of proteins [101], since lubricating protein motions appears to be the main role of water. The quaternary domain is the combination of two or more monomeric chains (or subunits), to form a complete (multimeric) unit. The interactions between the monomeric chains are not different from those in the tertiary domain, but are distinguished only by being interchain rather than intrachain. Some (multimeric) proteins are composed of identical subunits while other proteins are composed of non-identical subunits (as in the case of insulin), which is made up of two chains, the α-chain and the β-chain, linked by two disulfide bridges. The three-dimensional spatial architecture of proteins is mainly determined by two classes of non-covalent interactions, viz. electrostatic and hydrophobic. Electrostatic interactions between polar and ionized groups include ion pairing, hydrogen bonds, weakly polar interactions and van der Waals forces. Hydrophobic interactions imply van der Waals forces and hydration effects of non-polar moieties. The physical nature of the latter was recently interpreted as being entropic and enthalpic due to significant contributions from van der Waals forces. Whether we talk about native or recombinant proteins, all have in common the existence of three protein domains in the case of monomeric proteins, and the existence of a fourth protein domain in the case of multimeric proteins. As we shall see later, it is the third and the fourth domains the ones directly involved in the stabilization of monomeric and multimeric proteins, respectively.

1.4.1   MOLECULAR STABILITY (BIO) THERMODYNAMICS

At the molecular level, a protein molecule may oscillate between many (slightly) different three-dimensional conformations. Each oscillation between two conformations is ruled by the second law of thermodynamics [106].

$$\Delta G = \Delta H - T\Delta S$$

where, ΔG is the difference between the Gibbs free energy of the two three-dimensional conformations, ΔH is the transition enthalpy, ΔS is the transition entropy, and T is the absolute temperature.

The conformational Gibbs energy depends on the contribution of conformational enthalpy and mainly on the enthalpy of intramolecular hydrogen bonds and on the van der Waals interactions between all the protein atoms [106]. In these transitions between configurations, with the protein molecule seeking the lowest energy state possible, if the value of ΔG is negative the protein molecule would denature. Application of the mathematical expression of the second law of thermodynamics assumes, naturally, that the native and the denatured states are unique states corresponding to unique three-dimensional configurations of the protein moiety. However, in their works, Sanchez-Ruiz, Ragoonanan and Aksan, Scharnagl and colleagues, Shenoy and Jayaram, Becktel and Schellman, Doster and Settles, and Shah and colleagues have shown that a protein can assume a very high number of slightly different configurations both in the native and in the denatured states, among which the protein molecule continuously oscillates through bond vibrations, throwing thermodynamics of protein transitions into a highly complex energy landscape. Equilibrium of a protein moiety in any of these states is dictated both by entropic and enthalpic factors, and therefore stability theories predict that proteins can denature not only by increasing the temperature but also by decreasing it [177, 102]. The total entropy of the protein molecule together with the associated water shell decreases as the protein folds, while the enthalpic contribution is mostly due to hydrogen bond formation.

Protein stability is, as mentioned above, quantitatively described by the standard free Gibbs energy change involved in unfolding the unique, three-dimensional structure, to a randomly-coiled polypeptide chain [103, 104]. According to Miyawaki and Tatsuno and Khechinashvili, the Gibbs energy of the hydration process of protein unfolding is a large and negative value. While crosslinks of inert polymers increase stability entropically by decreasing the entropy gain on unfolding or dissociation, disulfide bonds in proteins can increase stability also by increasing the transition enthalpy. In their studies, Zhou et al. [105] showed that molecular mobility directly determines the storage stability of amorphous pharmaceutical compounds. Although proteins are relatively large molecules, they are not infinite systems. There is, therefore, a direct relationship between molecular motions of the protein moiety and the molecular motions of its immediate neighborhood, which implies that stabilization can eventually be achieved by dampening any motions of the micro-neighborhood. Since the protein moiety is constantly exploring a multitude of configurational landscapes while, at the same time, interacting with the solvent molecules and other surrounding solutes, it can adopt configurations that favor intermolecular interactions, virtually resulting in the formation of hydrophobic interactions among protein moieties (aggregation) [106]. Enzymes are thus devices, which select by construction a small fraction of events out of a large number of fast structural fluctuations, selecting the most stable three-dimensional architecture, which is the one with the lowest energy.

Denaturation of a protein moiety involves unfolding of its three-dimensional architecture, in a way that biological or biochemical activity is lost. Chemical agents such as chaotropes can also lead to denaturation [107, 108]. In high-viscosity, low-molecular mobility systems, molecular mobility and denaturation kinetics have been shown to be coupled [109]. All these processes may affect the stability of protein entities during both preservation processes and storage. According to Khechinashvili and coworkers, the thermal stability of proteins does not correlate with the energy of intramolecular interactions, and so the mechanism of thermal stabilization of a compact protein with an ordered structure is largely of an entropic nature. During the denaturation process of a protein molecule induced by temperature, the protein molecule changes from a rather well organized structure into a random coil structure in which the hydrophobic amino acid moieties come into intimate contact with water. As a consequence of this intimate contact, water molecules form locally ordered structures around the hydrophobic amino acid moieties, which are characterized by both a low entropy and a low enthalpy, due to the well-ordered hydrogen bonds.

1.4.2   THERMODYNAMIC STABILITY

Thermodynamic stability relates to protein stability to reversible changes of structure in non-native in vitro conditions. The biological function of a protein is guaranteed if equilibrium is established between the native and the unfolded states of the protein. One can judge the stability of any protein structure by studying its disruption and, since a protein molecule is a macroscopic system, the disruption of its structure can be thought of as a change of the macroscopic state of the system. The stability of a protein molecule is usually expressed as Gibbs energy, since $\Delta N \, DG$ is the work required for disruption of the native protein structure [108–112]. Therefore, this difference between the Gibbs energies of the native and completely random states serves as a measure of protein stability, and the larger this value the more stable the protein is. While the Gibbs energy of hydration of nonpolar groups, though a relatively small value, shifts the imbalance of forces towards the formation of the native structure. According to Privalov, the native state of a protein is most stable at the temperature where the entropy difference of the native and denatured states is equal to zero, and it is stabilized only by the enthalpy difference of these states. Understanding how the biopolymers of life adapt to their microenvironments is central to devise thermodynamic stability. The rationale is that protein stability can be partially understood by examining the α-helices in the three-dimensional architecture and their constituent amino acid residues. One of the principal factors that controls α-helix stability is the presence of intra-helical, non-covalent bonding interactions, such as hydrogen bonds. Such interactions exist in thermodynamic equilibrium and, therefore, their strength will be strongly influenced by physical factors of the environment. Imprisonment of a protein entity leads to thermodynamic stability, which can be correlated with a change in the thermodynamic conditions of the microenvironment surrounding each bioparticle, since the movements of (aqueous) solvent molecules in their micro-neighborhood become seriously reduced by the effect of being contained within the matrix's core. The result is that the protein entity's rotational, translational and vibrational viscosity becomes enhanced, leading to a more rigid three-dimensional architecture with concomitant decrease of entropy and producing stabilization.

1.4.3 KINETIC STABILITY

Long-term or operational stability characterizes a protein's ability to resist irreversible changes of structure in in vitro non-native conditions [113, 114]. Kinetic stability is a measure of how quickly a protein unfolds, hence measuring the resistance to irreversible inactivation [113, 114], often being expressed as the protein's half-life (t1/2). It can be considered a "long-term stability." In the case of irreversible unfolding, it is the kinetic stability or the rate of unfolding that assumes particular importance. When irreversible alterations of non-native states occur very fast, the rate of irreversible denaturation is given by the rate of unfolding [113, 114]. A kinetically stable protein entity will unfold slower than a kinetically unstable one. In a kinetically stable protein entity, a higher free energy barrier (Ea) is necessary for the unfolding process, and the factors that affect stability are the relative free energies of the folded (GF) and transition state (GTS) conformations. A protein entity can denature irreversibly if its unfolded form undergoes a permanent quick change such as aggregation and proteolytic degradation. For the former change, it is the patches of contiguous hydrophobic groups in the folding–unfolding intermediates that initiate the aggregation process, because mutual attraction of hydrophobic moieties and/or patches minimizes the area of unfavorable protein–solvent interface. Proteins aggregate to minimize thermodynamically unfavorable interactions between solvent and exposed hydrophobic residues. Hence, it is not the difference between the free energies of folded and unfolded states that is important; such difference only affects equilibrium.

Truly important is, as stated, the difference between free energies of folded and transition states (Ea), since it is the magnitude of this difference that determines the unfolding rate. Besides the biotechnological implications for biopharmaceutical industry, there is a general interest in understanding protein kinetic stability since some emerging molecular approaches to the inhibition of amyloidogenesis focus on the increase of kinetic stability of protein native states [183].

1.4.4 STRUCTURAL AND FUNCTIONAL STABILIZATION OF MONOMERIC AND MULTIMERIC ENTITIES

The term "protein stabilization" refers to preservation of the unique chemical and three-dimensional structure of a polypeptide chain under extremes of physical conditions. Stabilization of multimeric enzymes and proteins in general represents a especially complex problem, if we realize that among the most interesting enzymes with potential biopharmaceutical applications many of them are of multimeric nature. As mentioned above, in monomeric proteins the first step in their inactivation involves generally alterations in the tertiary structure [115]. However, for multimeric problems, inactivation begins generally either with dissociation of the enzyme subunits or with alterations in their correct three-dimensional assembly [116–117]. But the multimeric nature of these protein entities does not translate at all into a low stability. On the contrary, the quaternary architecture resulting from the multiple subunits assembly provides a decrease in the surface area accessible, which in turn has been associated with extreme thermophilicity [118–121]. Quaternary interactions and closer packings are typical characteristics of proteins from thermophiles [122]. In multimeric proteins, subunit–subunit multi-interactions also contribute to improve protein rigidity due to a lower mobility of the moieties involved in these interactions. Notwithstanding the fact that multimeric proteins are intrinsically more stable than monomeric ones due to their rigidity when compared to their monomeric counterparts, under certain experimental conditions such subunit–subunit multi-interactions may be weakened leading to dissociation of subunits and concomitant rapid inactivation of the multimeric protein. Hence, prevention of subunit dissociation is the first goal when stabilization of such highly interesting protein entities is sought [123–128]. When we speak of enzymes, and particularly multimeric enzymes, their structural and functional stabilization assumes special relevance through immobilization and post-immobilization techniques [129–131], because as well as being necessary to ensure rigidification of the molecule it is also necessary to stabilize it against the action of denaturing organic solvents [132]. Protein stability inversely correlates with its flexibility, with the flexible patches in proteins typically being the labile areas [133]. A folded protein in an aqueous environment (solution) has hydrophobic regions sequestered from, and hydrophilic areas in contact with, the aqueous environment. But, when the polarity of an aqueous solvent decreases via addition of a nonaqueous solvent, protein hydrophobic cores tend to dissipate in contact with the latter which leads to disruption of the protein hydration shell, with concomitant destabilization and unfolding of the macromolecule. In terms of structural stabilization, its rigidification occurs via intense multi-subunit covalent immobilization, but maintenance of its functional stability occurs only after cross-linking all of its subunits [134], which promotes additionally its stability in the presence of organic solvents by maintaining the three-dimensional architecture of the enzyme perfectly hydrated due to creation of a hyperhydrophilic nanoenvironment. When structural rigidification is sought, multipoint covalent immobilization into a highly activated support is the preferred immobilization strategy. It is possible to structurally stabilize proteins by sacrificing activity to a certain degree [135]. An ideal support should be characterized by a subtle balance between the affinity towards the protein entity and the absence of critical structural distortions [136]. Stabilization of proteins and similar molecules translates into preservation of both structure and functionality during storage and/or targeting, with such stabilization being attained mainly through establishment of a thermodynamic equilibrium with the environment. Protein stabilization is based on elimination of molecular motion(s) and, therefore, on the elimination of conformational transitions while the biomolecule is still in its native (folded) three-dimensional architecture. A bioactive molecule encapsulated in the core of a nanovesicle becomes an independent phase within a food/pharmaceutical formulation, with the advantage of being protected and having its useful active life extended. Encapsulated biomolecules can therefore be considered as being localized in a given defined region of space, limited by an imaginary or physical barrier that allows physical separation between the biomolecule and the surrounding food/pharmaceutical formulation, and between the biomolecule and the immune/digestive system. Immobilization of a biomolecule can then be achieved via engineering the microenvironment or engineering the macroenvironment of the biomolecule [137]. Structural stabilization is related to three-dimensional rigidification of the biomolecule, while functional stabilization involves maintenance of the hydration shell of said biomolecules and, in the case of multimeric proteins, preservation of the quaternary assembly [138]. Hence, to achieve full stabilization of a protein entity, irrespective of its number of subunits, this will always involve two levels of stabilization: structural stabilization and functional stabilization. These will be tackled in the following subsections in detail.

1.4.1.1   Stabilization of Protein Entities via Engineering at the Level of the Microenvironment

Engineering at the level of the biomolecule's microenvironment includes immobilization by physical containment in a barrier or establishment of covalent bonds to a macroscopic support [139]. However, the immobilization of an enzyme does not guarantee per se the stabilization of the enzyme structure [140]. Thermal stability generally results from the molecular rigidification introduced by attachment of the protein molecule onto a solid support with concomitant creation of a protected microenvironment. Crosslinking proteins in solution to form oligomeric microstructures without any macroscopic carrier are a strategy used to increase protein activity per unit area, providing enhanced solvent- and thermal stability. Cross-linked enzyme aggregates allow engineering of the microenvironment. Through the co-aggregation of enzymes and polymers (for example, to reduce solvent interactions or to reduce oxygen dissolution). However, cross-linked enzyme aggregates are mechanically fragile and present extense diffusional limitations in mass transfer, hence they should then be encapsulated within LentiKats™, so as to gain mechanical resistance [141]. Several different enzymes have been stabilized using this approach, as reviewed by Talbert and Goddard, who extensively reviewed all the changes occurring upon contact of enzymes with material surfaces Protein stabilization can also be achieved via covalent modification, thus improving some properties of the protein surface. Such properties are especially important for in vivo stability, by determining, for example,

- protein distribution between tissues,
- ability to penetrate into different cellular compartments.

Several approaches can be used to enhance protein stability in vitro by covalently modifying surface functional groups, viz.

- protein surface modification with bifunctional reagents to crosslink surface functional groups (e.g., with glutaraldehyde, diimidates, or disulfonyl chlorides),
- chemical modification of the protein with nonpolar reagents to enhance hydrophobic interactions,
- protein modification with hydrophilic reagents to promote formation of additional hydrogen or ionic bonds, since the more hydrophilic groups are introduced in the molecule the greater the stabilizing effect.

Other approaches for the chemical modification of proteins include grafting to either polysaccharides. Fuentes and colleagues [142] reported on the successful immobilization of IgG anti-horseradish peroxidase onto magnetic nanoparticles previously coated with aldehyde-aspartic-dextran, aiming at structural stabilization of the antibodies with proper orientation on the support.

1.4.1.1.1 *Stabilization Via Physical Containment*

The thermodynamic stabilization of protein molecules is particularly important when it comes to nanocontainment as a way to convey this type of biomolecules. Within this context, nanoencapsulation procedures within lipid nanovesicles have started to gain momentum and assume now special relevance. By providing a hydrophilic core for the imprisonment of the protein entities, lipid nanovesicles make their rigidification possible since they greatly limit the molecular movements around the trapped bioentities. They limit, in particular, the movements of solvent molecules co-entrapped in their core. Lipid nanovesicles are, therefore, ideal for the immobilization by physical containment of protein entities, promoting their thermodynamic stabilization. In the process of their production, when forming multiple emulsions, the poloxamer used promotes formation of a three-dimensional network favorable to the maintenance of imprisonment of the protein entities.

The advantages of protein nanoconfinement are several:

- localization of the biomolecules in a given, defined, portion of space;
- enhanced thermal and chemical resistance;
- ease of application by the improved resistance transmitted to the biomolecules;
- potential for stabilization at room temperature.

The stability of protein entities following their nanoconfinement can be described both in thermodynamic and kinetic terms. Thermodynamically, protein stability comes from two large but opposing forces, viz. enthalpic and entropic. Both of these forces are temperature dependent. The enthalpic forces stabilizing, and the entropic forces are destabilizing. Thermodynamic stability can be correlated with a change in the thermodynamic conditions of the microenvironment surrounding each bioparticle, since the movements of (aqueous) solvent molecules in their micro-neighborhood become seriously reduced by the effect of being contained within the matrix's core. Since there is a direct relationship between the molecular movements of the biomolecule and the molecular motions of its immediate neighborhood, when we encapsulate a biomolecule we are eliminating the motions of the solvent molecules in its immediate neighborhood. The Reynolds number (Re) is a dimensionless value that gives us a measure of the ratio between inertial and viscous forces, thus quantifying the relative importance of these two types of forces under certain flow conditions. We live in a world where Reynolds numbers are very high due to the presence of high inertial forces and very low viscous forces. However, at the micro-scale, viscous forces are by far the dominant forces, and therefore Reynolds numbers at the microscale are very low. Hence, at the micro-scale, fluid mixing occurs by simple molecular diffusion due to the absence of both molecular turbulence and inertial forces. In the interior of vesicles, Re is null, while in their exterior Re is higher than zero. Since the encapsulated protein entities are to be used in some kind of formulation, with flow properties, Re values can also account for the stabilization produced by the encapsulation procedure due to the absence of turbulent flow in their inner core. Confinement of a protein entity in a nanoporous matrix has a similar effect as that of osmolytes with respect to changing the water activity and thus modifying biochemical reaction rates. Within such nanoenvironments, the motions of water molecules are so much restricted that they do not crystallize even at temperatures close to the absolute zero but instead, transition into a highly viscous state, stabilizing the protein entities by hyper-increasing its rotational, translational and vibrational viscosity [143]. Confining multimeric proteins within liposomes have been proposed as a general strategy to stabilize such protein entities [144]. Additionally, reverse micelles can be seen as a two-phase system, with enzymes encapsulated therein exhibiting very high activities when compared with their native counterparts.

1.4.1.1.2 *Stabilization Via Physical Adsorption onto Macroscopic Supports*

Enzyme immobilization by physical adsorption traditionally refers to binding of the enzyme via weak attractive forces to an inert carrier, which has not been chemically derivatized. Since the carrier is directly involved in binding to the enzyme, its chemical nature, particle size or thickness and pore size distribution play important roles [145]. In general, physical adsorption of an enzyme is achieved by simply contacting the buffered enzyme solution with the carrier, which may require minimal pretreatment such as wetting, washing, and presoaking in the buffered solution and usually requires minimal post-treatment such as washing out the excess solution containing nonadsorbed enzyme. Adsorption of proteins at solid/liquid interfaces is of great technical significance. Particularly in the case of lipases, hyper-activation of these enzymes has been observed upon contact with hydrophobic interfaces that generally mimic the presence of oily substrates, due to the so-called "interfacial activation" phenomenon. The use of physical supports previously coated with very large and flexible ionic polymers may allow the immobilization of multimeric enzymes via multi-subunit interactions, providing that the enzyme molecule can penetrate into such flexible coating and interact with the polymers. Additionally, several researchers have proven that the selective physical adsorption of a multimeric protein, viz.

glutamate dehydrogenase from thermus thermophilus, onto a very lowly activated support followed by covalent immobilization could promote immobilization of the enzyme by the maximum amount of subunits and attain rigidification of its subunits involved in the immobilization. These authors found out that glutamate dehydrogenase immobilized on a support possessing at the same time a low density of amine groups and a high density of epoxy groups, after incubation at pH 8.5 during 3 days did not release a single enzyme subunit. Vaidya and Singhal [146] used epichlorohydrin to convert the free hydroxyl groups in insoluble yeast β-glucan into activated epoxy groups capable of forming irreversible covalent bonds with various groups on the surface of Candida rugosa lipase, thus considerably improving its structural and functional stability with concomitant operational stability in non-aqueous medium. In the same line, Boscolo and colleagues reported on the structural and functional stabilization of a lipase from Pseudomonas fluorescents following its physical adsorption onto cyclodextrin nanosponges, while Serno and colleagues [147] comprehensively reviewed the use of cyclodextrins for the stabilization of proteins both in the liquid and dried states.

1.4.1.1.3 Stabilization Via Chemical Bonding onto Highly Activated Supports

Ideal supports for the immobilization and concomitant rigidification of multimeric protein entities involve large concentrations of available glyoxyl groups in such supports. This makes it possible to establish multipoint covalent attachment between protein entities and the support. If stabilization is performed in solid phase there are intrinsic added advantages in the sense that there is a greater control of the chemical modification because undesirable protein–protein interactions are minimized. Immobilization of proteins improves both thermodynamic and operational stability, since these are mainly determined by the ability to withstand protein unfolding. The most important feature of protein unfolding is the exposure (to the solvent) of its hydrophobic core, with a volume change negative and very large in magnitude. Therefore, an increase in both thermodynamic stability and operational stability should correlate with an increase in rigidity promoted by immobilization, which in turn correlates with the number of bonds established between protein and support. However, the thermodynamic behavior of proteins in the process of "order–disorder" cannot be explained if the hydration of hydrophobic amino acid moieties from the protein molecule is not taken into account. In the chemistry of protein immobilization there are two fundamental steps: multipoint covalent attachment followed by reduction of the immobilized protein derivative. This allows to structurally stabilize the protein through its permanent rigidification, without major distortions in its structure. A higher stabilization may be explained because of an optimal geometrical congruence among protein and support surfaces, together with a high surface density of reactive groups on the support. Two types of active groups especially important for this purpose are the glutaraldehyde groups and the glyoxyl groups. This explains the high enzyme stabilization usually achieved by using this immobilization technique. López-Gallego and colleagues also reported on the immobilization of several enzymes via adsorption onto a cationic support followed by crosslinking with glutaraldehyde, which allowed a remarkable stabilization of such covalently immobilized enzymes. The use of enzymes stabilized via multipoint or multi-subunit immobilization also limits the extent of their inactivation that may arise from chemical modification processes. Cowan and Fernández-Lafuente recently described the immobilization/stabilization of a series of thermophilic enzymes, viz. thermophilic amylases and xylanases, thermophilic sugar isomerases, thermophilic redox enzymes, thermophilic glycosidases, and thermophilic lipases and esterases. Of all stabilization strategies, immobilization and post-immobilization techniques possess additional advantages due to allowing the enzyme to be reused for several cycles, thus increasing enzyme productivity. Similarly, Serra and colleagues reported on the immobilization of homodimeric thymidine phosphorylase from *Escherichia coli* via ionic adsorption onto amine-functionalized Sepabeads™ coated with polyethyleneimine, followed by cross-linking with aldehyde-dextran.

1.4.1.1.4 *Multipoint Covalent Immobilization*

Protein unfolding can be inhibited mechanically by covalently attaching it to a macroscopic support through multiple covalent bonds, thus promoting its rigidification. Agarose highly activated with glyoxyl groups has been described as a suitable support for the multipoint covalent immobilization of both monomeric and multimeric enzymes, allowing their full structural stabilization via attainment of a high degree of rigidity. Multipoint immobilization of multimeric proteins may prevent subunit dissociation by inter-subunit crosslinking while simultaneously reducing conformational inactivation by intra-subunit crosslinking [148]. The highly reactive aldehydes from glyoxyl react with amino groups from the surface of the proteins to form reversible imino bonds. Proteins become immobilized on glyoxyl supports through a simultaneous multipoint bonding, and this occurs via the richest Lys region of the protein surface at alkaline pH. Despite this, proteins displaying several terminal amine groups with a pK in the range of pH 7–8 will become immobilized on glyoxyl agarose even at pH 7 via multiple subunits, as long as those amine groups are located in such a position that their simultaneous interaction with a flat surface becomes possible. Having in mind the extraordinary features of glyoxyl supports so stabilize proteins, immobilization of multimeric proteins on these supports at pH 7 can be coupled to immobilization and stabilization via multi-subunit and multipoint covalent attachment. Hence, following the first attachment at pH 7 via the terminal amine groups of multiple subunits located in the same plane, the immobilized molecules can be subsequently incubated at pH 10 for longer periods of time to allow an intense multipoint covalent immobilization onto the support. After this time period, the protein derivative may be reduced with sodium borohydride so as to re-convert the unreacted aldehyde groups back to inert hydroxyl groups while, at the same time, fully reducing the reversible Schiff's bases to irreversible covalent bonds yielding irreversible secondary amine bonds between the protein or the support, and the polyfunctional aldehyde-dextran polymer thus reducing the aldehyde groups present in the alde-hyde-dextran to a highly hydrophilic and inert polyhydroxyl. However, some decrease in activity may be observed during the reduction process, which might be ascribed to the change of the reversible bonds formed during immobilization to irreversible ones thus fixing the position of the protein residues relative to the support surface. In the copolymerization method of enzyme immobilization, the enzyme is first modified with a compound having a double bond, and then is copolymerized with a monomer and bifunctional crosslinking reagent producing a three-dimensional polymeric network. Immobilization of therapeutical enzymes by copolymerization is extensively used for biomedical uses. However, some of the disadvantages of enzyme immobilization are related to diffusional problems and activity losses due to the somewhat harsh conditions prevailing during the immobilization process. Additionally, López-Gallego and colleagues reported on the multipoint covalent attachment of glutaryl-7-aminocephalosporanic acid acylase onto amino-epoxy Sepabeads™, allowing achieving a high degree of stabilization of the enzyme since in addition to Lysine moieties also tyrosines have reacted with the support. In the same way, López-Gallego and colleagues reported on the structural and functional stabilization of several different multimeric alcohol oxidases via covalent immobilization onto glyoxyl-agarose followed by post-immobilization via inter-subunit cross-linking with aldehyde-dextran to prevent subunit dissociation. Similarly, Bernal and colleagues reported on the multipoint covalent immobilization of beta-galactosidase from bacillus circulans onto hierarchical macro-mesoporous silica previously modified with glyoxyl groups, aiming at thermal stabilization of said enzyme for the production of prebiotic galacto-oligosaccharides. Those researchers found out in their studies that the degree of thermal stabilization attained was affected by the concentration of glyoxyl groups on the surface of the porous silica particles, in a directly proportional fashion probably because the enzyme rigidification attained is insufficient at low concentrations of glyoxyl groups for producing thermal stabilization. However, the excessive rigidification of the protein molecule also leads to destabilization of the tertiary structure of the enzyme. According to Grazu and colleagues, a lowering in the number of the attachments protein-support may hamper an optimal rigidification of the protein tertiary structure, avoiding a high final stability of the immobilized protein.

1.4.1.1.5 Co-Immobilization With Polycationic Moieties

Through coimmobilization with polyfunctional macromolecules, full stabilization of any protein entity will be possible from the point of view of its structure or of its functionality by modifying the microenvironment of such immobilized molecules, which allows a tremendous increase of their stability in organic media [149]. Additionally, co-immobilization of flexible polycationic polymers directly in the support allows a strong and non-distortive immobilization of the biomolecule, with the added advantage of facilitating the removal of the biomolecule and substitution for another one. Fernández-Lafuente and colleagues [150] attached Penicillin G acylase to glyoxyl agarose via multipoint covalent attachment followed by modification with ethylenediamine to increase free amine groups on the surface and intramolecular crosslinking of the amine groups with bifunctional glutaraldehyde, thus attaining increased stability to urea. Furthermore, Grazu and colleagues succeeded in fully stabilizing Penicillin G acylase from *E. coli* by site-directed immobilization onto commercial monofunctional epoxy-acrylic supports, viz. Eupergit C®, a type of support able to form very stable covalent linkages with different reactive groups on protein surfaces. Polymer conjugation may also be used to change the surface of an enzyme. Modifications typically use polymers or small molecules to attach to the surface of an enzyme with the most common being carbohydrates and PEG. Such molecules are believed to provide additional points for hydrogen bonding with the enzyme surface, decreasing dehydration or providing thermodynamic barriers to unfolding. Minimization of any interactions between the protein entity and solvent or gas bubbles can proceed via grafting a polyhydrophilic (macro)polymer over the surface of the protein entity, thus creating a stable nanohydrophilic environment around the protein. López-Gallego and colleagues also employed another stabilizing immobilization protocol for multimeric alcohol oxidase enzymes, viz. the ionic adsorption on agarose coated with a polymeric bed of 600 kDa polyethyleneimine, with the adsorption of such large proteins onto the large polymeric bed allowing full stabilization of the quaternary structure of said multimeric enzymes.

1.4.1.1.6 *Further Inter-Subunit Crosslinking With Polyfunctional Hydrophilic Macromolecules*

Fernández-Lafuente and Cowan and Fernández Lafuente provided an extense and specific review on stabilization of multimeric enzymes, in particular with regard to strategies aiming at preventing subunit dissociation. Chemical modification is used to generate covalent cross-linking bonds between different groups on the enzyme surface; whether such crosslinking occurs between different structural elements of a protein, it will typically amplify the structural rigidity and hence increase protein stability against agents that induce conformational changes. Both physical and chemical crosslinking with poly-ionic macromolecules have been reported as a means to fully stabilize multimeric proteins. In physical crosslinking, the multimeric protein is coated with apoly-ionic macromolecule, such as polyethyleneimine [151], followed by further treatment with glutaraldehyde. Although preventing both subunit dissociation and other phenomena such as oxidation and aggregation, this strategy does not however promote rigidification of the multimeric structure. In inter-subunit chemical crosslinking, aldehyde-dextran has proved to be suitable for fully stabilizing multimeric enzymes from the functional point of view [152]. Such polymeric multifunctional reagents have effectively prevented dissociation of multimeric proteins or multiprotein complexes [152]. When, due to geometrical constraints, it is not possible to immobilize all enzyme subunits onto a planar support, as is the case for example of tetrameric L-asparaginase from *E. coli* or hexameric bovine liver glutamate dehydrogenase, hexameric alpha-galactosidase from Thermussp. T2, glutamate dehydrogenase from T. thermophilus, or multimeric microbial alcohol oxidases from *Candida boidinii, Hansenula sp., or Pichia pastoris*, coupling multisubunit and multipoint covalent attachment to subsequent intersubunit crosslinking between covalently immobilized and non-immobilized subunits using polyfunctional hydrophilic macromolecules such as aldehyde-dextran, allows to prevent dissociation of the whole set of subunits composing the multimeric protein, thus allowing to stabilize the quaternary structure of these enzymes. Such rigidification of the multimeric structure implies a full degree of structural stabilization of the protein. It should, however, be noted that once subunit dissociation is no longer a problem, inactivation will proceed via distortion of the protein three-dimensional structure by altering either the assembly of monomeric subunits or the structure of the individual subunits. Pessela and colleagues have reported on the successful structural and functional stabilization of a multimeric enzyme via covalent immobilization onto cyanogen-bromide agarose followed by chemical modification with aldehyde-dextran aiming at inter-subunit crosslinking. Also, Kotzia and Labrou [153] reported on the successful stabilization of L-asparaginase from *Erwinia chrysanthemi* onto epoxy-activated Sepharose™ CL-6B using 1,4-butanediol diglycidyl ether as a coupling agent. Similarly, Bolivar and colleagues successfully immobilized a multimeric alcohol dehydrogenase from Baker's yeast on glyoxyl-agarose in the presence of acetyl cysteine, followed by inter-subunit crosslinking with polyethyleneimine. Polyethyleneimine is advantageous because due to its polymeric nature, it is likely to interact with areas on the protein surface located in different enzyme subunits, and is therefore desirable in the case of multimeric enzymes. A new and effective methodology to covalently immobilize multimeric and sensitive enzymes at close to neutral pH values has been described, involving the use of thiolated compounds during the immobilization process, providing the opportunity to immobilize highly sensitive enzymes via highly reactive glyoxyl groups but under very mild processing conditions. However, immobilization via glyoxyl chemistry at neutral pH does not stabilize the quaternary structure as it does the tertiary structure. Thus, polyethyleneimine is highly suitable for use in the post-immobilization stabilization of the quaternary structure of multimeric enzymes that are very sensitive to covalent rigidifications, by allowing rigidification of the tertiary structure.

1.4.1.2 Stabilization of Protein Entities Via Engineering at the Level of the Macroenvironment

Engineering on the level of the macroenvironment of the protein entity may proceed via modification of the reaction medium aiming at modifying the surrounding environment. The three-dimensional architecture and molecular motions of a protein entity are determined to a large extent by the properties of its macroenvironment. Intrinsic characteristics of the macroenvironment where the protein entity is embedded, such as pH, pressure, temperature, presence of salts and surfactants, all interfere with the structural and functional stability of the protein entity. Hence, stabilization methods via engineering at the level of the macroenvironment aim at modifying the thermodynamic state in the immediate vicinity of protein, either by cooling, freezing, vitrification, or removing the medium. From the thermodynamic point of view, the contribution of salts to a biocatalytic process is almost entropic and related to the degrees of conformational freedom along the transition states, with no enthalpic role. According to the generally accepted Lumry–Eyring mechanistic model [154], enzyme inactivation is considered a two-step process:

$$N \leftrightarrow K$$
$$U \rightarrow k$$

where, I, where N represents the native form of the enzyme; U represents the reversibly unfolded form of the enzyme; I represent the irreversibly inactivated form of the enzyme; K represents an equilibrium constant of the reversible conformational change; k represents the rate constant of the irreversible inactivation reaction.

Since it is easier to characterize the reversible conformational change of the protein, due to being highly sensitive to changes in pH, ionic strength, chaotropes, among other additions and/or changes in the reaction medium, one can vary the value of K by changing the medium composition, which will directly affect the value of k. In this way, the operational stability of the protein can be tailored.

1.4.1.2.1 Medium Engineering

The structural stability of protein entities is extensively controlled by the interactions between the protein molecules and the surrounding solvent molecules. Various substances were found to stabilize the native structure as a reflection of their effect on the water structure around the protein molecule. Since it is almost impossible for the protein entity to retain its native configuration while its micro- and/or macroenvironment is modified during any process that involves cooling, freezing or desiccation, certain polyol moieties are added into the solution [115]. All these chemicals share a common feature, in that they modify both the structure and motions of water molecules in the immediate vicinity of the protein entities. Irrespective of the exact mechanism of protection offered by such chemical moieties added to the solution, the immediate vicinity of the protein entity is modified, which reflects in the physical, thermodynamic and chemical properties of the surrounding solution. However, most cryoprotectant agents used are extremely cytotoxic at high temperatures, and so their immediate removal should be pursued after thawing.

1.4.1.2.1.1   Sugar and Salt Addition

Hydrophobic interactions, which can be regarded as the reluctance of non-polar moieties to be exposed to water, are considered to be the major driving force for both protein folding and aggregation. Addition of polyols and sugars to aqueous solutions of proteins promotes strengthening of the hydrophobic interactions among non-polar amino acid residues, leading to protein rigidification and enhancing thermostability. According to Miyawaki and Tatsuno, Haque et al. and Kumar et al. [115], the effect of polyols on the activity of water seems to be the main factor that governs the stabilization effect produced by such polyhydroxy moieties. There are two postulated explanations commonly employed to describe the mechanism of action of chemical moieties added to the reaction medium, through which both the molecular motions and structure of protein entities are affected:

- the "water replacement" hypothesis;
- the "preferential exclusion" hypothesis.

The special ability of sugar moieties to bind protectively to the surface of biological molecular structures has been ascribed to their ability also to form hydrogen bonds [156]. Since the three-dimensional architecture of virtually all protein entities in solution depends on stabilization of said architecture by a shell of water molecules hydrogen-bonded to their surfaces, the former hypothesis aims at explaining the protective effect of certain sugars against damage promoted by freezing and desiccation [157], predicting that the stabilization moieties added replace the water molecules that are removed from the hydration shells of the protein entities, thus stabilizing their native state; the later hypothesis states that the protective moieties are excluded from the surfaces of the protein entities and thus the available water molecules in solution can interact preferentially with the protein entity, thus stabilizing its native configuration [158]. Due to the preferential exclusion of the protective sugar moieties from the (immediate domain) hydration shell of protein entities, sugar moieties shape a protective and stabilizing shield around those biomolecules [159]. The basis of this phenomenon is the difference in size between molecules of water and those of the sugar moieties. Essentially, a shell is formed around the protein at the radius of closest approach between the protein and the sugar moiety, a shell that is impenetrable to the sugar moieties but is penetrable to water, resulting in an excess of water in the vicinity of the protein. This is termed preferential hydration. Preferential hydration of proteins is favored due to stronger interactions between sugar and water molecules compared to those between sugar and protein molecules. Protein hydrophobicity increases with a decrease in pH due to the protonation of COO- groups, and so the increased exclusion of sugar moieties from the immediate vicinity of the unfolded protein molecule as compared with its native counterpart results in a larger stabilization effect at low pH values. Xie and Timasheff and Kaushik and Bhat [161] arrived at the same conclusion for ribonuclease-A in the presence of trehalose. Such preferential exclusion increases the chemical potential of the protein molecule proportionally to the solvent exposed surface area and, according to the Le Chatelier's principle, sugar osmolytes favor the more compact state over the structurally expanded state, which leads to an increase of the Gibbs free energy change associated with the denaturation process in the presence of osmolytes. Such increase in the chemical potential of the protein is thermodynamically unfavorable, meaning that protein interaction with the osmolyte is unfavorable relative to water, and so the protein is preferentially hydrated and the osmolyte is preferentially excluded from the protein surface. Typically, this enhancement of stability increases with both increasing sugar concentration and increasing sugar molecular size. According to Arakawa and colleagues, Wong and Tayyab, Timasheff and Arakawa, and Hall and Minton, the generalized exclusion of carbohydrate molecules from the protein surface might be explained by the molecular crowding concept, with steric exclusion, increase of the surface tension of water by the added moieties, repulsion by charged loci on the protein and solvophobicity. The interactions of carbohydrate moieties with protein molecules progressively inhibit, with increasing extent of crowding, any conformational change of a protein that increases its effective volume, such as protein unfolding. The sugar's concentration near the protein surface is lower than in the bulk, and so any process that increases the protein's solvent-exposed surface area is disfavored because exclusion of the sugar from a larger surface area entails a correspondingly larger free energy cost. In particular, in the case of multimeric entities, these become thermodynamically stabilized against stress-induced dissociation. Since protein unfolding increases the total surface area, in the presence of preferentially excluded moieties the free energy of unfolding is increased, and so preferentially excluded moieties can also stabilize quaternary structures because the combined surface area of the dissociated subunits is greater than that for the fully assembled protein entity. The unfolding process of a protein in solution can be described by the Tanford's equation,

$$N + \Delta iW + \Delta jY \rightarrow U$$

where, N is the native form of the protein; W is the water; Y is the co-solute moiety to bind protein upon unfolding; Δi is the change in hydration number per protein molecule; Δj is the change in the bound co-solute molecules per protein molecule; U is the unfolded form of the protein.

Osmolytes stabilize proteins since they do not directly interact with the protein molecules, being excluded from their surface. This is a sine quanoncondition for maintenance/stabilization of biological function. Such qualitative correlation between preferential osmolyte exclusion and protein stabilization is expected from the thermodynamic consequences arising from preferential interactions. Changing the microenvironment of a protein via addition of sugars, polyols, salts, polymers, among others, leads to inhibition of protein aggregation and concomitant stabilization of these macromolecules via preferential interactions. In this way, added polymers might inhibit protein aggregation via several mechanisms, viz. surface activity, preferential exclusion, steric hindrance of protein-protein interactions and increasing viscosity thus limiting protein structural movements. Lavelle and Fresco found, in the same way, that protein stabilization by high concentrations of salts follows the Hofmeister series. Also, certain surfactants may also increase the viscosity of a protein solution, restraining the motions of the protein backbone to inhibit protein aggregation and induce thermal stabilization. However, it should be stressed that these thermodynamic arguments hold true only when osmolyte interactions with the protein are affected by increased surface area upon unfolding. Since the unfolded state of protein molecules has a greater surface area than the native state, it should exclude more co-solute molecules leading to a more negative preferential interaction parameter for the unfolded state than for the native state and a greater increase in the values of ΔG and in turn to stabilization of the protein molecules by the co-solutes. Excipients are generally required to increase long-term stability of protein entities following processing and storage. The thermodynamic interaction of osmolytes with protein molecules therefore stabilizes the equilibrium state of the (more folded) native or native-like protein structure. These molecules stabilize proteins not by interacting with them directly, but by altering the solvent properties of the surrounding water and therefore the protein–solvent interactions. Trehalose is a non-reducing (polyhydroxy) disaccharide composed of two residues of D-glucopyranose, and is able to increase the stability of the folded conformation of proteins. Sugars in general protect proteins against dehydration by hydrogen bonding to the dried protein by serving as water substitute, increase the thermal unfolding temperature and inhibit irreversible aggregation of protein molecules. The interaction of sugar moieties with the protein reduces the contact area of hydrophobic fragments in the protein with water, leading to a decrease in free energy of the system and concomitant stabilization of the protein molecule. Magnesium sulfate acts as a stabilizer, and its stabilization property is determined by a competition phenomenon between

- salt exclusion effect; and
- salt binding effect.

The stabilizing salts seem to increase surface tension at water–protein interface and strengthen hydrophobic interactions by keeping hydrophobic moieties away from water molecules, inducing preferential hydration of proteins. When in presence of a stabilizing osmolyte, a protein entity prefers to interact with water molecules, with the osmolyte being preferentially excluded from the protein domain hence proportionally more water molecules and less osmolyte molecules are found at the protein surface than in the bulk solution. By increasing the surface tension of solvent, salt exclusion of sulfate anion results, whereas the salt binding effect refers to the magnesium cation affinity for ionic residues and peptide bonds.

According to Wimmer and colleagues, the most widespread mechanism behind exclusion of polyols from the protein surface lies in its enhancing effect on the surface tension of water. Arakawa and colleagues reviewed in detail the factors affecting short-term and long-term stabilities of proteins. The long-term stability of firefly luciferase in the presence of additives showed that trehalose has a positive effect on storage stability of this enzyme at refrigeration temperature, whereas magnesium sulfate has a positive effect on its short-term stability (i.e., thermal stability). Similarly, Miyawaki and Miyawaki and Tatsuno studied the contribution of the water activity of several sugar solutions for the stabilization of proteins. In comparison to sucrose and trehalose, maltose, glucose and ribose are reducing sugars, interacting with proteins by forming a Shiff-base between the reducing terminal aldehyde of the sugar and the amino group of proteins at high temperature, explaining at least partially the incomplete precipitation of proteins in the presence of such sugars. Haque and colleagues studied the effect of several polyol osmolytes on the stabilization of *RNAse*-A and egg white lysozyme, having concluded that the secondary and tertiary structures of the native proteins remained unchanged upon addition of the aforementioned polyols. In contrast to magnesium sulfate, trehalose cannot dissociate into ion species, but it can increase the solvent viscosity which in turn enhances the preferential hydration of the enzyme and can affect the structural dynamics and decrease the molecular collisions of the protein molecules. Such high viscosity causes a motional inhibition that hinders processes leading to the loss of structure and denaturation.

Glycerol, on the other hand, lowers the surface tension of water and has been hypothesized to preferentially hydrate protein molecules by enhancing the solvent ordering around the hydrophobic groups of the protein molecules. Such ordering results in a decrease in the entropy of water present in the immediate hydration shell surrounding protein molecules, shifting the equilibrium towards the native state and producing stabilization of the protein molecule. Any increase in the hydrophobic surface area of protein entities upon unfolding would thus be rendered even more unfavorable in the presence of glycerol. Liao and colleagues reported on the stabilization of lysozyme following freeze-drying in the presence of trehalose, sucrose, or dextran. On the contrary, due to a high frequency of protein molecular collisions in the presence of magnesium sulfate, aggregation occurs at higher temperatures, due to the exposure of hydrophobic patches to solvent. Miyawaki and Tatsuno and Kaushik and Bhat have studied the mechanism of trehalose mediated thermal stabilization of a set of globular proteins and concluded that trehalose in addition to imparting thermodynamic stability to proteins also helps in the retention of activity of enzymes during storage at high temperatures. According to Xie and Timasheff, Kumar et al., Anjum et al. and Kaushik and Bhat, trehalose stabilizes proteins by shifting the equilibrium constant (Native⇔Denatured) in favor of the native state, and that it is the smaller preferential binding to the unfolded protein than to the native one which gives rise to the stabilization. O'Connor and colleagues found out that, at the same molar concentration, the disaccharide sucrose stabilizes proteins against thermal denaturation to a greater extent than does the monosaccharide fructose, suggesting that geometric differences between carbohydrates play a predominant role in determining their relative stabilizing ability, because the larger sugar makes a greater contribution to the unfavorable entropy of interface formation between protein and solvent. According to O'Connor and colleagues, the increase in protein stability observed in the presence of sugars is primarily due to the increase in the free energy of creating the protein–solvent interface. Poddar and colleagues also studied the thermal denaturation of ribonuclease-A in the presence of glucose, fructose, galactose, sucrose, raffinose and stachyose, and concluded that an oligosaccharide has more stabilizing effect than the individual monosaccharide constituents, and also that on the same molar scale the order of stabilization is stachyose Nraffinose Nsucrose Nglucose = fructose = galactose. Another postulate, the "water entrapment hypothesis," states that sugars concentrate residual water molecules close to the surface of the biomolecule, thus enabling the biomolecule to preserve its solvation/ hydration layer and native properties. In their studies, Liao and colleagues found out that the combination of either trehalose or dextran with sucrose as lyoprotectants resulted in an equivalent protective capacity to sucrose alone and was better than either trehalose or dextran alone, in the stabilization of lysozyme. Since water plays a crucial role in the establishment of hydrogen bonds between protein molecules and the medium, ions mimicking the water role in accepting or donating hydrogen bonds facilitate conformational changes and functioning of bioactive proteins such as enzymes. Additionally, the effect of kosmotropic ions on the water surrounding hydrophobic residues of the protein has been reported to be associated with an increase in the order of water molecules while, at the same time, enhancing the strength of the hydrophobic interactions providing a more closed protein shell structure with concomitant increase in stability. Eriksson and colleagues studied the effect of both inulin and trehalose in the stabilization of bovine intestine alkaline phosphatase in tablets, and concluded that inulin was by far superior to trehalose as stabilizer of bovine intestine alkaline, with the poor stabilizing capabilities of trehalose after compaction being explained by both crystallization of trehalose induced by the compaction process and moisture in the material. According to Wang, when selecting sugars for protein stabilization reducing sugars should be avoided, since these sugars have the potential to react with the amino groups in proteins via the Maillard pathway. According to Arakawa and colleagues and Eriksson and colleagues, stabilization of proteins via addition of a sugar followed by lyophilization or spray-drying is attained because the protein is incorporated in a matrix consisting of amorphous sugar in its glassy state, which imparts a strong reduction in the mobility of the protein molecule, and this has in fact been referred as the main factor for protein stabilization. Wong and Tayyab reported on the stabilization of bovine serum albumin following addition of a simulated honey sugar cocktail. According to these researchers, honey may be a potential natural stabilizer for proteins due to the presence of a high sugar content in it.

1.4.1.2.1.2   Addition of Glycerol, DMSO, PEG, and Other Synthetic Polymers

Stabilizing agents are inert substances that modify the physicochemical properties of aqueous solutions used to preserve biomolecules. DMSO and glycerol are preferentially excluded from contact with the protein surface, which means that the protein molecule is preferentially hydrated, thus accounting for their cryoprotective effect. The stabilizing effect of PEG addition increases with both increasing PEG concentration and PEG chain length, which is related with the promotion of increased solvent ordering in the medium. The stabilizing effects of polymer addition into the reaction medium may be explained by the fact that the polymer promotes exclusion of the protein molecule from part of the solvent and, in this way, prevents detrimental effects of the environment upon the protein molecule. Shulgin and Ruckenstein reported also that upon addition of PEG into the medium, protein molecules become preferentially hydrated, which could be explained by a steric exclusion mechanism due to the large difference in sizes between water and PEG molecules, with PEG being excluded from the vicinity of the protein molecule. It is thus established the ability of a cosolvent to stabilize a protein molecule, because the preferential hydration of a protein in an aqueous solution containing an organic compound is related to the ability of the latter to stabilize the structure of the protein. These authors also concluded that the excess number of water molecules around a protein molecule increases monotonically with increasing PEG molecular weight, which is in agreement with the aforementioned steric exclusion mechanism.

1.5 GENERAL STABILITY ISSUES

One of the primary differences between the conventional drug entities and peptide/protein drugs is their stability profile. The chemical and physical instability presents peculiar difficulties in the purification, separation, formulation, storage, and delivery of these compounds. Physical instability involves alterations in the secondary, tertiary and quaternary structure of the molecule. These changes are manifested as denaturation, adsorption, aggregation and precipitation. Chemical instability results in the generation of a new chemical entity, by bond formation or cleavage. Changes brought about by physical and chemical changes almost always lead to a loss of biological activity.

1.5.1 PHYSICAL INSTABILITY

• *Denaturation*: Non-proteolytic modification of a unique structure of a native protein that affects definite change in physical, chemical and biological properties. Several examples of denaturating agents are urea, alcohol, acetic acid, sodium dodecyl sulfate, polyethylene glycol.

- *Adsorption*: Ampiphilic nature of protein causes adsorption at various interfaces like air-water and air-solid.
- *Aggregation and Precipitation*: The denatured, unfolded protein may rearrange in such a manner that hydrophobic amino acid residue of various molecules associate together to form the aggregates. If aggregation is on macroscopic scale, precipitation occurs.

1.5.1.1   Denaturation

Peptides and proteins are comprised of both polar amino acids and nonpolar amino acid residues. The hydrophobic, nonpolar amino acid residues fold upon themselves in an aqueous environment to form globular molecules. The hydrophobic, polar amino acid residues of these molecules are exposed to the aqueous environment. On changing the aqueous environment to nonaqueous, they start unfolding and thereby exposing their hydrophobic residues to the hydrophobic environment. This leads to rearrangement and loss of quaternary and tertiary structure. On unfolding hydrophobic and hydrogen bonds are broken. The term denaturation is used to describe any nonproteolytic modification of the unique structure of a native protein that gives rise to define changes in physical, chemical, and biological properties. Conditions that denature protein include:

- Solvent change from an aqueous to a mixed solvent as alcohol and water;
- PH change alters the ionization of the carboxylic acid and amino groups and there by the charges by the molecules. With an increase in the number of like charges on an individual molecule, there is a charge repulsion with in the molecule and as a consequence it unfolds;
- Alteration in ionic strength affects the charge carried by molecules as well. The molecule would be surrounded by ions of the opposite charge to that of the molecule itself, thereby leading to a net reduction in its effective charge;
- Temperature increase leads to an increment in the thermal energy of the molecules, which may suffice to break the hydrogen bonds that stabilize the secondary, tertiary and quaternary structure of these entities. Even chemical bonds may break at very high temperature;

- With an arrangement in structure the inherent biological activity may be lost with an emergence of totally new range of activity. This is usually accounted for the concealing of particular residues and exposure of newer ones respectively.

Denaturation can either be reversible or irreversible. In irreversible denaturation the conformational changes in the molecules are reversed by coming back to the original state. For example, a protein that has been denatured by temperature in crement may revert to its native structure when the temperature is restored. However in the case irreversible denaturation the proteins are unable to restore to their original structure. This may be due to the fact that protein undergoes has undergone some physical or chemical process that inhibits the original pattern of folding, or even the proteins may be misfolded that disallow their proper renaturation.

1.5.1.2 Adsorption

Proteins and peptides pose both polar and nonpolar residues and are thereby amphiphilic in their disposition. This property imparts them the tendency to be absorbed at interfaces such as air-water and air-solid. True to its nature, polar amino acid residues have the inbuilt preferences for the aqueous environment and the nonpolar ones for the hydrophobic environments on the surface of container or the air. So a conformational rearrangement and denaturation can be induced once they are adsorbed at the interface. Once the protein molecules are absorbed, they are absorbed they are capable of forming short range bonds (van der walls, hydrophobic, electrostatic, hydrogen, ion pair bonds with the surface and this may lead to further denaturation of protein acetous moieties. Despite the rapid absorption of peptides and proteins at the interfaces, the rates of conformational changes are relatively much slower. It is unlikely that the surface proteins may disrobe in their original state. Rather they will be rapidly desorb with their hydrophobic residues exposed and in all probability this triggers off the unfavorable interactions with water and formation of aggregates and precipitates.

On adsorption there may be a loss or change in biological activity as the molecular structure is rearranged. For instance as blood proteins are absorbed on surface, the antigenic moiety may become exposed. Also if peptide and protein drug entities are absorbed at interfaces there may be a reduction in the concentration of drug available to elicit its function. Such loss of these drug moieties may occur during purification, formulation, storage and/or delivery.

1.5.1.3   Aggregation and Precipitation

The denatured unfolded protein may rearrange in such a manner that hydrophobic amino acid residues of various molecules associate together leading to the formation of aggregates. If the aggregation is on a macroscopic scale, precipitation occurs. As discussed in the preceding paragraphs interfacial adsorption may be followed by the aggregation and absorption. The extent to which the aggregation and precipitation occurs is defined by the relative hydrophilicity of the surfaces in contact with the polypeptide/protein solution. The acceleration of aggregation and precipitation occurs by hydrophobic surfaces is a class example of this. Insulin form finely divided precipitates on the wall of the container, referred to as the frosting. The presence of large air-water interface accelerates this process. Agitation of polypeptide and protein solutions introduced air bubbles, thereby increasing the hydrophobic air interface. In this process, aggregation and precipitation of polypeptides/proteins is augmented. Another factor contributing to this may be the increase in thermal motion of the molecules due to agitation.

1.5.2   CHEMICAL INSTABILITY

The stability of peptides and proteins against a chemical reagent is decided by temperature, length of exposure, and amino acid composition, sequence and conformation of the peptide/protein. Conformation of the moiety is of particular importance since the reacting groups may be buried and unavailable for reaction with the reagent. The conformational stability may be vital, for example, chemical reaction may elevate the conformational energy of a peptide in an unstrained conformation or may lower the conformational energy of the peptide in a highly strained conformation. Some of the major reaction mechanisms are discussed in brief below:

1.5.2.1 Deamidation

The hydrolysis of the side chain amide linkage of an amino acid residue leading to the formation of a free carboxylic acid. This reaction involves the hydrolysis of the side chain amide linkage of an amino acid residue leading to the formation of a free carboxylic acid. Asparagine and glutamine are particularly susceptible to deamidation. In vivo deamidation could be enzymatic or nonenzymatic. In vitro deamidation is observed with human growth hormone, prolactin and insulin. Factors that favor the rate of deamidation include increased pH, temperature and ionic strength. The tertiary structure of the protein also affects its stability, as observed with trypsin in which the tertiary structure prevents deamidation. Lowering in biological profile after deamidation has been reported with procine adrenal corticotropin hormone.

1.5.2.2 Oxidation and Reduction

Oxidation occurs during isolation, synthesis and storage of proteins. Temperature, pH, trace amount of metal ion and buffers influence these reactions. Glucagon is an exception as it retains biological activity even after oxidation. Oxidation of susceptible amino acids is the major degradation pathway for peptides and proteins that is very common during isolation, synthesis and storage of the same. Oxidation my occur in the side chins of histidine, lysine, methionine, tryptophan, and thronine residue of proteins. The thioether group of methinonin is particularly susceptible for oxidation. Under acidic conditions Met residues can be oxidized by atmospheric oxygen. Oxidizing agents like hydrogen peroxide, dimethylsulfoxide and iodine can oxidize methionine to methionine sulfoxides. The thiol group of cysteine can be oxidized to sulfonic group, oxidation by iodine and hydrogen peroxide is catalyzed by metal ions and may occur spontaneously by atmospheric oxygen. These acid or base catalyzed oxidations may be blocked by oxidation scavengers. Air oxidation can be prevented by avoiding vigorous stirring and exclusion of air by degassing solvents. Usually the oxidation of amino acid residues is followed by significant loss in biological activity as observed after oxidation of methionine residues in calcitonin, corticotrophin and gastrin. Glucagon is an exception as it retains its biological activity even after oxidation. Restoration of the lost biological activity can be achieved by reduction. By reduction of Methionine sulfoxide to Methionine nearly all the lost biological activity is restored to lysozyme and ribonuclease.

1.5.2.3 Proteolysis

It may occur on exposing the proteins to harsh conditions like prolonged exposure to extreme of pH or high temperature or proteolytic enzyme. The hydrolysis of peptide bonds within the polypeptide or protein destroys or at least reduces its activity. The vulnerability of peptide bonds to cleavage is dependent on the residues involved. In comparison to other residues, asparagine residues are more unstable and in particular the Asp-Pro bond. Proteolysis may occur on exposing the proteins to harsh conditions, such as prolonged exposure to extremes of pH or high temperature or proteolytic enzymes. Bacterial contamination is the most common source for introduction of proteases. This can be avoided by storing the protein in the cold under sterile conditions. Proteases may also gain access during the isolation, purification and recovery of recombinant proteins from cell extracts or culture fluid. The particular protease co-purified with the recombinant proteins decides the fate of proteolysis. The problem can be minimized by the manipulation of the solution conditions during the stage of purification and/or addition of protease inhibitors. Some proteins even have autoproteolytic activity. This property aids in controlling the level or function of protein in vivo but is highly unwarranted for a drug moiety if cleavage leads to a loss in biological activity.

1.5.2.4 Disulfide Exchange

A peptide chain with more than one disulfide can enter in to this reaction and thereby change in conformation. Disulfide bonds may break and reform incorrectly and thereby an alteration in the three-dimensional structure is followed by change in activity. This reaction can be catalyzed in neutral as well as alkaline media by thiols, which may arise as a result of hydrolytic cleavage of disulfides. In the presence of HSR,' a disulfide R-S-S-R, interchanges mercaptan groups to give a mixed disulfide, R-S-S-R.'A peptide chain with more than one disulfide can enter into disulfide exchange reactions, leading to scrambling of disulfide bridges and thereby a change in conformation. Another possibility is the addition of peptide change.

1.5.2.5 Racemization

It is alteration of L-amino acids to D, L-mixtures. Racemization form peptide bonds that are sensitive to proteolytic enzymes. With the exception of glycine all the amino acids are chiral at the carbon bearing side chain and are susceptible to base catalyzed racemization. Racemization is the chemical alteration of L and D amino acids. This reaction can be catalyzed in neutral and alkaline media by thiols, which may arise as a result of hydrolytic cleavage of disulfides. The thiolate ions carry out nucleophile attack on a sulfur atom of the disulfide. Addition of the thiol scavengers such as p-mercuribenzoate, N-ethylmaleimide and copper ions, may prevent susceptible sulfur of disulfide.

1.5.2.6 β-Elimination

It proceeds through a carbanion intermediate. Protein residues susceptible to it under alkaline conditions include Cys, Lys, Phe, Sre and Thr.

1.6 FACTORS AFFECTING DELIVERY OF PROTEIN-BASED DRUGS

- Some protein, peptide or antibodies only act extracellularly.
- Rapid elimination of proteins by enzymatic degradation and RES.
- Proteolytic degradation.
- Non covalent complexation with high or low molecular weight compound.
- Low permeability of cell membranes.
- Dissociation of subunit of proteins, for example, coenzyme require cofactor for biochemical transformation.

- Conformation of proteins at different pH and temperature: conformation changes at different pH and temperature, which causes activation and deactivation of proteins.
- Chemical modification, for example, oxidation, acetylation, deamination, sulfation.
- Association of proteins-peptide: such association causes the inactivation of therapeutic protein.
- Antigenic determinants.

1.7   PHYSICOCHEMICAL PROPERTIES OF PEPTIDES AND PROTEINS

1.7.1   SOLUBILITY AND PARTITION COEFFICIENT

Aqueous solubility of peptide is strongly dependent upon pH, presence of metallic ion, ionic strength and temperature. At isoelectric point the aqueous solubility of peptide is minimal where the drug is neutral or has no net charge. Peptides are very hydrophilic with a very low octanol-water partition coefficient, so to improve the absorption of peptides by passive diffusion, their lipophilicity should be increased.

1.7.2   AGGREGATION, SELF ASSOCIATION AND HYDROGEN BONDING

Self-aggregation tendency of peptides modifies their intrinsic properties. Human insulin was found to be more self-aggregating than bovine insulin. Additions of additive like non ionic surfactants stabilize the peptide formulation against self aggregation. Intermolecular hydrogen bonding with water decreases the permeability of protein in lipid membrane.

A protein drug has to face with a number of lipophilic and hydrophilic barriers to cross. The drug must first dissolve in the contents of the intestinal lumen if it is not already in solution; then there is a mucus layer and a water layer protecting the surface of the epithelial cells. The protein or peptide drug must have sufficient water and lipid-solubility to pass through these layers. The epithelial tissue represents the next barrier. There are cases where a protein can be absorbed in to these cells by endocytosis, and then transported to the basement membrane on its way to the capillaries.

1.8 BARRIER TO PEPTIDE AND PROTEIN DELIVERY

1.8.1 ENZYMATIC BARRIER

Enzymatic barrier is the most important barrier that limits absorption of protein/peptide drugs from G.I. tract. The enzymatic degradation is brought about mainly in two ways by:

- Hydrolytic cleavage of peptide bonds by proteases, such as insulin degrading enzymes, enkephalinase, angiotensin covering enzyme and rennin. Proteolysis is an irreversible reaction and is short of absolute specificity thereby potentiating the probabilities of damage of the peptide and protein drug.
- Chemical modification of protein such as phosphorylation by kinases, oxidation by xanthine oxidase or glucose oxidase, carbamylation by pronase, chymotrypsin and trypsin, denaturation and ubiquitination. These chemical changes in the substrate protein are reported to affect the rate and site of hydrolysis catalyzed by proteases, thus the peptides/proteins become more susceptible to proteolytic attack.

The enzymatic barrier predominantly has three essential features:

- The proteolytic enzymes are ubiquitous and hence the protein/peptides are prone to degradation in multiple anatomical sites. Thus they have to be guarded against degradation in all the anatomical sites to ensure their reaching site of action in intact form.
- The anatomical site where the peptide/protein is located is likely to have the presence of all the protease capable of degrading the same. Thus the peptide or protein has to be protected from all the enzymes before it can elicit the pharmacological affect.
- A particular peptide/protein is prone to degradation at more than one linkage within the backbone, each locus being mediated by a specific protease. Thus all the vulnerable linkages call for protection or modification.

1.8.2 INTESTINAL EPITHELIAL BARRIER

There are several mechanisms that are involved in the transport of peptide/protein drugs across the intestinal epithelium. These are discussed below.

1.8.2.1   Passive and Carrier Mediated Transport

The extensive absorption of di- and tri-peptides from the small intestine is well documented. Active transport appears to be the predominant mechanism. Also, there is little evidence that peptides with more than three or four amino acid residues are transported across the intestinal mucosa by the peptide transport system. Stereoisomerism, side chain length and N and C terminal substitution are reported to affect dipeptide absorption. For, for example, L-Ala-L-Phe and L-leu-L-leu were found to be more rapidly absorbed than their D isomers. Likewise, the presence of a larger side chain in either the N-terminal or the C-terminal favors dipeptide absorption. Acid and basic dipeptides have lower affinity than neutral dipeptides for peptide transport. Methylation and acetylation of N-terminal amino groups, esterification of C-terminal carboxyl groups, presence of t-linkage and incorporation of β-amino acids are known to decrease the affinity of dipeptides for the peptide transport system.

1.8.2.2   Endocytosis and Transcytosis

Cellular internalization of peptides/proteins may occur by endocytosis whereby peptides/proteins, which are too large to be absorbed by carrier mediated transport are taken up. The two different pathways of endocytosis are:

1.8.2.2.1   Fluid-Phase Type (Non Specific Endocytosis, Pinocytosis)

In fluid-phase endocytosis (FPE) the macromolecules dissolved in the extracellular fluid are incorporated by bulk transport into the fluid phase of endocytic vesicles (Figure 1.3).

1.8.2.2.2   Adsorptive or Receptor-Mediated Type (Specific Endocytosis)

In this process the macromolecules are bound to the plasma membrane before they are incorporated into endocytic vesicles. When the membrane-binding site is a specific receptor for the macromolecule involved, the

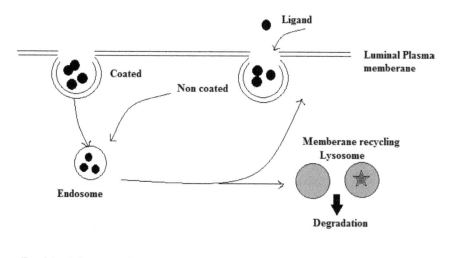

FIGURE 1.3 Cellular uptake by fluid phase endocytosis.

process is referred to as receptor-mediated endocytosis (RME) as shown in Figure 1.4.

The small intestine epithelial mucosa serves as a barrier to the permeation of macromolecules. However, endocytic uptake provides ways to circumvent this barrier. By RME the absorptive cells of the intestinal epithelium are able to select and transport specific molecules while excluding undesirable/harmful ones like bacterial enterotoxins and endotoxins. Some peptides and proteins that are reported to enter intestinal mucosal cells include, nerve growth factor, epidermal growth factor and IgG.

1.8.2.2.3 Transcytosis

The process of transcytosis combines elements of membrane protein sorting and endocytosis. In polarized cell, such as the intestinal epithelial cells, some endosomes carrying the ligands or the receptor-ligand complex bypass the lysosomes and migrate toward the basolateral membrane. Thereby the ligand is released in the extracellular space bound by the basolateral membrane (Figure 1.5). This process is known as transcytosis.

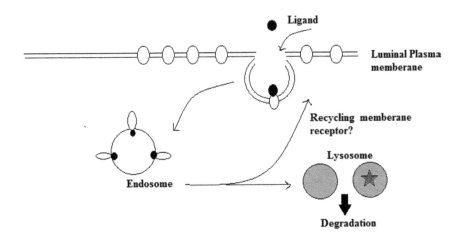

FIGURE 1.4   Cellular uptake by receptor mediated endocytosis.

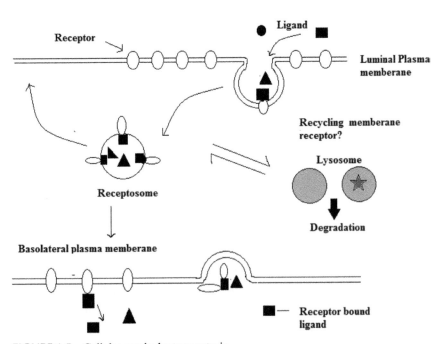

FIGURE 1.5   Cellular uptake by transcytosis.

This pathway presents an important route for mucosal transport of peptides and proteins.

1.8.2.3 Paracellular Movement

This involves the movement between the tight junctions between cells and/or across the spaces formed when cells are extruded into the intestinal lumen. Paracellular movement plays an important role in the absorption of water from the intestinal lumen. Presumably, the passage of water across the tight junction is capable of carrying the dissolved drug(s) or facilitates the transport of macromolecules, which are otherwise not capable to travel across the apical membrane.

1.8.3 CAPILLARY ENDOTHELIAL BARRIER

The Lumina' surface of capillaries consists of the plasma membrane surface of a monolayer of endothelial cells which are joined together by more or less continuous tight or occluding junctions. The structural features of capillary endothelium are displayed in Figure 1.6.

The structural and permeability properties of capillaries vary enormously between tissues. Capillaries can be broadly classified into three major classes:

- sinusoidal or discontinuous;
- fenestrated; and
- continuous.

To cross the capillary endothelium the peptides/proteins must pass between the cells or alternatively traverse the endothelial cells themselves. Solutes that traverse the endothelial cell membrane may be modified or metabolized by cytoplasmic enzymes. Thus, the endothelial passage poses metabolic or enzymatic barrier to solute passage. The failure of circulating dopamine, ammonia and fatty acids to enter brain can be accounted for by considering the cellular metabolism. Endothelial tight junctions also serve as the major extracellular barrier to solute exchange. The transcapillary movement of macromolecular tracers injected into plasma or interstitial fluid is hampered by these junctions.

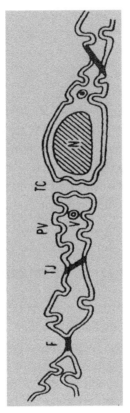

FIGURE 1.6 Structural features of capillary endothelium (F: fenestrated diaphragm; TJ: tight junctions; PV: plasmalemmal vesicles; V: vesicles; TC: transmembrane channel; N: cell nucleus).

1.8.3.1   Mechanisms of Solute Transit

1.8.3.1.1   Passive, Non-Facilitated

Continuous capillaries of the peripheral tissues are permeable to proteins that are <70 kD or even larger. The continuous capillaries contain two sets of rigid water-filled pores: 120A in diameter and 600A in diameter, at densities of 12 and 0.05 per square micrometer of capillary surface respectively. The primary mediators of macromolecular exchange between plasma and interstitial fluid are plasmalemmal vesicles. The sequence of events involves the uptake into vesicles at the luminal plasma membrane of the capillary endothelial cells, followed by migration across the endothelial cells and release of the tracer on the abluminal side. This process is referred to as capillary-transcytosis and appears to function bidirectionally unlike other bulk-phase pinocytosis or receptor-mediated endocytosis, transcytosis does not appear to utilize coated pits or culminate in vesicle fusion with lysosomes. Instead, the vesicles appear to be composed of a pool of membrane that does not mix with the plasma membrane pool.

1.8.3.1.2 Carrier-Mediated

Specific solute transport pathways allow the movement of solutes across capillary endothelium that otherwise would fail to penetrate. In brain capillary endothelium eight independent nutrient transport systems have been identified. They are: (1) a hexose carrier; (2) a mono-carboxylic carrier; (3) a neutral amino acid carrier, which facilitates transport of 14 neutral amino acids; (4) a carrier for lysine, arginine and ornithine; (5) a choline carrier; (6) an adenine/guanine carrier; (7) a porter of purine and uracil nucleosides; and (8) a carrier for aspartic and glutamic acids. Of all the carriers mentioned above, the neutral amino acid carrier holds greatest interest with regard to drug delivery. This carrier holds great similarity to the Na-independent leucine-preferring L system of the peripheral tissues. Its low Km renders it highly sensitive to competition effects.

1.8.3.1.3 Receptor-Mediated

Endothelial receptor mediated transcytosis pathways exist for various polypeptides including insulin, transferrin, P-lipotropin and insulin-like growth factor I.

Basement membrane of the capillaries also imposes a barrier to solute transport. The basement membrane is present everywhere except the sinusoidal gaps of discontinuous endothelium. For instance, in the glomerular capillaries, a thick basement membrane is present and this is reported to limit the capillary permeation by macromolecules like dextrans. The filtration properties are extremely sensitive to charge owing to the anionic disposition of the basement membrane. Basement membranes of the intestinal capillaries are also reported to be limiting barrier. The endothelial barrier is modulated by several physiological parameters. Some of these are outlined below:

- Glucocorticoids tighten the blood-brain barrier by decreasing vesicular transport.
- Angiotensin, bradykinin, histamine and serotonin increase vascular permeability by opening large gaps in the endothelial junctions of post capillary venules.
- Inflammatory agents, vasopressor agents, certain hormones and hypertension relax the endothelial barrier.

1.8.3 BLOOD BRAIN BARRIER (BBB)

The impermeable nature of BBB has been sincerely appreciated and strategies have been evolved to alleviate and obviate the barrier potential to facilitate the targeting of drug to brain compartment. It has been gradually accepted that BBB response particularly in context with permeability as recorded for some vital dyes may not hold the same for bioactive molecules. Recent studies which plunge upon transport kinetics, metabolic, cellular or molecular strategies related to BBB have given major impetus and provided close and better understanding of BBB functioning. It has been demonstrated that various types of metabolic substrates, neuroactive and regulatory peptides and centrally active pharmacotherapeutics could effectively utilize special shuttle services at the BBB. Recently, it has been realized that microvascular endothelium in conjunction with its accessory components, for example, astrocytes, pericytes and microglia, regulates the homeostasis of neural milieu, rather than simply impeding solute exchange. The neurons, glial cells and brain extracellular fluid are separated from blood by BBB, however, the BBB should not be misconceived as an absolute restriction to blood borne molecular moieties, however, it could be realized as a multiple regulatory unit located at sites within the brain.

The morphology of BBB though represented by continuous capillaries of the cerebral micro-circulation, where the endothelial cells are sealed by tight junctions to form a complete composite cellular layer. The functional character of BBB depends upon, geometrical relations amongst the glial and nerve cells and brain extracellular fluid on one hand and cellular fluid as well as humoral extra-brain signals on the other (Figure 1.7). The endothelial layer needs to be viewed as two separate membranes; one is referred as luminal located on the inner sides of the vessels and other, abluminal on the outer side of the vessels. These two layers are separated by thick cytoplasm of nearly 300–500 nm thickness. Furthermore, with regard to surface area, it is about 100 cm² per gram of brain tissue, in capillary endothelium. Thus total surface area in human brain is about 12 m² with the length of capillaries nearly 650 km, the diameter of capillary in human brain 6 um, where the capillaries are separated apart with a distance of 40 um. Numerous functional proteins, that are involved in various transports, receptor signal transduction, cell mediated responses of the brain are expressed by the endothelial cells. It appears that different regulatory molecules, many

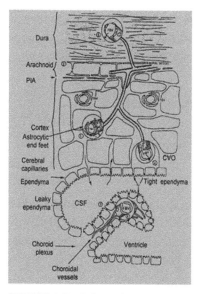
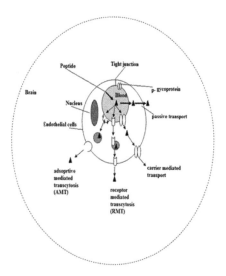

FIGURE 1.7 Schematic presentations of components of blood brain barrier.

of them being peptidal in nature, mainly control the diversified BBB mechanisms. These peptides could be secreted by astrocytes and some neural endings, acting similar to paracrine, being operative at abluminal site of BBB. The circulatory molecules could use a number of different mechanisms for their transport across BBB. These include:

- lipid mediated transport of small but lipophilic molecules;
- carrier mediated transport of hydrophilic nutrients and their drug analogs;
- plasma protein mediated transport of acidic drugs, peptides and highly lipophilic drugs;
- bulk flow transcytosis.

The latter could be pinocytosis or tubulocanalecular transport, independent of molecular size. This transport system however, remains minimal under physiological conditions but may turn out to be significant under certain pathological circumstances.

Peptides which range from two amino acid size to a protein of molecular weight of a Lac or more, require special attention when they are considered for permeability across a given barrier. Multiple biologic actions in brain positively suggest about their utility as neuropharmaceuticals in the treatment of diseases of the central nervous system. The role of Optides in the CNS is complex and largely operates at the level of physiologic phenomenon of the brain, such as:

- regulation of neuroendocrine axis.
- regulation of expression of specific proteins at BBB.
- regulation of cerebral blood flow.
- permeability to nutrient modulation.
- neurotransmission.
- maintenance of integrity of BBB.

Therefore, peptides register important behavioral effects on higher integrative functions of brain as well as on the vegetative functions of the CNS. For years it was thought that BBB remains, as a composite barrier for peptides and proteins not allowing their access to CNS, however, of late through the set of experiments based on brain perfusion models it has been established that intravenously perfused peptides could be detected in CSF in measurable quantities. It was further propounded that these problem molecules utilize various special transport mechanisms, for their transbarrier localization. It was largely attributed to the passive diffusion of molecules where the concentrations obtained in CSF were sub-therapeutic. Barrier factors which have been considered to be responsible for peptide permeability include, lipophilicity, molecular size and charge. Various strategies which have been suggested for improvisation of permeability profile include:

- through cytoporter systems;
- peptide lipidization;
- chimeric protein formation;
- cationization;
- antibody mediated transport.

Chemically modified drug could be permeable in brain capillaries; however it may be converted back into an impermeable salt within the brain tissues. Unfortunately, the lipidization approach appears not to be much useful for peptides with molecular weight greater than 1,000 Da. Another approach suggests inclusion of peptides in detergent based liposomes or poloxamer coated nanoparticles. Specific nutrient transport system could be utilized for brain capillary wall permeation in order to deliver the therapeutic agent selectively to the CNS. Most classical example is the systemic treatment of Parkinson's disease using L-3,4-dihydroxyphenylalanine (DOPA), a metabolic precursor of dopamine. This utilizes neutral amino acid transport system. The same system could be utilized wherein phenylalanine and analogs could be utilized as ligand modules. In some cases the permeability of brain capillaries for proteins could be increased. The change in permeability is suggested to be transient in nature. This could be achieved by intra-arterial injection of hyper osmolar solution, which selectively disrupts inter endothelial tight junctions. This approach has been used with success to deliver antibodies to the brain. Similarly, cationization using hexamethylenediamine or anionization by succinylation could enhance the uptake of proteins in the brain. Another interesting strategy describes chemical conjugations of peptides or proteins to antitransferrin receptor antibody as a ligand whereas, transferrin receptor as a shuttle port for permeation enhancement. The approach classically depends upon receptor-mediated transcytosis of transferrin receptor complexes by brain endothelial cells.

1.9   DELIVERY OF PROTEIN AND PEPTIDE DRUGS

1.9.1   PARENTERAL SYSTEMIC DELIVERY

For the systemic delivery of therapeutic peptides and proteins, parenteral administration is currently believed to be the most efficient route and also the delivery method of choice to achieve therapeutic activity. Parenteral delivery consists of three major routes: intravenous, intramuscular, subcutaneous. Among them intravenous administration is currently the method of choice for systemic delivery of proteins and peptides. Insulin, interferons, and gamma globulins have been reportedly metabolized and/or bound to tissue at injection sites following IM administration, and as a result, the systemic bioavailability of these protein drugs following IM administration is often less than that obtained by IV injection. The bioavailability of tissue plasminogen activator following IM administration was facilitated by the co-administration of hydroxyl amine or by electrical stimulation of muscles, which results in prompt attainment of therapeutic blood levels and achievement of coronary thrombolysis. For SC administration, insulin is best example for the treatment of diabetes. The controlled delivery of peptide or protein based pharmaceutical from subcutaneously implanted polymeric devices was reported by Davis. He used a gel formulation of cross linked polyacrylamide-polyvinylpyrrolidone to achieve the prolonged release of immunoglobulin, luteinizing hormone, bovine serum albumin, insulin, and prostaglandin.

1.9.1.1   Biomedical Applications

- *LHRH and Analogs*: Luteinizing hormone-releasing hormone is a naturally occurring decapeptide hormone with a molecular weight of 1182 Daltons. They are poorly absorbed when taken orally, so it requires parenteral delivery. There is a development of a biodegradable subdermal implant for the SC controlled delivery of goserelin (zoladex), a potent synthetic analog of LHRH for the treatment of prostate carcinoma. Another example is SC delivery of nafarelin, a potent LHRH agonist.

- *Insulin:* It is a protein molecule, with a molecular weight of 6000 Daltons. Three long acting Zn-insulin preparation, Semilente, Lente, and Ultralente insulin, have been formulated for the treatment of diabetes. For delivery of insulin by SC route require sprinkler type of needle and Novolin pen. Various types of infusion pumps like continuous, subcutaneous insulin infusion (CSII), piezoelectrically-controlled micropump (P-CRM) are use for programed delivery of insulin.
- *Vasopressin:* Vasopressin, a nonpeptide with a molecular weight of 1084 Daltons, is an antidiuretic hormone. Using a device prepared by covering a section of microporous polypropylene (Accurel) tubing with collodion, a long lasting and constant in-vitro release of vasopressin was achieved. By using the Alzet pump, subcutaneous delivery of vasopressin was achieved.
- *Systemic delivery of peptides and proteins across absorptive mucosae:* As therapeutic peptides and proteins become readily available through rapid advances in recombinant technology, and because rapid presystemic elimination renders them ineffective when administered orally, pharmaceutical scientists are faced with the challenge of delivering these macromolecules systemically; therefore, alternative routes of delivery need to be investigated. Transmucosal delivery through absorptive mucosae represents one of these alternatives. This route has the advantage of being noninvasive and of bypassing hepatogastrointestinal clearance. The absorptive mucosae that have been investigated for delivery of peptides and proteins include buccal, nasal, pulmonary, rectal, and vaginal. Nasal delivery has been studied extensively and has been the most successful–nasal sprays for buserelin, desmopressin, oxytocin, and calcitonin are already available commercially. In general, enzyme inhibitors and permeation enhancers need to be co-administered for successful delivery of these biopharmaceuticals. Classes of enhancers used for transmucosal delivery include bile salts, dihydrofusidates, cyclodextrins, surfactants, and chelating agents. Each of these agents exerts its enhancing effects by a different mechanism, and each has been associated with adverse effects. This article

discusses the physiology of each of the mucosae used, the fundamentals of transmucosal delivery, and recent progress in systemic delivery of therapeutic peptides and proteins across each of the mucosae; in an effort to highlight principles of transmucosal delivery, it also discusses the transmucosal delivery of enkephalin, calcitonin, and insulin as case studies.

• *Protein and peptide parenteral controlled delivery:* Protein and peptide delivery has been a challenge due to their limited stability during preparation of formulation, storage and in vitro and in vivo release. These biopolymers have traditionally been administered via intramuscular or subcutaneous routes. Recent efforts have been made to develop formulations for non-invasive routes of administration, including oral, intranasal, transdermal and transmucosal delivery. Despite these efforts, invasive delivery remains the main method of administering peptide and protein drugs.

1.9.1.2   Parenteral Drug Delivery Systems

These systems include those intended for intravenous, intramuscular, intraarterial, subcutaneous, intraperitoneal and intrathecal use. The drug carrier systems used for defined and controlled delivery of drug through this route can be:

1.9.1.2.1   Particulates

(a) Microspheres: These are solid spherical particles in the particle size range of few tenths of a micrometer up to several hundreds micrometer. They contain dispersed drug in either solution or microcrystalline form. They are prepared by various polymerization and encapsulation processes. Microspheres have immense potential in controlled release and targeting of drugs. Biodegradable microspheres of 1:1 copolymer of lactic acid and glycolic acid containing ACTH, poly (d, 1-lactide-co-glycolide) microspheres of LHRH, poly (lactide-co-glycolide) microspheres of human serum albumin and poly (d, 1-lactide) microspheres of insulin have been tested successfully with promising results. Similar results were obtained

with growth hormone and for vaccines based on entrapped antigens and immunomodulators in crystallized carbohydrate spheres.

Microspheres can be targeted to a particular organ, a specific part of the organ or to a selective intracellular site. Passive targeting can be achieved by occlusion, cellular uptake or local injection. A classical example is targeting of microspheres to the RES (1–71 nm particles) and to the lung capillaries (7–12 nm particles). The microspheres conjugating receptor specific moieties, such as monoclonal antibodies, (immunomicrospheres) or incorporating magnetic particles, or based on a combination of the two, (magnetic immunomicrospheres) could be used for active peptide(s) or protein targeting.

Advantages:

- Can be prepared cheaply if the correct encapsulation method is optimized and chosen.
- Can be administered subcutaneously, intramuscularly or intraperitoneally and thus implantation is not necessary.

Disadvantages:

- High-molecular weight compounds have limited and restricted loading and their release may be difficult;
- May successfully pass through biological barriers (like blood, endothelium, RES) and cellular barriers before and they can be effective.
- May interact or complex with the blood components.

(b) Microcapsules: This carrier system holds immense potential for controlled release of peptide moieties from mammalian cells and tissues. The microcapsules are polymeric in nature and prepared employing interfacial polymerization, or interfacial coacervate phase separation of capsule wall forming polymers. The capsule membrane serves as a permeability barrier. The polymers conventionally used include polyvinyl alcohol, polyvinyl acetate, nylon, polyurethane, gelatin, polyacrylonitrile, etc. The proteinaceous molecules are effectively entrapped and remain encapsulated within the microcapsules.

(c) Nanoparticles: They are very much similar to microspheres but for their particle size which is in the nanometer range (10–1000 nm). They can also be used for targeted delivery. Owing to their small size they

can pass through the sinusoidal spaces in the bone-marrow and spleen. Targeting moieties monoclonal antibodies can be attached to nanoparticles to enhance their specificity. Polycyanoacrylate and glycolic acid co-lactate based nanoparticles have been discussed as an effective adjuvant demonstrated effective adjuvanticity. The immunogenicity of nanoparticulate based antigen(s) was t many fold higher than conventional plain form. The better adjuvanticity is attributed to the effective of antigens by nanoparticles. The typical constitutive polymers include cyanoacrylate, polymer polystyrene poly co-glyco-lactide, albumin and acrylic resins. The methods employed for pre conventional solvation, desolvation, in situ micellar polymerization, etc.

(d) Aquasomes: Aquasomes are self-assembling nanoconstructs comprising of a solid ceramic core and glassy polyhydroxyl oligomeric surface coating. The system has been studied for the immobilization of bioactive molecules. These include insulin, antigens for Epstein-Barr virus, HIV, Mussel adhesive protein, hemoglobin, etc. Water is a vital requirement for maintaining structural conformation of proteins and their biological activities. Nevertheless, it alone cannot sub-serve as a medium which can resist denaturation of the protein molecules. A variety of environment changes such as pH, temperature, tonicity and solvents can cause protein inactivation when in aqueous state leading to irreversible protein inactivation. This has led to the designing of aquasomes which are referred to as water bodies being with water-like interactive properties. The latter appear to enable them to preserve biological molecules as well as to act as delivery vehicle. Proteins are more stable in the solid state, however, dehydration, the loss of water molecules is critical in maintaining the molecular shape and activity. It follows that in order to maintain structural integrity as well as the activity a well balanced mini environment should be contrived so that even on drying a minimum required aqueous domain in immediate vicinity is maintained. Aquasomes are special systems which while dry and in the solid-state, exhibit water-like properties that enable the molecules to stabilize the 'aqueous conformation.' Since they are based on ceramic materials coated with polyhydroxyl-oligomeric substances with inherent aqueous properties (bound water) can successfully be utilized for immobilization of susceptible protein and peptide molecules into their well-defined interior or through adsorption on to their surface through some keying agent

or simple adsorption phenomenon. Biologically active molecules when adsorbed on to aquasomes and administered intravenously in rabbits, a remarkable in vivo activity was observed. For example, aquasome-delivered insulin showed 149.31–156.99 mg/dL fall in blood glucose concentration. This has been attributed to the steric hindrance and hydrodynamic barrier to opsonic which may deter the contents from RES recognition.

(e) Liposomes: Liposomes are spherical vesicles formed when phospholipids are allowed to hydrate in an aqueous media. They consist of one or more concentric bilayers surrounding aqueous phases. Liposomes are of particular interest since their structure bears great resemblance to that of cell membranes. Proteins and proteinaceous drug(s) are incorporated in liposomes using dehydration rehydration vesicle and reverse phase evaporation methods. Some interesting reports discuss modification of protein in order to lend them an amphiphilic character. The hydrophobic and hydrophilic segment allows them to be intercalated in to the membrane packing. However, the integration could be parallel as shown in Figure 1.8 (c) or antiparallel (b) or perpendicular depending on the membrane potential. The membrane potential dependent orientation of α and β helix of a protein is a voltage responsive phenomenon as does operate in the case of natural protein constituents of biomembranes. Nevertheless, antiparallel and

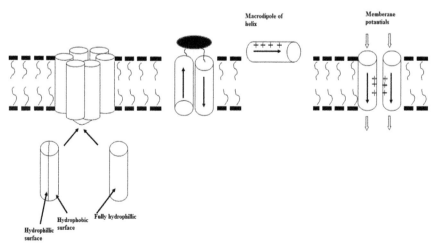

FIGURE 1.8 Topological presentation of protein incorporation into lipid bilayer.

β helix orientation has found to be in a stable state. The protein(s) bearing liposomes can further be surface modified in order to endow them with long circulatory character. The surface modifier(s), for example, PEG, pullulan coating, Gm coating etc., delay them from their selective RES uptake probably through steric hindrance to opsonin adsorption and as a consequence recognition by reticuloendothelial cells (RES). The modified version has long circulatory half-life thus allowing contained bioactive protein(s) to function as circulatory bioreactive units or slow leaching of therapeutically active protein may result in to its protracted activity profile stretched over a defined period of time segment.

Some of the distinct applications of liposomes in parenteral peptide and protein delivery are:

- After subcutaneous, intramuscular and intraperitoneal injection, liposomes can act as a 'depot,' the drug being released slowly with enzymatic degradation and by neutrophils. Intramuscular or subcutaneous administration of liposomes encapsulated insulin and growth hormone immobilized in a collagen matrix resulted in their retention at the site. Similar results were obtained after intramuscular injection of liposomal interferon. Liposomes can be designed to release drug preferentially at a higher temperature attainable by mild local hyperthermia or by application of a magnetic field or by microwave. Various engineered liposomes for selective and programed delivery of bioactives.
- Liposomes protect the entrapped peptides from enzymatic degradation after intravenous administration Insulin encapsulated in multilamellar liposomes was reported to be more stable against proteolytic enzymes cyclosporin entrapped in liposome was observed to be low in nephrotoxicity. Similarly 13-fructofuranosidase, amyloglucosidase and neuraminidase have been successfully delivered in liposomes.
- Liposomes can be exploited for passive targeting of the peptide to a particular site of action. This object can be achieved by overcoming the two limiting factors, for example, localizing the liposomes in macrophage gaining access to target cells across blood vessel walls. Immunopotentiating agents such as muramyl (MDP) can be delivered to macrophages in liposomes. Calcitonin entrapped in liposome has also been administered successfully. Covalent coupling of antibodies to liposomes can incorporate the specificity to specific cell types or subset of cells within the vascular system.

Potency of antibody-directed liposomes is proportional to their relative binding and endocytosis by the cells and to the reactivity of the particular antibody with the cell.

The advantages of liposomes include:

- Flexibility in size, shape and structure.
- Relatively nontoxic disposition.
- Ability to encapsulate both hydrophilic and lipophilic peptides and proteins. For instance, the tertiary and quaternary structures of protein can be damaged irreversibly by dehydration. The presence of aqueous phase makes them an excellent delivery system for this class of pharmaceuticals, for example, those soluble in aqueous media.

Drawbacks with particular reference to peptide and protein drug includes:

- The constituent phospholipids have an inherent tendency to interact with peptides and proteins.
- This can affect the release kinetics of the peptides/proteins and also the shelf life of the liposomal preparation. Their capacity as adjuvants.
- Proteins susceptible to aggregation may affect the liposomal stability by initial fusion owing to hydrophobic interaction followed by aggregation.

(f) Emulsions: Colloid-sized emulsion droplets can be utilized for parenteral delivery of peptides. This delivery system can be of great significance and utility in protecting hydrophilic or hydrophobic drugs from direct contact with body fluids and also in delivering the drug over a prolonged period of time. Emulsion droplets are usually taken up by the cells of the RES. Multiple emulsion can further prolong the release of drug. After delivery of influenza vaccine and diphtheria toxoid in emulsion, prolonged and higher antibody levels were observed. Subcutaneous administration of muramyl dipeptide in w/o emulsion significantly prolonged the effect. Therapeutic levels of cyclosporin were achieved on its intramuscular administration in intralipid emulsion. A typically ordered however biphasic emulsion system having one phase as dispersed component where other serves for dispersion vehicle can be a simple, multiple or fat emulsion. The fat emulsion is based on tri-glyceride(s) of unsaturated fatty

acid(s) being used as an internal phase. The protein and peptide(s) could be embedded in to the matrix of quasi solidified (viscous) fat droplets. The system offers prolongation of half-life to the contained circulatory protein. The stability and fate of parentally administered emulsion droplets is decided by the nature of the emulsifying agent used. Drug release from emulsions is controlled by the droplet size of the dispersed phase and the emulsion viscosity. The advantages of emulsion systems are their clinical acceptability and the amenability to large-scale production. The use of this system is rather limited but investigations are in progress to utilize their potential for delivery of peptides.

(g) Cellular carriers: Enzymes and/or other proteinaceous pharmaceuticals can be encapsulated in erythrocytes to achieve a prolonged release or targeting of the same. Some of the methods of encapsulation include hemolysis, dialysis, and electric field breakdown. The carrier erythrocytes usually display a bimodal survival curve with a rapid loss of the cells in the first 24 h following injection and a much slower loss of cells afterward. Cell damage during in vitro handling is responsible for an early loss. Once damaged erythrocytes are removed by phagocytosis, these carriers can be taken up by spleen or liver. The release of drug occurs by simple diffusion or by a specific transport system following phagocytosis. Some of the enzymes investigated for delivery using erythrocytes include arginase in hyperargininemia L-asparaginase for leukemia and aminolevulinic acid dehydratase in porphyria's. Advantages of erythrocytes include:

- biodegradability;
- nonimmunogenicity;
- large circulation life;
- easy availability;
- large quantities of material can be entrapped in small volume of cells extracellular concentration;
- afford enzymatic and immunological protection.

Drawbacks include:

- long-term storage is problematic;
- permeable to a large number of drugs;
- only those drugs that are not susceptible to irreversible denaturation under hypotonic conditions can be used.

An alternative approach is the utilization of mammalian cells. Ricin toxin has been administered in this delivery vehicle. Immobilized enzymes bound covalently to polymers are reported to have increased stability in circulation and decreased ability to provoke undesirable complications. Another approach that prevents immune rejection of the cells and also aids in controlling the release is to encapsulate the cell within semi-permeable aqueous microcapsule. Microencapsulated pancreatic islet of Langerhans cells are reported to provide a long-term insulin delivery in comparison to pumps. Replication defective viruses: Retroviral gene delivery systems have been developed to assist in entry of genes into the cells. This system consists of an RNA copy of a gene package into a viral particle. This is an excellent alternative to physical methods of gene delivery, such as direct injection into the nucleus. The basic concept of gene therapy is that functionally active genes are delivered into the somatic cells of a patient with genetic defect. This mode of treatment can either be gene replacement or gene augmentation. The former involves specific correction of a mutant gene sequence while in the latter a correctly functioning gene is inserted into non-specific sites in a genome and thereby leaving the malfunctioning gene alone. In a virus, an RNA-protein core is encapsulated in a lipid envelope. The viral glycoproteins bind with specific receptors on target cells, and the viral envelope then fuses with the cell membrane thereby introducing RNA genome into the target cell cytoplasm. Non replicating viral vectors that infect human cells have been developed. An RNA copy of a replacement gene is packed into a nonreplicating viral particle, when the viral gene delivery systems infect target cells, both the viral and exogenous genes are expressed by these cells. However, problems may be posed if sequences are deleted during replication, recombination with endogenous viral sequences occur to produce infectious recombinant viruses, by activation of cellular oncogenes, by-introduction of viral oncogenes and by the inactivation of genes.

1.9.1.2.2 *Soluble Carriers (Macromolecules)*

Soluble carrier systems include conjugates, chemically modified drugs and hybrid proteins. The drug can be conjugated with a polymer/macromolecule. This not only decreases immunogenicity, but Ebb improves protease stability and helps in achieving selective or targeted drug delivery. Immunological properties of bovine serum albumin can be reduced by conjugation with polyethylene glycol. Likewise, asparaginase was chemically modified by a DL-alanine-N-carboxyanhydride polymerization technique. The modified enzyme not only retained most of its catalytic activity but also had better protease stability and demonstrated a 7–10 fold prolongation in plasma clearance properties. It is hypothesized that polymers serve as a steric barrier, which hampers the interaction of macromolecules with the native enzyme, while macromolecular substances can still interact with the enzyme. Polymer modifications bring with it the following changes:

- masking of antigenic determinants;
- masking of protease-susceptible sites;
- masking of immunogenic recognition signals;
- masking of clearance recognition signals;
- allow free access to low-molecular weight substrates; and
- alter optimum pH by changing microenvironment.

Antibody, protease, clearance or immunogenic recognition sites: Model explaining the enzymatic and biological properties of poly-DL-asparaginases protein molecules can themselves act as carriers for targeting of other peptides/proteins. The four basic mechanisms involved in targeting are schematically presented in Figure 1.9.

- Interaction with surface receptor.
- Facilitated uptake and interaction of intact molecules with intracellular target site.
- Facilitated uptake and intracellular release of warhead moiety.
- Facilitated uptake and release of a suicide warhead, which is activated only after interacting with its target.

Naturally occurring multi-subunit proteins of toxins also act as conjugates. The subunits of these moieties have different roles. For instance, if one subunit mediates binding, the other is the pharmacologically active entity. A classic example of this type of carriers is the B chain of ricin toxin that has been used with interferon and insulin. The specific binding and the antiviral activity of the conjugate was inhibited in the presence of galactose, an inhibitor of ricin B chain binding. The ricin A chain can also

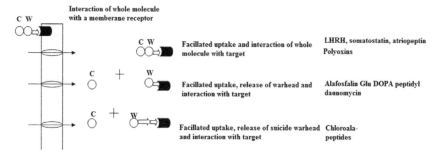

FIGURE 1.9   Schematic representation of mechanisms for targeting peptide drugs.

be conjugated to binding subunit of other proteins or monoclonal antibody to tumor antigen for cytotoxic effect in tumor cells. Synthetic polypeptide carriers have been used as ligands such as muramyl dipeptide. With these carriers the conjugation is multiple and results in to an increased receptor affinity, increased potency, enhanced resistance to enzymatic degradation and prolonged duration of action.

One of the major drawbacks of this technique is the limited ability of these conjugates to cross capillary endothelium and their susceptibility to removal by the RES.

1.9.1.2.3   Others

Some of the other sophisticated systems designed and engineered for parenteral controlled and targeted delivery of proteins and peptides include:

(a) On-demand systems
In certain cases it is beneficial to have externally augmented delivery on-demand as in delivery of insulin to patients with diabetes mellitus. Magnetically modulated systems have been designed to achieve this end. The release rate is influenced by the position, orientation and strength of the embedded magnets, the amplitude and frequency of the applied magnetic field and the mechanical properties of polymer matrix. Another approach for external modulation of release is the application of ultrasound.

(b) Self-regulated systems

The self-regulated systems are of great importance to deliver insulin in response to blood glucose concentration' for diabetic patients. One of the polymer-based systems utilizes a cationic hydrogel polymer with immobilized glucose oxidase. As glucose diffuses into this glucose-sensitive polymer, the glucose oxidase catalyzes its conversion to gluconic acid. Thereby, the pH of the microenvironment within the membrane is lowered and the amine groups in the membrane are protonated. Due to this, the membrane swells and its permeability to the insulin held in a continuous reservoir increases. However, this approach holds promise only for small peptide molecules. Another approach with a slight modification is based on the increase in solubility of a modified insulin derivative with a decrease in pH in the membrane and corresponding increase in release. A biochemical approach based on competitive binding between glucose and glycosylated insulin to concanavalin A has also been reported. The functioning of system is classically based on affinity binding principle where high affinity ligand occupies the site of binding while low affinity ligand as a consequence released in to the environment where it acts as a therapeutic entity (insulin). The strategy-judiciously operates through affinity exclusion as higher affinity ligand that is sequestered and removed from circulation otherwise is a toxic molecule, for example, (glucose). With an increase in glucose level, the influx of the glucose to the pouch increases and thereby G-insulin is displaced from the concanavalin A substrate). The increase in displaced G-insulin in the pouch results in efflux of G-insulin from the system body.

(c) Temperature sensitive systems

Some polymers like polyacrylamide derivatives have inherent sensitive swelling behavior. This leads to a temperature-dependent release pattern that can be explored as pulsatile delivery of peptides/proteins. It was observed that the temperature sensitivity of insulin pen through a poly-N-isopropylacrylamide-butyl methacrylate copolymer membrane varies with the hydro component of the copolymer. With increasing hydrophobic co-monomer component in a copolymer, however gradual drop in temperature sensitivity was recorded. The sensitivity implies polymer gate opening character to allow efflux of entrapped protein. The thermosensitive perm demonstrated reversibility without noticeable lag times. Similar temperature-dependent release pattern due swelling was observed for myoglobin in N-isopropylacrylamide gel and bovine serum albumin in ethylene acetate matrix. Schematic diagram of a self-regulating insulin delivery system illustrated in Figure 1.10.

(d) Pumps

A pump is different from other diffusion-based system in that the primary driving force for delivery by a pump is pressure difference and not the concentration difference of the drug between the formulation and the surroundings. The pressure difference can be generated by pressurizing a drug reservoir, osmotic action or by direct mechanical actuation. The pump can either be implantable or externally portable.

(i) Mechanical pumps: Most of the portable pumps for insulin delivery are syringe driven, either lead screw or direct drive. Another widely used principle is roller peristaltic. This utilizes a disposable bag and silicone outlet tube. The tube is stretched around the roller mechanism, and as the roller turns, the contents of the tube are expelled from the cannula. However, this type of drive requires careful filling to avoid air inclusion. Infusaid® implantable pumps sensitive to ambient temperature and pressure have been developed. Portable infusion devices for open-loop insulin delivery and for pulsatile therapy of LHRH have also been designed and developed. For practical reason,

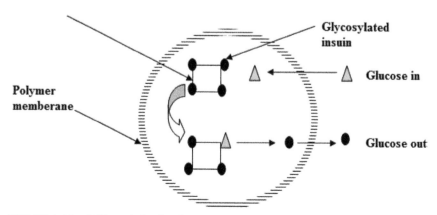

FIGURE 1.10 Self-regulating insulin delivery system.

the use of infusion pump is restricted to subcutaneous insulin delivery as the routine use of the intravenous and intraperitoneal routes is too hazardous. Infusion pumps have also been used for growth hormone, vasopressin, calcitonin, glucagon and somatostatin. The advantages of infusion pumps include:

- More flexibility and freedom for the patient.
- The potential for achieving physiological levels of the drug.

The disadvantages of pump therapy may or may not be pump-specific. They are:

- Possibility for mechanical or electrical failure.
- Inconvenient and difficult to use.
- Costly.
- Chances of dermatological complications.
- Aggregation of peptide/protein drug may be problematic since many of the pumps rely on a linear relationship between volumetric flow rate and actual drug delivery rate.

(ii) Osmotic Pumps: Osmotic pumps have been used extensively for delivery of a large number of peptides and protein drugs in animals. Some of the representative examples include insulin, ACTH, calcitonin, LHRH, growth hormone, neurotensin, vasopressin. Osmotic pumps are simple in principle, small and reliable. The critical features of this pump are stability of the drug at 37°C within the pump for the entire infusion period and the compatibility of the drug with the internal pump components. The osmoregulatory mini pump is structurally consisted of a central drug and osmogen core coated with a semipermeable polymeric rigid lamella. The osmogen facilitates the penetration of biofluid(s) into the pump however the dimensional variation in pump is not allowed thus forcing out drug solution under osmotic gradient through appropriates openings or orifices. There is less flexibility in the delivery program and reservoir volume is low, and hence requires for frequent replacement.

(iii) Controlled-release micropumps: This is a novel addition to the implantable pumps (Figure 1.11). The concentration difference between the drug reservoir and the delivery site causes diffusion of the drug to the delivery site to provide basal delivery. Thus, no external power

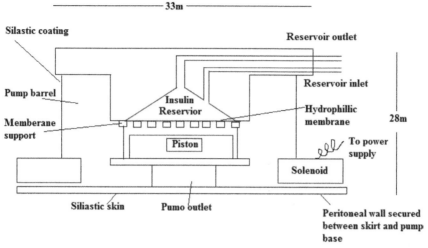

FIGURE 1.11   Schematic illustration of prototype VIII of controlled-release micropump.

source is required. A piezoelectric controlled micropump has also been developed. This comprises of piezoelectric disk bender and a cellulose acetate micro-porous membrane located at the opposite sides of the insulin reservoir. A considerable augmentation in delivery was obtained on applying a square-wave voltage to the piezoelectric bender.

The exceptional advantages of controlled-release micropumps include:

- Less susceptible to limiting problems such as catheter blockage and mechanical failure due to absence of an outlet catheter, valves and motors.
- In case of failure no chances of accidental overdosage due to presence of the membrane and lack of a valve system.

Some of the limitations are:

- Long term stability of the membrane to repeated compression is to be considered.
- The location of implant is limited by the need to minimize diffusion resistance and the biocompatibility of the pump exterior plays an important role in this.

Some approaches based on electrochemical principles have been proposed for implantable pumps. One suck technique is electroosmosis through a cation-exchange membrane. The membrane is placed between two Ag/AgCl electrodes and an aqueous NaCl solution. On application of electric potential, hydrated Na- ions pen through the membrane accompanied by the movement of volume of fluid: A to-and-fro transport of fluid, the membrane was effected due to the reversing of the direction of the current.

Some of the identified advantages of this pump are:

- Low power requirement.
- High reliability.
- Very low flow rates (in pliday).

Drawbacks include:

- Possibility of the consumption of electrodes.
- Necessary valve and reservoir arrangements have not yet been devised.

1.9.1.3 Pharmaceutical Approaches Related to Systemic Delivery of Protein and Peptide Drugs

In new era of medication, recently under development of biotechnology and genetic engineering, the industry is capable of producing number of therapeutic peptides and proteins commercially. The research efforts to formulate the immediate and sustained release formulation of protein and peptide based pharmaceuticals which is given to body directly or using polymeric carriers. Disease related to CVS, CNS, GIT, hormones and cancer can be effectively treated by this new class of therapeutic agents. We have primary aimed to improve bioavailability, half-life, penetrability and stability. Generally, protein and peptide drugs are administered by parenteral route. Many peptide based pharmaceuticals can't accomplish their full range of therapeutic benefits when delivered by parenteral route because they are limited by extremely short duration of their biological function. Alternate nonparenteral routes like oral, nasal, pulmonary, ocular, buccal, rectal, vaginal and transdermal have been tried with varying degree of success. Various problems associated with administration of protein and peptide drugs are needed to overcome by different pharmaceutical approaches. Several approaches available for maximizing pharmacokinetic and pharmacodynamic properties are chemical modification, formulation vehicles, mucoadhesive polymeric system, use of enzyme inhibitors, absorption enhancers, penetration enhancers, etc. This review summarizes recent approaches in protein and peptide drug delivery system

Proteins and peptides are the most abundant components of biological cells [1]. They exist functioning moieties such as enzymes, hormones, structural element and immunoglobulin. Also, take part in metabolic process, immunogenic defense mechanisms and other biological activities [2]. Each peptide or protein molecule is a polymer chain with α-amino acid linked together by peptide bonds in a sequential manner, formed by interaction between the α-carboxyl and α-amino groups of the adjacent amino acids, resulting polymers are generally called peptides [3]. A polypeptide containing 50–2500 or more units of amino acids in a peptide chain is called protein. For contribution of three-dimensional structure of the protein, regularly repeating backbone with distinctive side chains of polypeptide interact with each other [4]. Basically, they are macromolecules with the high molecular weight. Molecules referred to as polypeptides generally have molecular weights below 10,000 Dalton and those called proteins have higher molecular weight [1]. Diffusion of drugs through the epithelial layer greatly affect by high molecular weight and size of protein [5, 6]. Difficulties arise in formulation and delivery of proteins due to its unique physical and chemical properties [7]. Proteins are very prone to physicochemical degradation and easily get denatured by heat or by agitation, so kept at refrigerated temperatures along with stabilizing agents for long-term storage [8]. The discovery of numerous hormones and peptides those have found applications as a biopharmaceuticals, role of regulatory proteins and peptides in pathophysiology of human disease, increasing importance of proteins and peptides [9]. Peptide based pharmaceuticals are well accepted in medical practice and research activities because proteins serve as significant role in the integration of life processes and act with high specificity and potency [10]. Due to tremendous growth in molecular biology and genetic engineering, the large-scale production of polypeptides is possible by using recombinant DNA and hybridoma techniques [11]. More than 200 proteins and peptides have received US Food and Drug Administration approval for treating a variety of human diseases [12]. The total global market for protein drugs was about $47.4 billion in 2006. The market will reach $55.7 billion by the end of 2011, an average annual growth rate of 3.3% [13]. The 2006 "Biotechnology Medicines in Development" identifies 418 new biotechnology medicines for more than 100 diseases, including cancer, infectious diseases, autoimmune diseases, AIDS and related conditions, which are in human clinical trials or under review by the Food and Drug Administration [14]. Protein and peptide drugs most commonly administered by parenteral route. Other routes such as oral, intranasal, transdermal, buccal, intraocular, rectal, vaginal, pulmonary route, etc. have been tried with varying degree of success [15–19]. Some examples of peptide based pharmaceuticals and their potential function/biomedical applications are mentioned below:

- β-endorphin: Relieves pain;
- Vasopressin: Treats diabetes insipidus;
- Pancreatic enzyme: Digestive supplement;
- Interferons: Enhance activity of killer cell;
- Insulin: Treats diabetes mellitus;
- Human growth hormone: Treats dwarfism;
- Gastrin antagonist: Reduce secretion of gastric acid;
- Cholecystokinin (CCK-8 or CCK-32): Suppress appetite;
- Bursin: Selective B cell differentiating hormone;
- Bradykinin: Vasodilation;
- Angiotensin II antagonist: Lowers blood pressure.

1.9.2 NON-PARENTERAL SYSTEMIC DELIVERY

1.9.2.1 Oral Route [160]

Till recent, injections remained the most common means for administering therapeutic proteins and peptides because of their poor oral bioavailability. However, oral route would be preferred to any other route because of its high levels of patient acceptance and long term compliance, which increases the therapeutic value of the drug. Designing and formulating a polypeptide drug delivery through the gastro intestinal tract has been a persistent challenge because of their unfavorable physicochemical properties, which includes enzymatic degradation, poor membrane permeability and large molecular size. The main challenge is to improve the oral bioavailability from less than 1% to at least 30–50%. Consequently, efforts have intensified over the past few decades, where every oral dosage form used for the conventional small molecule drugs has been used to explore oral protein and peptide delivery. Various strategies currently under investigation include chemical modification, formulation vehicles and use of enzyme inhibitors, absorption enhancers and mucoadhesive polymers. This review summarizes different pharmaceutical approaches which overcome various physiological barriers that help to improve oral bioavailability that ultimately achieve formulation goals for oral delivery.

Due to rapid progress in biotechnology, as well as gene technology, the industry is capable of producing a large number of potential therapeutic peptides and proteins in commercial quantities. Endogenous proteins and peptides play an important role in the regulation and integration of life processes and act with high specificity and potency. For example, in the form of enzymes, hormones, antibodies and globulins, they catalyze, regulate and protect the body chemistry, while in the form of hemoglobin, myoglobin and various lipoproteins, they affect the transport of oxygen and other chemical substances within the body. In the form of skin, hair, cartilage and muscles, proteins hold together, protect and provide structure to the body of a multicellular organism.

The increasing importance of proteins and peptides can be attributed to three main developments. First, improved analytical methods have promoted the discovery of numerous hormones and peptides that have found applications as biopharmaceuticals. Second, molecular biology and genetic engineering have enabled the large-scale production of polypeptides previously available only in small quantities. Lastly, there is a better understanding of the role of regulatory proteins/peptides in the pathophysiology of human diseases. Simultaneously, pharmaceutical companies around the world have endeavored to develop the processes for producing therapeutically active entities at commercial scales.

Till recently, injections (i.e., intravenous, intramuscular or subcutaneous route) remain the most common means for administering these protein and peptide drugs. Patient compliance with drug administration regimens by any of these parenteral routes is generally poor and severely restricts the therapeutic value of the drug, particularly for disease such as diabetes.

Among the alternate routes that have been tried with varying degrees of success are the oral, buccal, intranasal, pulmonary, transdermal, ocular and rectal. Among these, oral route remains the most convenient way of delivering drugs. Oral administration presents a series of attractive advantages towards other drug delivery. These advantages are particularly relevant for the treatment of pediatric patients and include the avoidance of pain and discomfort associated with injections and the elimination of possible infections caused by inappropriate use or reuse of needles. Moreover, oral formulations are less expensive to produce, because they do not need to be manufactured under sterile conditions. In addition, a growing body of data suggests that for certain polypeptides such as insulin; the oral delivery route is more physiological.

Designing oral peptide and protein delivery systems has been a persistent challenge to pharmaceutical scientists because of their several unfavorable physicochemical properties including large molecular size, susceptibility to enzymatic degradation, short plasma half-life, ion permeability, immunogenicity, and the tendency to undergo aggregation, adsorption, and denaturation. Consequently, the absolute oral bioavailability levels of most peptides and proteins are less than 1%. The challenge here is to improve the oral bioavailability from less than 1% to at least 30–50%.

Designing and formulating a protein and peptide drug for delivery though G.I. tract requires a multitude of strategies. The dosage form must initially stabilize the drug making it easy to take orally [4]. It must then protect the drug from the extreme acidity and action of pepsin in the stomach. In the intestine, the drug should be protected from the plethora of enzymes that are present in the intestinal lumen. In addition, the formulation must facilitate both aqueous solubility at near-neutral pH and lipid layer penetration in order for the protein to cross the intestinal membrane and then basal membrane for entry into the bloodstream.

The purpose of this article is to review the general approaches that have been studied for improving oral protein and peptide bioavailability by overcoming various physiological barriers associated with therapeutic proteins and peptides.

Oral route is the most popular, convenient and acceptable route of delivery from the patients point of view. However, successful oral therapy for peptide and protein drugs has largely eluded solution. Their oral administration is severely prohibited by strong acidic environment, enzymatic and cellular barriers of the intestinal tract. The macromolecular peptide/protein drug(s) also have a low permeability across the gastro-intestinal mucosa. Figure 1.12 illustrates the various barriers to oral absorption. This is the reason why the oral bioavailability of peptide and protein drug is often less than 1%. Molecules that are absorbed via the lumen of the small intestine can either be taken up into a blood capillary or alternatively enter lymphatic lacteal located within the submucosal space. Molecules entering the blood are directed through the hepatic-portal system to the liver and are normally substantially metabolized before they reach the systemic circulation. However, in some cases there is an underlying physiological advantage in directing a peptide or protein to the liver prior to their entry in the, systemic circulation. A classic example of such entity is insulin. On referring to its physiology we learn that insulin released from the pancreas travels directly to the liver and first acts on the glucose reserves in respective organ before it acts peripherally. The ease of administration and higher degree of patient compliance with oral dosage forms are the major reasons for preferring to deliver proteins and peptides by mouth.

In conclusion, delivering proteins and peptides by the oral route is extremely challenging. The very nature of digestive system is designed to breakdown these polypeptides into amino acids prior to absorption. The low bioavailability of drugs remains to be an active area of research. Several sites in the GIT have been investigated by researchers, but no major breakthrough

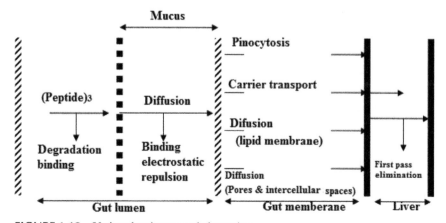

FIGURE 1.12   Various barriers to oral absorption.

with broad applicability to diverse proteins and peptides has been achieved. Considerable progress has been made over past few years in developing innovative technologies for promoting absorption across G.I. and numbers of these approaches are demonstrating potential in clinical studies. Chemical modification and use of mucoadhesive polymeric system for site-specific drug delivery seen to be promising candidates for protein and peptide drug delivery.

1.9.2.1.1 Strategies for Oral Delivery

Formulating for delivery through the gastrointestinal tract requires a multitude of strategies. One strategy for overcoming the body's natural processes is to alter the environment for maximum solubility and enzyme stability of the protein by using formulation excipients such as buffers, surfactants and protease inhibitors. If the enzyme attack can be defeated or delayed, the proteins can be presented for absorption. Proteins and peptides could be derivatized with polyethylene glycol to achieve properties such as retention of activity, prevention of immunogenicity and prevention of excessive enzymatic degradation. The use of oligomers with both hydrophilic and lipophilic properties, confer the enzymatic stability necessary for proteins to survive the digestive processes in the gut. Another strategy for oral delivery is to promote absorption through the intestinal epithelium. Absorption may be enhanced when the products formulated with acceptable safe excipients. A typical transport mechanism for proteins across the epithelial boundary is paracellular transport. There are tight junctions between each of the cells in the epithelium that prevent water and aqueous soluble compounds from moving past these cells. A number of absorption enhancers are available that will cause these tight junctions to open transiently, allowing water-soluble proteins to pass. Fatty acids, surface-active agents, EDTA, glycerides and bile salts have all been shown to be effective in opening these tight junctions. Most of the data showing the successful oral delivery of model proteins, such as insulin and calcitonin, has been generated in animal studies.

The various pharmaceutical approaches and their outcomes are:

Chemical modification
- Amino acid modification: Improves enzymatic stability.
- Hydrophobization: Improve membrane penetration.

Use of enzyme inhibitors
- Resist degradation by enzymes present in stomach and intestine.

Use of absorption enhancers
- Increases membrane permeability.

Formulation vehicles
- *Emulsions*: Protects drug from acid and luminal proteases in the GIT. Enhance permeation through intestinal mucosa.
- *Liposomes*: Improves physical stability. Increases membrane permeability.
- *Microspheres*: Prevents proteolytic degradation in stomach and upper portion of small intestine. Restricts release of drug to favorable area of GIT.
- *Mucoadhesive polymeric system*: Achieve site-specific drug delivery. Improves membrane permeation.
- *Nanoparticles*: Prevent enzymatic degradation. Increases intestinal epithelial absorption.

1.9.2.1.2 Nobex Technology

The Nobex technology involves the bonding of polyethylene glycol and alkyl groups or fatty acid radicals to produce desired amphiphilic oligomers. These oligomers are conjugated to proteins or peptides to obtain desired amphiphilic products that can traverse the aqueous and lipid layers of the mucosa, and can resist excessive degradation of protein or peptide drugs. In this, an amphiphilic protein conjugate is prepared. This technology reduces self-association, increases penetration and increases compatibility with formulation ingredients than parent drug.

1.9.2.1.3 Oral Delivery of Insulin

This conjugated protein is known as hexyl insulin monoconjugate [2], and is the first successfully delivered oral dosage form of insulin to show good oral bioefficacy in humans. The oral form of insulin goes directly to the liver and is believed to stimulate normal biochemical pathways, including glucose control. Human clinical results with conjugated insulin are a clear demonstration that a protein can be developed into a therapeutically viable product. With the conjugated form of insulin and an appropriate formulation, a product can be developed which has the attributes required for oral delivery. Oral delivery of insulin with sodium deoxycholate on ascending colon of rat causes 50% reduction in blood glucose level. Stefanov et al. investigated the feasibility of delivering insulin systemically by oral route using a liposomes prepared from phosphatidylcholine and cholesterol as delivery system and reported no change in blood glucose level is noted in normal animals, but significant reduction is obtained with diabetic rats. Oral delivery of arginine, lysine vasopressin, synthetic analog, 1-deamino-8-D-arginine vasopressin was studied on rats and rapid anti diuretic response achieved. Coating of peptides with polymers with azoaromatic groups and the cross linking of azopolymers to form an impervious film to protect orally administered proteins or peptides from degradation and metabolism in the stomach and small intestine.

1.9.2.1.4 *Potential Problem Associated With Oral Protein Delivery*

The oral administration of peptide and protein drugs faces two formidable problems. The first is protection against the metabolic barrier in GIT. The whole GIT and liver tend to metabolize proteins and peptides into smaller fragments of 2–10 amino acids with the help of a variety of proteolytic enzyme (proteases), which are of four major types; aspartic protease (pepsin, rennin), cysteinyl proteases (papain, endopeptidase), metallo proteases and serinyl proteases. The second problem is the absence of a carrier system for absorption of peptides with more than three amino acids.

1.9.2.1.5 *Prodrug Approach*

Proteins are labile due to susceptibility of the peptide backbone to proteolytic cleavage, as well as their molecular size and complex secondary, tertiary and sometimes even quaternary structures. Therefore proteins can be modified chemically to give more stable prodrugs with increased plasma half-lives. Some strategies for prodrug formation include olefinic substitution, d-amino acid substitution, dehydro amino acid substitution, carboxyl reduction, retro inversion modification, polyethylene glycol (PEG) attachment to amino group and thio-methylene modification.

1.9.2.1.6 Modifications by Chemical Synthesis of Prodrugs and Analogs

This strategy of chemically altering the peptide/protein structure is aimed to modify physicochemical pro of drugs, such as lipophilicity, charge, molecular size, solubility, configuration, isoelectric point, stability, enzyme lability and affinity to carriers. This modification aids in manipulating the pharmacokinetic parameters and as a result to improve the therapeutic value of the parent drug by facilitating permeation and providing stability against degradation and thereby altering bioavailability. Modification succinylation, acylation, and deamination conjugation with polymers such as albumin, de polyvinylpyrrolidone, DL-poly(aminoacid) and poly(ethylene glycol) and lipoamino acids and their home-oligomers has been tried out to increase lipophilicity, the blood circulating life and/or to reduce immunogenicity. Some of the enzymes are specifically located at the luminal and subluminal sites of the epithelial cells of the G.I. tract. The strategy interestingly discusses the chemical modification process, where protein is modified in its prodrug. The later serves as substrate with defined propensity for a membrane localized enzyme. The pro-drug is converted into the parent drug via enzymatic reaction and, generated parent drug gets absorbed. The physicochemical characteristics help preventing its possible precipitation at the absorption site. The prodrug synthesis approach has been utilized with success to improvise the absorption of water insoluble peptide renin inhibitor where mucosal membrane peptidases were used as conversion sites. The approach offered better solubility and high permeability. The approach offers an increase in polarity of drug and simultaneously the membrane permeability. As a result of a systematic study conducted with renin inhibitors, it was inferred that substantially improvised delivery of peptidomimetics is possible through structural modification approach. The modification is so carefully conducted that pharmacological effects of the drug remains to be unchanged. It was further proposed that carrier-mediated transport via peptide cytotransporters is one of the important uptake mechanisms that operates in the case of brush border membrane. Taken collectively it could be concluded that oral systemic delivery of peptide and peptidomimetics is an achievable aim. It also suggests that nutrient and other natural native cytoportal systems can judiciously be exploited for effective transportation and as a result systemic absorption of protein and peptide drugs. Lipid conjugates by employing phospholipids have been proposed. Phospholipids provide unique interface between lipophilic acyl region and hydrophilic biomolecules. Lipid conjugates have contributed significantly towards promising prodrugs analogs with improved stability, bioavailability and barrier permeation characteristics. Peptides, proteins and carbohydrates are appropriate drug target for conjugation. A variety of approaches including derivatization with phospholipids, fatty acids, cholesterol and long chain alcohols are used to generate pharmaceutical lipid conjugates.

Conventional methods, for example, amidation, cross-linking and active anchoring using carbodiimide are utilized for chemical derivatization. The developed prodrugs are amphiphilic thus via supramolecular aggregation state may form lymphotropic system. The latter has appreciable potential for oral administration of protein and peptide drugs, via lymphatic route. Thyrotropin-releasing hormone (TRH) is employed clinically to exert control over pituitary functions. However, after oral administration, low levels are observed in the brain. TRH is reported to be resistant to proteolytic degradation in G.I. tract. The nominal oral activity seems to be due to poor absorption and rapid clearance in the blood stream. To circumvent these problems, analogs have been synthesized. A dimethyl analog of TRH, p-glu-his-RX 77368 was synthesized and on clinical examination reported to be long-lasting and more potent. This property was attributed to its stability to protease action. Another analog MK-771 was synthesized having a sulfur atom in the pyrrolidone ring at position 2 of the proline residue. This analog was more potent than the parent and also possessed much slower clearance rate but was equipotent in causing the release of TSH. The oral bioavailability of MK-771 was found to be only 2% in humans and 3% in monkeys.

Another approach is that of synthesis of tripeptide TRH and acetylation on the N-terminus. These conjugates were found to have improved oral uptake and stable over a considerable period of time than the native TRH. On the similar lines LHRH has been conjugated to lipoamino acids and lipopeptides, and are reported to promising results in vivo. Enkephalins, the endogenous pentapeptides, occur naturally in two types, leucine (YGGFL) and methionine (YGGFM) enkephalins. Their oral delivery is precluded due to their susceptibility to hydrolysis by intestinal peptidases. Pentapeptides structurally related to YGGFM have been synthesized by oxidizing YAGFM-ol to sulfoxide[YAGFM(0)-ol] and N-methylating YAGFM(0)-ol [YAGF (Me)M(0)-ol]. These analogs produced significant analgesic activity after oral administration. Several analogs of vasopressin, the antidiuretic hormone, have been synthesized. These include deamino (4-valine, 8-D-arginine)-vasopressin (DVDAVP), deamino (4-threonine, 8-D-arginine)-vasopressin (DTDAVP). 1-deamino-(8-D-arginine)-vaso-pressin (desmopressin, DDAVP) and (1-deamino penicillamine, 2–0-methyl-tyrosine)-arginine vasopressin. DDAVP differs structurally from vasopressin in two positions. At position I. instead of hemicystine, 13-mercaptopropionic acid (deaminohemicystine) is present and at position 8 instead of L-arginine, D-arginine is present. For the oral delivery of the proteins and peptides an interesting approach has been reported based on protein polymer complex. In this approach mainly a hydrophobic polymer, for example, co-polystyrene maleic anhydride\maleic acid is conjugated with protein or peptide utilizing their terminal amino active groups. So formed complex is dissolved in triglyceride (medium chain length) containing 5% polyglycerine trioleate. This facilitates solubilization and incorporation of the protein-polymer complex into the fat phase of the emulsion and so formed emulsion could be used for oral administration.

1.9.2.1.6.1   Chemical Modification

A chemical modification of peptide and protein drugs improves their enzymatic stability and/or membrane penetration of peptides and proteins. It can also be used for minimizing immunogenicity. Protein modification can be done either by direct modification of exposed side-chain amino acid groups of proteins or through the carbohydrate part of glycoproteins and glycoenzymes.

Modifications of individual amino acids combined with the substitution of one more L-amino acid with D-amino acids can significantly alter physiological properties. This was demonstrated by vasopressin analogs 1-deamino-8-D-arginine vasopressin (DDAVP) and [Val4, D-Arg8], arginine-vasopressin (dVDAVP), hereafter called desmopressin and deamino-vasopressin, respectively. While the former involves deamination of the first amino acid and replacement of the last L-arginine with D-arginine, the latter also has the fourth amino acid changed to valine. While the natural vasopressin is orally active in the water-loaded rat at large doses, desmopressin is twice as active at the 75th fraction of the dose, which is attributed to enhanced membrane permeation and enzymatic stability. Desmopressin absorption was shown to be passive and by the paracellular route across the rat jejunum and site dependent in rabbits. Whether the chemical modification alters the transport pathway, however, remains to be unknown.

Increasing the hydrophobicity of a peptide or protein by surface modification using lipophilic moieties may be of particular benefit to transcellular passive or active absorption by membrane penetration or attachment, respectively; or it may simply aid in the increased stability of the protein.

Nobex corporation has developed a proprietary insulin compound modified with small polymers, in which a single amphiphilic oligomer is covalently linked to the free amino group on the Lys-β29 residue of recombinant human insulin via an amide bond, that is intended, on delivery by mouth, to resist degradation by enzymes of the stomach and intestine and to be efficiently absorbed into the bloodstream. It is believed that once delivered by mouth to the intestine and into the bloodstream, Nobex oral insulin can follow the same pathway as insulin released by the pancreas, into a blood vessel called the portal vein and then directly to the liver. Since the liver is a significant participant in the control of blood glucose, it is believed that successfully activating the liver with oral insulin may provide a mechanism to potentially reestablish normal glucose control in the diabetic patient and turn on a number of metabolic activities that can help mitigate complications of diabetes.

Another example of hydrophobization to increase lipophilicity of insulin is palmitoylation. Insulin was conjugated to 1,3-dilpalmitoylglycerol at the free amino groups of glycine, phenylalanine, and lysine to form mono and dipalmitoyl insulin. This facilitated the transfer of insulin across the mucosal membranes of the large intestine and improved its stability against intestinal enzymatic degradation. To decrease binding to albumin, Brader et al. recently synthesized octanoyl-N-Lys β-29, co-crystallized with human insulin, and determined pharmacokinetic and insulin release profiles after subcutaneous injection in beagle dogs. However, these derivatives were not very effective after oral administration.

1.9.2.1.7 Site-Specific Delivery

There are reports that absorption of certain peptides and proteins are limited to a certain specific region within the G.I. tract, for example, an optimal site exists for their absorption. For example, dipeptide like angiotensin converting enzyme (ACE) inhibitor is absorbed only from the upper small intestine in humans. The presence of proteolytic enzymes also plays a vital role in hampering the favorable absorption of peptides/proteins from G.I. tract. This is the reason for better absorption of peptides like. Met-Met that are better absorbed from jejunum in comparison to otherwise well absorptive ileum. A number of techniques have been developed for delineating the absorption characteristics of drugs from different regions of the G.I. tract. Some of these are discussed below:

1.9.2.1.7.1 High Frequency Capsule

The release of contents from this at a predetermined location is achieved by means of an external radio signal. This device thus not only allows for targeting to the possible absorption windows but also offers protection from the enzymatic environments of the G.I. tract. Theophylline and isosorbide-5-mononitrate absorption have been studied following their administration as high frequency capsules.

1.9.2.1.7.2 Ileum Delivery

The absorption capacity of the ileum is lower than the duodenum but is reported to contain highly specialized mechanisms for absorption of specific macromolecules. For instance, a specialized transport mechanism exists for absorption of bile acids from ileum. For active transport acidic side chain at the I7-position of the ring system of the bile acids is indispensable. C_3 derivatives are also reported to undergo active transport. Conjugation of p-toluene-sulfonic acid with cholic acid improved the absorption of the former from the ileum. Ileum also provides a carrier system for cyanocobalamin (Vitamin B_2) that binds with high affinity to three distinct transport proteins viz. transcobalamin II, intrinsic factor (IF) and the nonintrinsic factor cobalamine-binding proteins (R-proteins). Cobalamine analogs have been studied for their binding affinity. It is highly selective and transcobalamin II is moderately selective. On the other hand R proteins are reported to have high affinity for wide range of cobamides and cobinamid. Cobamides are vitamin B_2 produced by bacteria and differ from cobalamine by the absence of the cyano and the 5,6-dimethylbenzimidazol groups. Studies on similar lines have led to the conclusion that vitamin B,2 analogs are absorbed from the ileum via a mechanism that is independent of the intrinsic factor. Thus, coupling of macromolecules to cobalamine can give an opportunity for transport into the enterocyte via the receptor-mediated transport process. Promising results have been obtained with LHRH-vitamin B,2 conjugate. Peyer's patches are aggregates of lymphoid nodules. Solitary lymphoid nodules are present along the entire intestine but are more numerous and concentrated in the ileum, where they are recognized in aggregate form as Peyer's patches. The migration pathway of Peyer's patch cells have been illustrated in Figure 1.13.

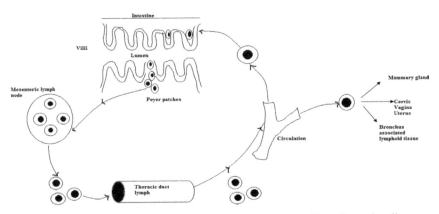

FIGURE 1.13   Schematic presentation of migration pathway of Peyer's patch cells.

Studies indicate that particles up to 5 pm can be transported within macrophages through Peyer's patches into mesenteric lymph nodes. Accumulation in Peyer's patches is governed not only by the particle size but the surface properties also play a vital role. For instance, better penetration is observed with particles having hydrophobic surfaces. These Peyer's patches (Figure 1.13) can be exploited as a port for oral immunization. B cells in the Peyer's patches can be induced to develop into IgA plasma cells by subjecting them to antigen challenge and exposure to bacterial lipopolysaccharides. Synthetic fragments of cholera toxin have been administered orally to induce the production of antibodies recognizing the intact toxin. Successful delivery of antigen by liposomes is reported following M-cell uptake. Another approach utilized coated liposomes composed of soya phosphatidylcholine with the reo-virus cell protein, for which receptors exist on the M cells.

1.9.2.1.7.3 Colon Delivery

Oral colon-specific drug delivery of protein and peptide drugs
With the advent of new technologies and radical growth in the field of biotechnology, dozens of protein and peptide drugs have been marketed. However, there are several challenges for successful delivery of such molecules. A number of routes have been used for the delivery of these fragile molecules by exploring various novel delivery technologies, including microspheres, liposomes, gel spheres, nano-spheres, niosomes, microemulsions, use of permeation enhancers, use of protease inhibitors, etc. But the route that has attracted the attention of worldwide drug delivery scientists is the oral route due to its various advantages. Even though the proteolytic activity is higher in a few segments of the gastrointestinal tract (GIT), this route has certain segments that have lower proteolytic activity, for example, the colon. The colon has captured attention as a site for the delivery of these molecules because of its greater responsiveness to absorption enhancers, protease inhibitors, and novel bioadhesive and biodegradable polymers. Although the success rate of these approaches, when used alone is pretty low, when used in combinations, these agents have demonstrated wonders in increasing the drug bioavailability. This review focuses on the challenges, pharmaceutical concepts, and approaches

involved in the delivery of these fragile molecules, specifically to the colon. This review also includes studies conducted on colonic targeting of such drugs. Further studies may lead to improvements in therapy using protein/peptide drugs and refinements in the technology of colon-specific drug delivery.

In comparison to small intestine, the colon has minimal enzyme concentration, especially peptidases and thus may be exploited for systemic peptide absorption. However, this site has several inherent limitations that have been discussed below:

- Colon has a flat epithelium with a small surface/volume ratio available for absorption, and therefore the permeability of polar compounds is relatively less when a comparison is drawn with the small intestinal epithelium.
- Longer residence time is supposed to counteract the above mentioned drawback but unpredictable colonic transit jeopardize a predictable absorption.
- There is a great variation in the colonic pH and this is all the more prominent in the presence of protein rich food, pH can be as high as 8.
- In the colon, free diffusion of drug across membranes is hampered due to the presence of surrounding solid fecal matter.
- Presence of higher concentration of bacteria, in particular the anaerobic ones, potentiates the drug degradation.

However, these limitations can be exploited to our advantage by utilizing the microbial enzymes as an aid to targeted delivery to colon. 5-aminosalicylic acid (5-ASA) has been linked to high molecular weight polymers via aromatic azo groups. This conjugate is capable of being absorbed only at the colon where the azo bond is reduced and thereby 5-ASA released. The same conjugation concept can be exploited for colonic delivery of protein or peptide drugs and their absorption thereafter. Another prodrug comprising of two 5-ASA molecules joined by an azo bond has been designed. An approach that has been used for delivery of insulin involves an azo cross-linked copolymer of styrene and hydroxyethylmethacrylate that coats the insulin. Thus, the drug remains protected from the vagaries of G.I. tract. The polymer disintegrates only at the ileocecal junction when the azoreductase reduces the $R-C6E14-N=N-C6H4-R$ bond. Thus, insulin is released at a site where it is least prone to degradation. The same concept has been used for the delivery of vasopressin. A system with Eudragit-S, an acrylic-based resin, coating has been designed. This resin is soluble only at weakly alkaline pH and thereby release of drug is achieved in colon by erosion of the coat.

1.9.2.1.8 Use of Enzyme Inhibitors

Enzymatic barriers are the deciding factors for the transport of unstable small peptides that could be effectively transported across the intestinal membrane but for their degradation by various proteases. Majority of the degradation events in the gut revolve around the resident proteases and peptidases. The catabolism of protein and peptide therapeutics can be limited by the use of protease and peptidase inhibitors. Specific catabolic protease and peptidase activities can be identified and a corresponding specific inhibitor can be employed to stabilize the sensitive protein/peptide.

The choice of protease inhibitors will depend on the structure of these therapeutic drugs, and the information on the specificity of proteases is essential to guarantee the stability of the drugs in the G.I. tract. The quantity of co-administered inhibitor(s) is essential for the intestinal permeability of a peptide or protein drug.

For example, enzyme degradation of insulin is known to be mediated by the serine proteases trypsin, α-chymotrypsin and thiol metalloproteinase insulin degrading enzymes. The stability of insulin has been evaluated in the presence of excipients that inhibit these enzymes. Representative inhibitors of trypsin and α-chymotrypsin include pancreatic inhibitor and soybean trypsin inhibitor, FK-448, Camostat mesylate and aprotinin. Inhibitors of insulin degrading enzymes include 1,10-phenanthroline, p-chloromeribenzoate and bacitracin reported the use of a combination of an enhancer, sodium cholate and a protease inhibitor to achieve a 10% increase in rat intestinal insulin absorption.

Thiomers are promising candidates within as enzyme inhibitors. Hutton ET AL. first reported the inhibitory properties of poly (acrylates) on intestinal proteases. They found a strong reduction of albumin degradation by a mixture of proteases in the presence of carbopol 934P. A subsequent study by Lueben et al. showed that polycarbophil and carbopol 934P were potent inhibitors of the proteolytic enzymes trypsin, α-chymotrypsin and carboxypeptidase A. As a result of the covalent attachment of cysteine to polycarbophil, the inhibitory effect of the polymer towards carboxypeptidase A, carboxypeptidase B and chymotrypsin could be significantly improved. This polycarbophil-cysteine conjugate also had a significantly greater inhibitory activity than unmodified polycarbophil on the activity of isolated aminopeptidase N and aminopeptidase N present on intact intestinal mucosa.

Another approach to enzyme inhibition is to manipulate the pH to inactivate local digestive enzymes. A sufficient amount of a pH-lowering buffer that lowers local intestinal pH to values below 4.5 can deactivate trypsin, chymotrypsin and elastase.

1.9.2.1.9   Use of Absorption Enhancers

As discussed earlier, the major pathways available for transport across membrane are the transcellular and the paracellular routes. The transport across this pathway is dependent on the lipophilicity of the membrane and molecule, the molecular and particular size and charge. Mechanisms involved in the improvement of intestinal membrane permeation through transcellular route. Absorption enhancers are presented in Figure 1.14 and enumerated below:

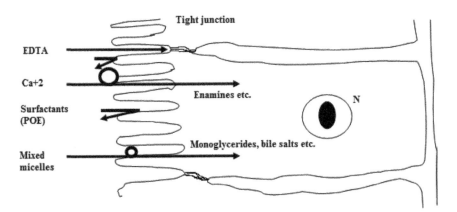

FIGURE 1.14   Schematic presentation of absorption enhancement strategies.

a) Interaction of absorption enhancers with membrane lipid protein leads to membrane perturbation follow an increase in permeability, for example, mixed micelles, salicylic acid, acyl carnitine, middle chain fatty enhance membrane permeability via membrane perturbation.

b) Disorder of membrane status by decrease in membrane nonprotein thiol, for example, diethyl maleate, salicylic The mechanisms involved in the improvement of intestinal membrane permeation by absorption e through paracellular route include:

- Chelation between enhancer and Ca^{++}/Mg^{++} around tight junctions which force water through by enhancing paracellular absorption of water-soluble drugs, for example, EDTA/bile acids, middle chain fatty.
- Activation of junctional actomyosin contraction, for example, glucose, amino acids, decanoyl carnitine, capric acid, mixed micelles.

In order for therapeutic agents to exert their pharmacological effects, they have to cross from the biological membranes into the systemic circulation and reach the site of action. Absorption enhancers are the formulation components that temporarily disrupt the intestinal barrier to improve the permeation of these drugs. Ideally, the action of absorption enhancers should be immediate and should coincide with the presence of the drug at the absorption site.

Numerous classes of compounds with diverse chemical properties, including detergents, surfactants, bile salts, Ca^{2+} chelating agents, fatty acids, medium chain glycerides, acyl carnitine, alkanoyl cholines, N-acetylated α-amino acids, N-acetylated non-α-amino acids, chitosans, mucoadhesive polymers, and phospholipids have been reported to enhance the intestinal absorption of large polypeptide drugs.

Many of these absorption enhancers act as detergents/surfactants to increase the transcellular transport of drugs by disrupting the structure of the lipid bilayer rendering the cell membrane more permeable and/or by increasing the solubility of insoluble drugs. The chelators are believe to exert their action by complex formation with calcium ions, thus rupturing the tight junctions (TJs) and facilitate paracellular transport of hydrophilic drugs. However, permeation enhancers often induce toxic side effects, for,

for example-Ca^{2+} depletion induces global changes in the cells, including disruption of actin filaments, disruption of adherent junctions, and diminished cell adhesion. Reports about some enhancers, including fatty acid sodium caprate and long chain acyl carnitines, have been shown to improve absorption without obvious harmful effects to the intestinal mucosa. But based on various studies, it would appear that a transient opening of TJs would seem less damaging than disruption of cell membrane structure. Several studies on sodium dodecyl sulfate, sodium caprate, and long-chain acylcarnitines shows increased permeability through the paracellular pathways. Tomita et al. and Lindmark et al. proposed that the mechanism of paracellular transport enhancement by sodium caprate was via phospholipase C activation and upregulation of intracellular Ca^{2+}, leading to contraction of calmodulin dependent actin-myosin filaments and opening of TJs. Dodecylphosphocholine and quillaja saponin, dipotassium glycyrrhizinate, 18 β-glycyrrhetinic acid, sodium caprate, and taurine also increases the permeability of hydrophilic compounds across Caco-2 cells.

Among the recent absorption enhancers displaying this principle and exhibiting the safest and most effective promising results in enhancing drug delivery is *Zonula Occludens* toxin or Zot. Zot is a single polypeptide chain of 44.8 kDa, 399 amino acids in length, with a predicted pI of 8.5, of bacteriophage origin, present in toxigenic stains of *V. cholerae* with the ability to reversibly alter intestinal epithelial TJs, allowing the passage of macromolecules through mucosal barriers. Zot possess multiple domains that allow a dual function as a morphogenetic phage protein and as an enterotoxin. After cleavage at amino acid residue, a carboxyl terminal fragment of 12 kDa is excreted, that is probably responsible for the biological effect of the toxin. The mechanism of action of ZOT has been constructed as protein kinase C-dependent actin reorganization through interaction with a specific receptor, whose surface expression on various cells may differ because the action of ZOT is not uniform throughout the G.I. tract34.

In vitro experiments in the rabbit ileum demonstrated that Zot reversibly increased intestinal absorption of insulin (MW 5733 Da) by 72% and immunoglobulin G (140–160 kDa) by 52% in a time dependent manner They further observed an encouraging 10-fold increase in insulin absorption in both rabbit jejunum and ileum in vivo with ZOT34. Karyekar et al.

has recently reported that Zot increases the permeability of molecular weight markers (sucrose, inulin) and chemotherapeutic agents (paclitaxel and doxorubicin) across the bovine brain microvessel endothelial cells in a reversible and concentration dependent manner and without affecting the transcellular pathway as indicated by the unaltered transport of propranolol in the presence of Zot. Extensive in vivo and in vitro studies have identified Zot receptors in the small intestine, the nasal epithelium, the heart and the brain endothelium. Moreover, toxicity studies have shown that Zot and its biologically active fragment ΔG do not compromise cell viability or cause membrane toxicity as compared to other absorption enhancers.

Another recently developed option for the use of absorption enhancers is to co-administer protein and peptide drugs with concentrated solutions of so-called "carrier" molecules. Emisphere Technologies has created a series of "transport carriers," designed to form a complex with the polypeptide, thereby altering the structure of the polypeptide to a 'transportable' conformation. These molecules promote protein and peptide drug absorption. The mechanism of action of these agents is still not clear, and efforts are being made to explore the same. Leone-Bay suggested that enhanced drug permeation across the G.I. tract is neither due to alteration in membrane structure (i.e., mucosal damage) nor a result of direct inhibition of degradation. Based on the structure-activity relationships, these authors concluded that more lipophilic compounds (i.e., high log P values) had better ability to promote protein (rhGH, sCT) absorption. They suggested that these delivery agents cause temporary stabilization of partially unfolded conformations of proteins, exposing their hydrophobic side chains. The altered lipid solubility permits them to gain access to pores of integral membrane transporter, and thus they are more absorbable through lipid bilayers [40]. Wu and Robinson used Caco-2 cell monolayers to show that interaction of rhGH with 4-(4-(2-hydroxybenzoyl) aminophenyl) butyric acid (IX) and N-(8-(2-hydroxybenzoyl) aminocaprylate (XI) makes the protein a better substrate for P-glycoprotein, thereby suggesting that the interaction causes the protein to be more lipophilic.

Kotze et al. have evaluated the transport enhancing effects of two chitosan salts, chitosan hydrochloride and chitosan glutamate (1.5% w/v), and the partially quaternized chitosan derivative, N-trimethyl chitosan chloride (TMC) (1.5 and 2.5% w/v), in vitro in Caco-2 cell monolayers.

The transport of the peptide drugs buserelin, 9-desglycinamide, 8-arginine vasopressin (DGAVP) and insulin was followed for 4 h at pH values between 4.40 and 6.20. They observed that all the chitosans (1.5%) were able to increase the transport of the peptide drugs significantly in the following order: chitosan hydrochloride > chitosan glutamate > TMC. Because of quaternary structure of TMC, it is better soluble than the chitosan salts and further increases peptide transport at higher concentrations (2.5%) of this polymer. The increases in peptide drug transport are in agreement with a lowering of the transepithelial electrical resistance (TEER) measured in the cell monolayers. No deleterious effect to the cell monolayers could be detected with the trypan blue exclusion technique. It is concluded from this study that chitosans are potent absorption enhancers, and that the charge, charge density and the structural features of chitosan salts and N-trimethyl chitosan chloride are important factors determining their potential use as absorption enhancers for peptide drugs.

1.9.2.1.10 Formulation Vehicles

A primary objective of oral delivery systems is to protect protein and peptide drugs from acid and luminal proteases in the GIT. To overcome these barriers, several formulation strategies are being investigated. Here, we discuss the use of enteric-coated dry emulsions, microspheres, liposomes and nanoparticles for oral delivery of peptides and proteins.

Emulsions protect drug from chemical and enzymatic breakdown in the intestinal lumen. Drug absorption enhancement is dependent on the type of emulsifying agent, particle size of the dispersed phase, pH, solubility of drug, type of lipid phase used, etc. The lipid phase of microemulsions is composed of medium chain fatty acids triglycerides increasing the bioavailability of muramyl dipeptides analog.

Torisaka et al. prepared a new type of oral dosage form of insulin, S/O/W emulsions, in which a surfactant-insulin complex is dispersed into the oil phase. This novel insulin formulation was designed to alleviate the previously mentioned two barriers: the solubilization into the oil phase can avoid degradation of protein and the noncovalent coating of insulin molecules with a lipophilic surfactant making it possible to enhance permeation through the intestinal mucosa without introducing a new chemical entity. The potential of the S/O/W emulsion was validated by hypoglycemic activity over several hours after oral administration to diabetic rats. However, a critical drawback of this formulation was physical-chemical instability in long-term storage and the requirement for storage at low temperatures. To overcome this drawback, it is formulated into dry emulsion. Dry emulsion formulations are typically prepared from O/W emulsions containing a soluble or an insoluble solid carrier in the aqueous phase by spray drying, lyophilization or evaporation. Dry emulsions are regarded as lipid-based powder formations from which an O/W emulsion can be reconstituted. From a pharmaceutical point of view, they are attractive due to their physical strength and ease of administration as capsules and tablets. In this study, Eiichi Torisaka et al. have developed a unique dry emulsion formulation in which the surfactant-insulin complex was entrapped in the oil phase of the solid formulation. Using a pH-responsive polymer, HPMCP, the dry emulsion was enteric-coated. The release behavior of encapsulated insulin was found to be responsive to external pH and the presence of lipase under the simulated G.I. conditions. Based on the results obtained in this study and the fact that any water-soluble drug can be complexed with surfactants, the new solid emulsion formulations could be extensively applicable to oral delivery of pharmaceutical peptides and proteins.

The influence of pH variability through the stomach to the intestine on the oral bioavailability of peptide and protein drugs may be overcome by protecting them from proteolytic degradation in the stomach and upper portion of the small intestine using pH-responsive microspheres as oral delivery vehicles. Lowman et al., loaded insulin into polymeric microspheres of poly (methacrylic-g-ethylene glycol) and observed oral bioavailability in healthy and diabetic rats. In the acidic environment of the stomach, the microspheres were unswollen as a result of the formation of intermolecular polymer complexes. The insulin remained in the microspheres and was protected from proteolytic degradation. While in the basic and neutral environments of the intestine, the complexes dissociated which resulted in rapid microspheres swelling and insulin release. Within 2 h of administration of the insulin-containing polymers, strong dose-dependent hypoglycemic effects were observed in both healthy and diabetic rats. Numerous pH-sensitive polymers have been investigated for a range of applications. These microspheres restrict the release of proteins to favorable area of GIT.

Recently, nanoparticles as particulate carriers are used to deliver protein and peptide drugs orally. It is stated that particles in the nanosize range are absorbed intact by the intestinal epithelium, especially, through peyer's patches and travel to sites such as the liver, the spleen and other tissues. The proteins and peptides encapsulated in the nanoparticles are less sensitive to enzyme degradation through their association with polymers. It is demonstrated that protein and peptide encapsulated in nanoparticles have better absorption through G.I. tract as compared to their native counterpart. The factors affecting uptake include the particle size of particulate, the surface charge of the particles, the influence of surface ligands and the dynamic nature of particle interaction in the gut.

Behrens studied the interaction of nanoparticles consisting of hydrophobic polystyrene, bioadhesive chitosans and (PLA-PEG) with two human intestinal cell lines and compared the in vivo uptake in rats. After intraduodenal administration of chitosans nanoparticles in rats, particles were detected in both epithelial cells and peyer's patches. In one example, insulin was encapsulated in nanospheres using phase inversion nanoencapsulation. The insulin released over a period of approximately 6 h, was shown to be orally active, and had 11.4% of the efficacy of intraperitoneally delivered insulin.

One problem using nanoparticles is the erratic nature of nanoparticles absorption. For example, proportion of intact particles reaching systemic circulation was estimated to be generally below 5%.

Liposomes are prone to the combined degrading effects of the acidic pH of the stomach, bile salts and pancreatic lipase upon oral administration. There are several reports on the intact liposomal uptake by cells in 'in vitro' and 'in situ' experiments. The results are, however, not convincing for the oral delivery of protein with a liposomal system. Attempts have been made to improve the stability of liposomes either by incorporating polymers at the liposome surface, or by using G.I.-resistant lipids.

In vitro release of insulin, a model peptide, from liposomes in the bile salts solution was markedly reduced by coating the surface with the sugar chain portion of mucin or polyethylene glycol. Encapsulation of insulin with the sugar chain portion of mucin and that of polyethylene glycol completely suppressed the degradation of insulin in the intestinal fluid, whereas uncoated liposomes suppressed it only partially. These results demonstrated that surface coating of liposomes with PEG or mucin gained resistance against digestion by bile salts and increased the stability in the G.I. tract. When insulin was orally administered to rats as a solution or non-charged liposome, no hypoglycemic effect was observed. Administration of insulin encapsulated in positively charged liposome caused the rapid decrease in the plasma glucose level that recovered to the control level within 3 h. In contrast, PEG containing liposomes and mucin containing liposomes caused a gradual decrease in the glucose level after administration. The hypoglycemic effect by PEG-Liposome lasted for much longer duration than that of uncoated liposomes. The slow release of insulin from the surface coated liposomes achieved longer duration of oral hypoglycemic activity. Consequently, the surface coating should be the potential way to add desirable functions to the liposome for oral drug delivery.

1.9.2.1.11 *Dosage Form Modifications*

This strategy is particularly applicable in the case of poorly absorbed peptides and proteins that are unstable in the G.I. lumen where their targeting to a specific tissue or organ is to be affected. The proper designing of the delivery system not only protects the drug from G.I. degrading components but also releases the drug at the site favorable for absorption, whether it is due to low protease content or enhanced permeability characteristics. Some of the delivery systems that have been explored are discussed in the following subsections.

1.9.2.1.11.1 *Lipid Vesicles and Emulsions*

These dosage forms have shown great potential in the delivery of proteins and peptides. Before a drug can exert its therapeutic effects, its penetration through the plasma membrane (a lipid bilayer) or its uptake through carrier systems is a mandatory step. The use of lipid vesicular carrier systems and emulsions have paved the way to circumvent membrane barriers and thereby promoting the uptake of this 'difficult' class of therapeutics. Solid nanospheres and fat emulsions have also been suggested as lymphotrophs to traffic 'problem' molecules effectively through lymphatic circulation utilizing typical endogenous lipid digestion and assimilation system specially the chylomicrons. Insulin delivered in liposomes produced better hypoglycemic effects than the free insulin following oral administration. Water-in-oil-in-water (w/o/w) emulsion also exhibited significant delivery potential when compared with the plain aqueous solutions.

1.9.2.1.11.2   Emulsomes

The emulsome drug delivery system as reported represents a new generation of colloidal drug carrier units. It is a lipiodol drug delivery vehicle which could be prepared using relatively higher concentration of (5–10%). The interesting feature of the system is that unlike oil phase of an emulsion (0/W), the internal e in the case of emulsomes remains to be in solid or quasi solid state at ambient temperature. Its preparation technique typically employs adequate emulsification at elevated temperature where an ordered bi-phasic dispersion is stabilized through the addition of lecithin in higher quantity. The internal phase may contain entrapped macromolecules which are proteinaceous in nature and could be cultivated as nanospheres typically surface stabilized resembling chylomicrons, the endogenous lymphotropic system. It follows that the system on oral administration could affect the translocation of its content via lymphatic route. The emulsomes can explicitly be distinguished from fat emulsions or lipid microspheres as they are distinctively sphere 'vesicular system' due to utilization of higher quantities of phosphatidylcholine both as emulsifying agent as well as surface modifier. Seemingly it forms phospholipid bilayer onto the surface of lipid solid core. The system holds promises for its effective utilization of oral administration of proteinaceous drug.

1.9.2.1.11.3   Particulate Carriers

Nano and microparticles can be employed as oral carriers for peptide and protein delivery. Intact uptake of particles up to 10 p.m. from intestinal wall have been reported. Another study claimed that there is no uptake of nanoparticles, but rather the contact time of the drug with the gut wall is increased by the particles and its degradation in the gut lumen alleviated. This unusual approach sounds promising, but it has been observed that the quantitative uptake into general circulation from the G.I. lumen is very small. Native surface properties and chemical composition of the carrier nanoparticles are crucial in determining the extent of uptake. For instance adsorption of poloxamer surfactants are reported to decrease total uptake while covalent attachment of tomato lectin decreases the extent of uptake. Insulin has been administered in 220 nm poly alkyl cyanoacrylate nano-particles which resulted in significant reduction in blood glucose level. Luteinizing hormone releasing hormone (LHRH) has also been administered as copolymerized peptide particle with successful results. In this straw a derivative of LHRH was prepared by conjugating it with vinyl acetic acid which was then co-polymerized with n-butylcyanoacrylate (n-BCA) and a radiolabel. The reaction conditions were manipulated to exploit the particle forming properties of n-BCA.

It is suggested that the uptake takes place via restricted but specialized mechanism involving the M-cells. This can be exploited in the development of oral vaccines. This strategy can be utilized if low bioavailability of protein/peptidal active species is not a restraint. Uptake also occurs through non-lymphoid epithelial tissues. Poloxamer and tomato lectin increases total uptake through this route only. Proteinoids are polymeric amino acid aggregation under acidic condition in the form of microspheres. These proteinoids serve to hold a cargo of drug(s) including proteins, peptides and immunogens, which on oral administration set improved absorption of these agents. Characteristically they are quite stable in G.I. tract. However, when discharged into the small intestine, they undergo spontaneous dissociation to release the contents. Apparently, the encapsulated peptides and proteins are protected from enzymatic degradation in stomach. Thus large amount ultimately reaches the absorption site in small intestine. For protein(s) and vaccine on oral administration in proteinoids an improved availability profile has been recorded.

1.9.2.1.11.4 Bioadhesive Systems

Bioadhesive systems are supposed to stick to the intestinal mucosa. This intimate contact between the delivery system and absorbing membrane and or the prolonged residence time at the site of absorption leads to an increment in absorption across the mucosa. Intestinal absorption of 9-desglycinamide 8-arginine vasopressin (DGAVP) with microspheres consisting of p-hydroxyethylmethacrylate having bioadhesive polycarbophil coating have been studied. The absorption of the studied drug has been reported to improve significantly.

1.9.2.1.11.5   Bioadhesion Technologies for the Delivery of Peptide and Protein Drugs to the Gastrointestinal Tract

For the efficient delivery of peptides, proteins, and other biopharmaceuticals by nonparenteral routes, in particular via the gastrointestinal, or G.I. tract, novel concepts are needed to overcome significant enzymatic and diffusional barriers. In this context, bioadhesion technologies offer some new perspectives. The original idea of oral bioadhesive drug delivery systems was to prolong and/or to intensify the contact between controlled-release dosage forms and the stomach or gut mucosa. However, the results obtained during the past decade using existing pharmaceutical polymers for such purposes were rather disappointing. The encountered difficulties were mainly related to the physiological peculiarities of G.I. mucus. Nevertheless, research in this area has also shed new light on the potential of mucoadhesive polymers. First, one important class of mucoadhesive polymers, poly(acrylic acid), could be identified as a potent inhibitor of proteolytic enzymes. Second, there is increasing evidence that the interaction between various types of bio(muco)adhesive polymers and epithelial cells has direct influence on the permeability of mucosal epithelia. Rather than being just adhesives, mucoadhesive polymers may therefore be considered as a novel class of multifunctional macromolecules with a number of desirable properties for their use as biologically active drug delivery adjuvants. To overcome the problems related to G.I. mucus and to allow longer lasting fixation within the G.I. lumen, bioadhesion probably may be better achieved using specific bioadhesive molecules. Ideally, these bind to surface structures of the epithelial cells themselves rather than to mucus by receptor-ligand-like interactions. Such compounds possibly can be found in the future among plant lectins, novel synthetic polymers, and bacterial or viral adhesion/invasion factors. Apart from the plain fixation of drug carriers within the G.I. lumen, direct bioadhesive contact to the apical cell membrane possibly can be used to induce active transport processes by membrane-derived vesicles. The nonspecific interaction between epithelia and some mucoadhesive polymers induces a temporary loosening of the tight intercellular junctions, which is suitable for the rapid absorption of smaller peptide drugs along the paracellular pathway. In contrast, specific endo- and transcytosis may ultimately allow the selectively enhanced transport of very large bioactive molecules or drug carriers across tight clusters of polarized epi- or endothelial cells, whereas the formidable barrier function of such tissues against all other solutes remains intact.

1.9.2.1.11.6   Mucoadhesive Polymeric Systems

Mucoadhesive polymeric systems are the most promising approach among several approaches. Mucoadhesive properties can provide an intimate contact with the mucosa at the site of drug uptake preventing a presystemic metabolism of peptides on the way to the absorption membrane in the gastrointestinal tract. Additionally, the residence time of the delivery system at the site of drug absorption is increased. Thus, we can achieve site-specific drug delivery by the use of mucoadhesive polymeric system. Mucoadhesive polymers are able to adhere to the mucin layer on the mucosal epithelium and thus results in the increase of oral drug bioavailability of protein and peptide drugs. These polymers decrease the drug clearance rate from the absorption site, thereby increasing the time available for absorption.

Most of the current synthetic bioadhesive polymers are either polyacrylic acid or cellulose derivatives. Examples of polyacrylic acid-based polymers are carbopol, polycarbophil, polyacrylic acid (PAAc), polyacrylate, poly (methylvinyl ether-co-methacrylic acid), poly (2-hydroxyethyl methacrylate), poly(methacrylate), poly(alkylcyanoacrylate), poly(isohexylcyanoacrylate) and poly(isobutylcyanoacrylate). Cellulose derivatives include carboxymethyl cellulose, hydroxyethyl cellulose, hydroxypropyl cellulose, sodium carboxymethyl cellulose, methylcellulose, and methylhydroxyethyl cellulose. In addition, seminatural bioadhesive polymers include chitosan and various gums such as guar, xanthan, poly(vinylpyrrolidone), and poly(vinyl alcohol).

A new gastrointestinal mucoadhesive patch system (GI-MAPS) has been designed for the oral delivery of protein drugs [60]. The system consists of four layered films contained in an enteric capsule. The backing layer is made of a water-insoluble polymer, ethyl cellulose (EC). The surface layer is made of an enteric pH-sensitive polymer such as hydroxypropyl methyl-cellulose phthalate, Eudragit L100 or S100 and was coated with an adhesive layer. The middle layer, drug-containing layer, made of cellulose membrane is attached to the EC backing layer by a heating press method. Both drug and pharmaceutical additives including an organic acid, citric acid, and a non-ionic surfactant, polyoxyethylated castor oil derivative were formulated in the middle layer. The surface layer was attached to the middle layer by an adhesive layer made of carboxyvinyl polymer. After oral administration, the surface layer dissolves at the targeted intestinal site and adheres to the small intestinal wall, where a closed space is created on the target site of the gastrointestinal mucosa by adhering to the mucosal membrane. As a result, both the drug and the absorption enhancer coexist in the closed space and a high-concentration gradient is formed between inside the system and the enterocytes, which contributes to the enhanced absorption of proteins because most drugs are absorbed by a passive-diffusion mechanism. As a result, the absorption enhancer makes full use of its capacity. As the GI-MAPS is a novel drug-delivery system preparation, the fabrication method is the second hurdle to overcome in the launch of an oral preparation of proteins. However, recent advances in microfabrication technology in the semiconductor industry have made it possible to produce many micron-size GI-MAPS. Several approaches to produce the micron-size GI-MAPS are described and the future of these technologies is discussed.

Carbopol polymers have been shown to inhibit luminal degradation of insulin, calcitonin, and insulin-like growth factor-I (IGF-I) by trypsin and chymotrypsin. Anionic polymers feature mucoadhesive properties via hydrogen bonding, van der Waal's interactions and chain entanglement with the mucus forces stronger than the electrical repulsion caused by electrostatic interactions. In contrast, cationic polymers adhere to the negatively charged mucus mainly due to electrostatic forces. As both anionic and cationic mucoadhesive polymers exhibit a high buffer capacity, a demanded microclimate regarding the pH can be adjusted and maintained over numerous hours within the polymeric network.

On the contrary, the strong mucoadhesive properties of thiomers are believed to be based on additional covalent bonds between thiol groups of the thiomer and cysteine-rich subdomains of mucus glycoproteins. This theory was confirmed by findings of mucoadhesion studies, where a higher amount of thiol groups on the polymer resulted in higher mucoadhesive properties.

Although thiomers show strongly improved mucoadhesive properties, the adhesion of delivery systems being based on such polymers is nevertheless limited by the natural mucus turnover. The mucus turnover in the human intestine, for instance, was determined to be in the range of 12–24 h. Consequently, at least within this time period, the adhesion of the delivery system will fail.

Hussain et al. have showed that surface conjugation of the bioadhesive molecule -tomato lectin increases the uptake of orally administered inert nanoparticles in rats. Improved intestinal absorption of 9-desglycinamide, 8-arginine vasopressin (DGAVP) was observed in rats in vitro as well as in vivo using the weakly cross-linked poly(acrylate) derivative polycarbophil dispersed in physiological saline (Haas and Lehr) [71]. Similarly, enhanced oral bioavailability of peptide and protein drugs was seen when these compounds were formulated with chitosan-EDTA conjugates. The authors suggested that chitosan-EDTA conjugates protect peptide and protein drugs from enzymatic degradation across the G.I. tract.

1.9.2.1.12 Drug Delivery Via the Mucous Membranes of the Oral Cavity

The identification of an increasing array of highly potent, endogenous peptide and protein factors termed cytokines, that can be efficiently synthesized using recombinant DNA technology, offers exciting new approaches for drug therapy. However, the physico-chemical and biological properties of these agents impose limitations in formulation and development of optimum drug delivery systems as well as on the routes of delivery. Oral mucosa, including the lining of the cheek, floor of mouth and underside of tongue and gingival mucosa, has received much attention in the last decade because it offers excellent accessibility, is not easily traumatized and avoids degradation of proteins and peptides that occurs as a result of oral administration, gastrointestinal absorption and first-pass hepatic metabolism. Peptide absorption occurs across oral mucosa by passive diffusion and it is unlikely that there is a carrier-mediated transport mechanism. The principal pathway is probably via the intercellular route where the major permeability barrier is represented by organized array of neutral lipids in the superficial layers of the epithelium. The relative role of aqueous as opposed to the lipid pathway in drug transport is still under investigation; penetration is not necessarily enhanced by simply increasing lipophilicity, for other effects, such as charge and molecular size, also play an important role in absorption of peptide and protein drugs. Depending on the pharmacodynamics of the peptides, various oral mucosal delivery systems can be designed. Delivery of peptide/protein drugs by conventional means such as solutions has some limitations. The possibility of excluding a major part of drug from absorption by involuntary swallowing and the continuous dilution due to salivary flow limits a controlled release. However these limitations can be overcome by adhesive dosage forms such as gels, films, tablets, and patches. They can localize the formulation and improve the contact with the mucosal surface to improve absorption of peptides and proteins. Addition of absorption promoters/permeabilizers in bioadhesive dosage forms will be essential for a successful peptide/protein delivery system.

The delivery of drugs via the mucous membranes lining the oral cavity, with consideration of both systemic delivery and local therapy, is reviewed in this paper. The structure and composition of the mucosae at different sites in the oral cavity, factors affecting mucosal permeability, penetration enhancement, selection of appropriate experimental systems for studying mucosal permeability, and formulation factors relevant to the design of systems for oral mucosal delivery are discussed. Sublingual delivery gives rapid absorption and good bioavailability for some small permeants, although this site is not well suited to sustained-delivery systems. The buccal mucosa, by comparison, is considerably less permeable, but is probably better suited to the development of sustained-delivery systems. For these reasons, the buccal mucosa may have potential for delivering some of the growing number of peptide drugs, particularly those of low molecular weight, high potency, and/or long biological half-life. Development of safe and effective penetration enhancers will further expand the utility of this route. Local delivery is a relatively poorly studied area; in general, it is governed by many of the same considerations that apply to systemic delivery.

1.9.2.1.13 Emerging Trends in Oral Delivery of Peptide and Protein Drugs

Most peptide and protein drugs are currently used as parenteral formulations because of their poor oral bioavailability. Development of an effective oral delivery system for these macromolecular drugs requires a thorough understanding of their physicochemical properties, such as molecular weight, hydrophobicity, ionization constants, and pH stability, as well as biological barriers that restrict protein and peptide absorption from the gastrointestinal (G.I.) tract, including pH variability, enzymatic degradation, and membrane efflux. Various strategies currently under investigation include amino acid backbone modifications, formulation approaches, chemical conjugation of hydrophobic or targeting ligand, and use of enzyme inhibitors, mucoadhesive polymers, and absorption enhancers. However, there is only limited success because of the hostile environment of the G.I. tract–e.g., strong pH extremes and abundant presence of potent luminal enzymes.

1.9.2.1.14 Oral Delivery of Peptide Drugs: Barriers and Developments

A wide variety of peptide drugs are now produced on a commercial scale as a result of advances in the biotechnology field. Most of these therapeutic peptides are still administered by the parenteral route because of insufficient absorption from the gastrointestinal tract. Peptide drugs are usually indicated for chronic conditions, and the use of injections on a daily basis during long-term treatment has obvious drawbacks. In contrast to this inconvenient and potentially problematic method of drug administration, the oral route offers the advantages of self-administration with a high degree of patient acceptability and compliance. The main reasons for the low oral bioavailability of peptide drugs are pre-systemic enzymatic degradation and poor penetration of the intestinal mucosa. A considerable amount of research has focused on overcoming the challenges presented by these intestinal absorption barriers to provide effective oral delivery of peptide and protein drugs. Attempts to improve the oral bioavailability of peptide drugs have ranged from changing the physicochemical properties of peptide molecules to the inclusion of functional excipients in specially adapted drug delivery systems. However, the progress in developing an effective peptide delivery system has been hampered by factors such as the inherent toxicities of absorption-enhancing excipients, variation in absorption between individuals, and potentially high manufacturing costs. This review focuses on the intestinal barriers that compromise the systemic absorption of intact peptide and protein molecules and on the advanced technologies that have been developed to overcome the barriers to peptide drug absorption.

1.9.2.1.15 Protein Drug Oral Delivery: The Recent Progress

Rapid development in molecular biology and recent advancement in recombinant technology increase identification and commercialization of potential protein drugs. Traditional forms of administrations for the peptide and protein drugs often rely on their parenteral injection, since the bioavailability of these therapeutic agents is poor when administered non-parenterally. Tremendous efforts by numerous investigators in the world have been put to improve protein formulations and as a result, a few successful formulations have been developed including sustained-release human growth hormone. For a promising protein delivery technology, efficacy and safety are the first requirement to meet. However, these systems still require periodic injection and increase the incidence of patient compliance. The development of an oral dosage form that improves the absorption of peptide and especially protein drugs is the most desirable formulation but one of the greatest challenges in the pharmaceutical field. The major barriers to developing oral formulations for peptides and proteins are metabolic enzymes and impermeable mucosal tissues in the intestine. Furthermore, chemical and conformational instability of protein drugs is not a small issue in protein pharmaceuticals. Conventional pharmaceutical approaches to address these barriers, which have been successful with traditional organic drug molecules, have not been effective for peptide and protein formulations. It is likely that effective oral formulations for peptides and proteins will remain highly compound specific. A number of innovative oral drug delivery approaches have been recently developed, including the drug entrapment within small vesicles or their passage through the intestinal paracellular pathway. This review provides a summary of the novel approaches currently in progress in the protein oral delivery followed by factors affecting protein oral absorption.

1.9.2.2 Nasal Route [161]

Until recently, the nasal route was limited to producing local action on the mucosa. But this route appears to hold immense promise for the delivery of peptides and proteins and has lately received considerable attention. This route is not only convenient but also has a large surface area available for absorption, the mucosa is high vascularized and first-pass metabolism can be avoided. Anatomically the length of nasal cavity is approximately 12 cm with a total volume of 15 ml. It covers about 150 cm^2 area divided into olfactory and vestibule regions. Most of the area is accounted for respiratory region. Major of nasal cavity is covered by epithelium Toe with microvilli and cilia forming a dense carpet like structure. Functionally, cilia remove mucus. The main contributory factors towards physiological environments include change in mucus secretion, pH and viscosity and ciliary motility and viability. These factors are mainly affected by the presence of foreign substances and appreciated to have an impact on drug absorption through this route. Figure 1.15 gives the cross-sectional view of the nasal cavity. The nasal mucosa is relatively more permeable to peptides as compared against routes like oral and transdermal. The bioavailability of peptides through this route is reported to be of the order of 1 to 20%. This depends on the physical properties and molecular weight of the peptide but can be extremely variable.

Some of the disadvantages of the nasal route include:

- Extent of absorption varies with the mucous secretion and turnover;
- Mucociliary clearance represents a physical and temporal barrier;
- Peptidase present in the nasal membrane serves as an enzymatic barrier in absorption;

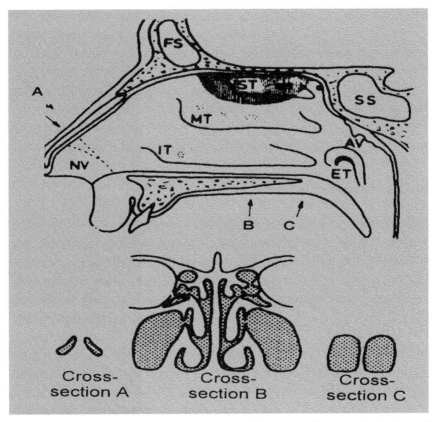

FIGURE 1.15   Lateral view and cross-sections of the nasal cavity through: A – internal ostium; B – middle of the nasal cavity; C – choanae; NV – nasal vestibule; IT – inferior turbinate; MT – middle turbinate; ST – superior turbinate; FS – frontal sinus; SS – sphenoidal sinus; AV – adenoid vegetation; ET – orifice of Eustachian tube.

- Alteration in absorption profile in diseased conditions like allergic and chronic rhinitis and upper respiratory tract infections;
- Penetration enhancers and preservatives may damage mucosal cell membrane and may even be ciliotoxic.

Peptidal and proteinaceous moieties like calcitonin, ACTH, insulin and interferon are reported to have appreciable absorption through nasal mucosa. Lypressin, a synthetic analog of vasopressin has been introduced commercially as an intranasal dosage form following encouraging clinical trials. Encouraging results have also been obtained with GnRH, LHRH, buserelin, enkephalins and GHRF. Pituitary hormones like vasopressin and oxytocin have been administered by the nasal route for many years.

With greater interest in delivery of protein and peptide-based drugs to the lungs for topical and systemic activity, a range of new devices and formulations are being investigated. Pulmonary protein delivery offers both local targeting for the treatment of respiratory diseases and increasingly appears to be a viable option for the delivery of proteins systemically. The lung is easy to access, has decreased proteolytic activity compared with the gut, and allows rapid absorption and avoidance of first-pass metabolism for systemically delivered drugs. Hundreds of proteins and peptides are undergoing clinical investigation for a range of clinical conditions. These include growth factors, hormones, monoclonal antibodies, cytokines, and anti- infective agents. For those being investigated for delivery via inhalation, the ultimate site of action may be the airway surface (e.g., DNase), the airway cells (e.g., cyclosporin), or the systemic circulation (e.g., insulin). Careful choice of carrier and device can facilitate delivery to a specific area of the lungs. Once delivered, a carrier can further influence the distribution and rate of clearance rom the site of action.. The only protein for inhalation currently available on the market is DNase, but a growing number of proteins/peptides are in various phases of clinical trials. Systemic inhaled insulin is in late phase 3 trials. Other proteins/peptides in phase 3 trials include leuprolide and gamma-interferon. Bile salts have been used to enhance the nasal absorption of peptide based pharmaceuticals.

1.9.2.2.1 Nasal Delivery of Proteins

Insulin: The nasal absorption of insulin is increased by coadministration of bile salts, surfactants. Gordon et al. reported that in case of sodium deoxycholate, a minimum concentration of 2.4 mM is required to enhance the transnasal permeation of insulin. Interferon: Merigan et al. studied the inhibition of respiratory virus infection by intranasal administration of human leukocyte interferon.

Nasal delivery of oligopeptides
Dipeptides: 1-tyrosyl-1-tyrosine and its methyl esters, 1-gycyl-1-tyrosine, 1-glycyl-1-tyrosinamide.

Tripeptides: Thyrotropin-releasing hormone(TRH) Pentapeptides: Leucin-enkephalin, et-enkephamide.

Nasal delivery of polypeptides

Nonapeptides: Vasopressin, OXYTOCIN

Decapeptides: LHRH

Undecapeptides: Substance P Glucagon, Calcitonin and Adrenal cortico-tropic hormone (ACTH) are another examples of polypeptides.

1.9.2.2.2   Advantages of Nasal Route

- Convenient, simple, practical way of drug administration
- The high vascularization permits better absorption.
- First pass metabolism can be avoided.
- Rapid onset of action.
- Disadvantages of nasal route:
- Long term use may lead to toxicity to mucosa.
- During disease states (e.g., common cold) some alteration in the nasal environment may occur.

1.9.2.2.3   Mechanism to Facilitate Nasal Peptide and Protein Absorption

Nasal absorption of drugs can be via passive diffusion or by special transport mechanisms. Nasal absorption of metkephamid for instance, is by passive diffusion. A prominent example of transport by special mechanism is sodium guaiazulene-3-sulfonate (GAS). It is believed to be absorbed mainly by a carrier-mediated transport mechanism. Large peptides and proteins are unable to cross the nasal membrane as efficiently as smaller peptides, therefore in order to facilitate their absorption usually adjuvants as absorption enhancer are required. Several approaches have been used to facilitate nasal peptide and protein absorption.

1.9.2.2.3.1   pH Modification

Peptides and proteins usually exhibit the lowest solubility at their iso-electric point. Thus, by adjusting the pH farther away from the isoelectric point of a particular peptide, its solubility can be increased. It has been demonstrated that insulin is capable of crossing the nasal membrane in an acidic medium. At pH 6.1 the nasal absorption recorded for insulin was the least. This pH was close to the isoelectric point of insulin.

1.9.2.2.3.2   Dissociation of Aggregation

Proteins are likely to form higher-order aggregates in solution. For instance, at pH 7.0, insulin exists in solution predominantly as hexameric aggregates. Insulin failed to cross the nasal membrane in the absence of an enhancer in the formulation. However, good nasal absorption of insulin was observed with sodium deoxycholate. Studies have suggested that sodium deoxycholate disrupts the formation of insulin hexamers and higher-order aggregates. On the basis of these reports, it is assumed that dissociation of insulin hexamers to dimers and monomers by sodium deoxycholate is partly responsible for enhancing the transport of insulin across the nasal epithelium.

1.9.2.2.3.3   Reverse micelle formation

Bile salts are known to promote the transmembrane movement of endogenous and exogenous lipids, and other polar substances within the G.I. tract by the virtue of their ability to affect the micellar properties of biomembranes. For this very reason, the bile salts present a lucrative option as an adjuvant for transmucosal delivery of drugs. It is reported that with an increase in the hydrophobicity of bile salts its adjuvant activity correspondingly increased. The adjuvant potency for nasal absorption of insulin correlates positively with increasing hydrophobicity of the bile salts. On the basis of the data it was concluded that the hydrophobicity of the steroid nucleus is the major determinant of adjuvant activity. This study also suggested that the insulin absorption commences at the critical micelle concentration of the bile salts and attains a maximal level when micelle formation approaches a well established level. In reverse micelles, the hydrophilic surfaces of the molecules face inward and the hydrophobic ones face outward from the lipid environment. Thus, reverse micelles can be utilized as transmembrane channels or mobile carriers for insulin to move down an aqueous concentration gradient through the nasal mucosal cells, into the intercellular space and finally into the blood stream.

1.9.2.2.3.4   Membrane Transport and Enzyme Inhibition

Penetration enhancers like bile salts, surface active agents and chelating agents are reported to increase nasal absorption of peptides and proteins. They increase the fluidity of the lipid bilayer membrane and open up aqueous pores as a result of calcium ion chelation. Peptidase inhibitors enhance the absorption of peptide/ protein by depressing peptidase activity in both the mucus and mucosal cells. Studies have been carried out to study the relationship between the absorption-promoting effect of surfactants like sodium lauryl sulfate and their effect on biomembrane in terms of hemolytic activity and protein releasing effect on the nasal mucosa. It was concluded that the effect of these surfactants on the permeability of the nasal mucosa to insulin may be due to the disturbance of the structural integrity of the nasal mucosa. However, bile salts exhibited less effect on biomembranes in comparison to other surfactants, in terms of hemolytic profile and protein release from the nasal mucosa. The addition of bile salts is reported to inhibit the enzymatic activity of leucine aminopeptidase and the enzymatic degradation of insulin. Thus, bile salts affect both the permeability of the nasal mucosa and the activity of proteolytic enzymes, and thereby enhance the absorption as a whole. However, reports indicate that penetration enhancers can damage the nasal mucosa. The membrane damage may be in terms of cell erosion, cell to cell separation and loss of cilia. Absorption promoters like linoleic acid and oleic acid seem to enhance absorption safely. The effect agents are readily reversible within 15–20 minutes after washing. For instance, sodium taurodihdrofusidate serves as an excellent nasal absorption enhancer of insulin and is nontoxic both systemically and local.

1.9.2.2.3.5   Increased Nasal Blood Flow

With an increase in local nasal blood flow an enhancement in nasal peptide absorption has been re can be attributed to an increase in the concentration gradient for passive peptide diffusion. vasoactive which are known to enhance nasal blood flow include histamine, prostaglandin E1 and beta-adrenergic. Studies indicate that in comparison to control, the combination of histamine and desmopressin, an led to an increase in nasal blood flow and a corresponding increase in antidiuretic activity. The duration of activity of desmopressin was found to be in line with the increased nasal absorption of the In practice, considering the long term use of nasal route, it appears to rank low due to the possible t mucosa and cilia. Alterations in the nasal environment during disease states also limits its application. It appears to hold immense potential for short-term delivery of peptides/proteins, small molecular peptides in particular.

1.9.2.3   Buccal Route

Oral mucosa, including the lining of the cheek (buccal mucosa), floor of mouth and underside of tongue (sublingual mucosa) and gingival mucosa, has received much attention in the last decade because it offers excellent accessibility and avoids degradation of proteins and peptides that occurs as a result of oral administration, gastrointestinal absorption and first-pass hepatic metabolism. Peptide absorption occurs across oral mucosa by passive diffusion and it is unlikely that there is a carrier-mediated transport mechanism. The penetration of macromolecules through oral epithelia has been studied by several investigators [162]. Merkle et al. [163] developed a self adhesive buccal patch and reported that it is feasible to deliver peptide base pharmaceuticals such as protirelin and buserelin through buccal mucosa. Various types of polymers like sodium carboxymethyl cellulose, hydroxypropylmethyl cellulose, polyvinyl pyrrolidone, acacia, calcium carbophil, gelatin, polyethylene glycol are used for delivery of proteins or peptides via buccal route. The anionic polyacrylate type hydrogel is the most commonly used polymer.

Peptides/proteins are an important class of drugs which are usually administered by parenteral route. In recent years, pharmaceutical research has been directed towards developing a nonparenteral route of delivery of peptide/protein drugs. These studies report that it may be possible to administer the peptides/proteins especially insulin, by nasal, buccal, rectal, or even transdermal route. Therefore, there is a great potential for future development of a nonparenteral route of delivery of peptide/protein drugs. The primary objective of this review is to report the present status of research involving nonparenteral administration of macromolecular peptides/proteins, with special emphasis on insulin. Buccal route is known to be the most popular route among all routes of nonparenteral administration of peptide and protein drugs.

For peptide and protein drug delivery buccal route offers some distinct benefits over other mucosal routes like nasal, vaginal, rectal, etc. Although the buccal route displays less efficiency in absorption but it is robust and comparatively much less sensitive even on long term treatment and this is of importance when penetration enhancers are to be employed. Absence of enzymatic barrier to peptide/protein absorption associated with other mucosal sites is an added advantage. Its close resemblance to oral route is expected to well acceptance by patients. Improved patient compliance is anticipated due to the easy accessibility and administration as dosage forms can be attached and removed without any pain or discomfort. The buccal membrane is linked by a stratified squamous epithelium that is keratinized in some areas. Drug that penetrates this membrane enter the systemic circulation via a network of capillaries and arteries. The lymphatic drainage almost runs parallel to the venous vascularization and ends up in the jugular ducts. A multitude of dosage forms are available that can be used to deliver peptide/protein. The conventional means include aqueous solutions and buccal or sublingual tablets and capsules. However, the inherent problem with these dosage forms is the risk of drug loss by accidental swallowing or by the salivary washout.' To overcome these drawbacks self-adhesive systems have been designed that are capable of being in intimate contact with the mucosa, either buccally, sublingually or on the gingiva. The various adhesive polymers include water-soluble and insoluble hydrocolloid polymers from both the ionic and the nonionic types. Some of the polymers are sodium carboxymethylcellulose (Sod. CMC), hydroxypropyl methylcellulose (HPMC), polyvinyl pyrrolidone (PVP), acacia, calcium carbophil, gelatin, polyethyleneglycol (PEG). The anionic polyacrylate-type hydrogel is the most commonly used polymer. The adhesion mechanism involves the entanglement and nonspecific or specific interaction between the polymer chains of polymer (used as the dosage form excipient) and the glycoprotein coat of the mucosal membrane. Drug release from soluble polymers is determined by polymer dissolution and drug diffusion. From the non-soluble hydrogels drug release is reported to follow fickian or non-fickian diffusion kinetics. The designing of dosage forms with different release rates can be performed on the basis of pharmacodynamics of the peptides. When fast and instantaneous release of peptides is warranted highly permeable or rapidly eroding carriers are desired. However, when sustained release is the motive then approaches like matrix diffusion control, polymer erosion control or membrane controlled transport of the peptide can be utilized.

1.9.2.3.1 Various Adhesive Dosage Forms

Adhesive tablets
Adhesive tablets for buccal administration were designed on the basis of eroding hydrocolloid/filler tablets. Adhesive tablets of nitroglycerin based on hydroxypropylcellulose have been designed and pharmacodynamic effects were observed for up to 5 hours. In stark contrast to conventional tablets, these adhesive tablets allow speaking and drinking without any major discomfort.

Adhesive gels
Viscous adhesive gels have been designed for local therapy using poly-acrylic acid and polymethacrylate as gel-forming polymers. Gels are reported to prolong residence time on the oral mucosa to a significant level. This not only improves absorption but also allows for sustained release of the active principle.

Adhesive patches
Adhesive patches is a relatively new addition to pharmaceutical technology (Figures 1.16 and 1.17). The various self adhesive set ups are illustrated in Figures 1.16 and 1.17. In these approaches the adhesive polymer may act as the drug carrier itself (case a and d), act as an adhesive link between a drug loaded layer and the mucosa (case c). Alternatively a drug containing disk may be fixed to the mucosa by using a self-adhesive shield (case b). In this approach drug loss to the saliva is decreased and the drug action and the effect of additives is confined to the site of application by creating a local microenvironment.

Encouraging results have been obtained with protirelin (a thyrotropin-releasing hormone) and buserelin (a synthetic LHRH derivative).

With the aid of absorption promoters
To augment the efficiency of buccal peptide administration some absorption promoters have been tried. For instance sodium lauryl sulfate, sodium myristate and bile acids have been tried to promote buccal absorption

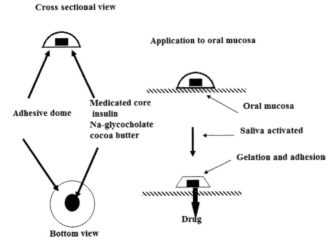

FIGURE 1.16   Diagrammatic illustration of the dome-shaped mucosal adhesive device and its application to oral mucosa.

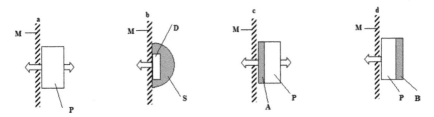

FIGURE 1.17   Schematic representation of various adhesive patches for transmucosal delivery, M – mucosa; P- polymer with peptide; S – adhesive shield; D – drug; A – adhesive layer; B – impermeable back layer.

calcitonin. Enhancement in buccal absorption of insulin was observed when coadministered with sodium glycocholate. Other absorption promoters that have been investigated include sodium 5-methoxysalicylate and citric acid. The challenge with this approach is the need to find biocompatible, non-toxic and effective absorption enhancers.

1.9.2.3.2   Factors and strategies for Improving Buccal Absorption of Peptides

Peptides and polypeptides have important pharmacological properties but only a limited number (e.g., insulin, oxytocin, vasopressin) have been exploited as therapeutics because of problems related to their delivery. The buccal mucosa offers an alternative route to conventional, parenteral administration. Peptides are generally not well absorbed through mucosae because of their molecular size, hydrophilicity and the low permeability of the membrane. Peptide transport across buccal mucosa occurs via passive diffusion and is often accompanied by varying degrees of metabolism. This review describes various approaches to improve the buccal absorption of peptides including the use of penetration enhancers to increase membrane permeability and/or the addition of enzyme inhibitors to increase their stability. Other strategies including molecular modification with bioreversible chemical groups or specific formulations such as bioadhesive delivery systems are also discussed.

1.9.2.3.2.1 *Various Strategies Employed for Buccal Delivery*

- Adhesive tablets, for example, adhesive tablet based on hydroxypropylcellulose.
- Adhesive gels, for example, by using polyacrylic acid and polymethacrylate as gel forming polymers.
- Adhesive patches, for example, protirelin in HEC patches and buserelin.
- Adhesive promoters, for example, sodium lauryl sulfate, sodium myristate, bile acids, sodium glycocholate, citric acid.
- Addition of absorption promoters/permeabilizers in bioadhesive dosage forms will be essential for a successful peptide/protein delivery system.
- Thyrotropin-releasing hormone, tripeptide, oxytocin, vasopressin, LHRH analogs, calcitonin, insulin have been easily applied through buccal route.

Advantages of buccal route

- It is robust, much less sensitive to irreversible irritation even on long term treatment. Absence of enzymatic barrier. Well acceptable to the patients. Easy accessibility administration as dosage forms. It is attached or removed without any pain or discomfort.

1.9.2.4 Ocular Route

The ocular route holds immense potential for peptides or proteins intended for pathological ophthalmologic conditions. The ocular route is the site of choice for the localized delivery of opthalmologically active peptides and proteins for the treatment of ocular disease that affect the anterior segment tissues of eye. Christie and Hanzal [164] reported the observation of a dose dependent reduction in blood glucose levels following the ocular administration of insulin to the rabbit. The use of nanoparticles, liposomes, gels, ocular inserts, bioadhesives or surfactants are necessary to enhance ocular absorption of proteins or peptides. The polypeptide antibiotics like cyclosporine, tyrothricin, gramicidin, tyrocidine, bacitracin and polymyxins have often been considered potential candidates for achieving local pharmacological actions in the eyes [165].

Proteins or peptides with opthalmological activities:

- Affect aqueous humor dynamics: calcitonin gene related factors, LHRH, vasopressin.
- Immunomodulating activities: cyclosporine, interferons.
- Act on inflammation: substance P, enkephalins.
- Affect wound healing: epidermal growth factor, fibronectin.

The systemic bioavailability achieved by this route is very low. Ocular tissues are sensitive to the presence of foreign substances and patient acceptance could be rather low in ocular route.

1.9.2.5   Rectal Route

The use of the rectum for the systemic delivery of organic and peptide based pharmaceuticals is a relatively recent ideas. The coadministration of an absorption promoting adjuvants such as sodium glycocholate, has been reported to enhance the rectal absorption of insulin. Touitou et al. [166] achieved hyooglycemia in rats by administering insulin via rectal route, in a dosage form that contained polyethylene glycols and a surfactant. It was recently reported that a solid dispersion of insulin with sodium salicylate can produce a rapid release of insulin from the suppositories achieve a significant reduction in plasma glucose levels in normal dogs, even at doses as low as 0.5 IU/kg [167]. Bile salts, such as sodium salts of cholic, deoxycholic and glycocholic acids, have also been shown to enhance the rectal absorption of insulin in rats [168, 169] and human volunteers [170].

Vasopressin and its analogs [171], pentagastrin and gastrin [172], calcitonin analogs [173, 174] and human albumin [175] have been investigated for rectal delivery of protein or peptide based pharmaceuticals.

Advantages of rectal delivery

- It is highly vascularized. It avoids first pass or presystemic metabolism. Drug can be targeted to the lymphic system. It is suitable for drugs that cause nausea/vomiting and irritate G.I. mucosa on oral administration. A large dose of drugs can be administered.

The large intestine is drained by the hepatic portal vein as well as by the lymphatics. It has been that in the lower colon the drainage is mostly lymphatic. With an increase in the molecular compound, its lymphatic uptake is also augmented. Compounds with molecular weight greater predominantly make an entry into the lymphatic fluid. Thus rectal delivery of peptides and proteins only the presystemic or first-pass metabolism but also the G.I. tract peptidases. The bypass of h system is attributed to the fact that the lower hemorrhoidal veins do not enter the hepatic portal. Thus, a considerable portion of the rectally absorbed drug enters the general circulation d' intercellular junctions of the columnar epithelium of the rectal mucosa limits the bioavailability of proteins. Rectal absorption of peptides and proteins has been reported. The literature is replete with examples delivery of insulin. Even relatively large polypeptides like lysozyme have been reported to be ab rectum. Encouraging results have been obtained with calcitonin, gastrin, pentagastrin and tetragastrin. The venous drainage of human rectum is highlighted in Figure 1.18.

1.9.2.6 Adjuvants to Enhance the Absorption

Most of the peptide/protein drugs require absorption enhancers to obtain reasonable extent of drug absorption. These include surface-active agents, bile acids, saponins, phospholipids, organic alcohols, acids, salts, and fats.

- *Surfactants*: Several surfactants including Tween 40, 60 and 61. Span 40, Cetomacrogol, various poly-oxycthylene ethers and esters, glycerylmonostearate, sodiumlaurylsulfate and dioctylsulfosuccinate are reported to enhance absorption. It appears that surfactants

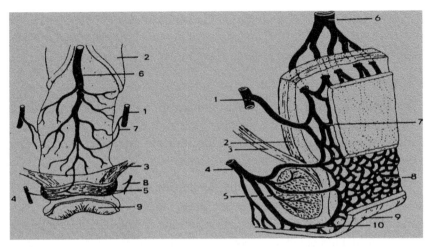

FIGURE 1.18   The venous drainage of human rectum: 1. Middle rectal vein, 2. Tunica muscularis; Statism longitudinal, 3. Levator ani, 4. Inferior rectal vein, 5. External anal sphincter, 6. Superior rectal vein, 7 and a Submucus venous plexus, 9. Skin, 10. Marginal vein and subcutaneous plexus.

interact with the lipoidal fraction of the membrane and in short term the effect is irreversible. It is postulated that surface-active agents enhance drug absorption through damage to the rectal mucosa.

- *Salicylates:* Sodium salicylate, 3-methoxysalicylate, 5-methoxysalicylate and homovanilate were found to be effective in enhancing the rectal absorption of a number of drugs. The effectiveness of sodium salicylate was augmented in the presence of sodium chloride and reduced in the presence of inhibitors like phlorizin and 4,4'-disothiocyano-stilbene-2,2'-disulfonic acid (DIDS). Small amounts (<0.5 mg/mL) of N-ethylamaleimide (NEM) or sodium p-chloromercuriphenyl sulfonic acid (p-CMP) on concurrent administration reduce the effectiveness of sodium salicylate. However, higher absorption of cefoxitin was observed with higher concentration of (>1 mg/mL) of the same. Ouabain and 2,4-dinitrophenol (DNP) are also reported to inhibit the effectiveness of salicylate as an absorption enhancer. It has been hypothesized that the effects of salicylate occur through a saturable process at the protein fraction of the rectal mucosa.
- *Ethyl/enediamine tetra acetic acid (EDTA):* EDTA has been reported to enhance the rectal absorption of salicylate and decrease the

absorption of m-and p-hydroxybenzoic acids. It has been found to enhance the absorption of sodium cefoxitin. Ouabain and DNP suppressed the enhancing effects of EDTA when EDTA was administered at low doses. At higher concentrations of EDTA, DNP had little effect and the effects of EDTA were partially suppressed by ouabain.

- *Fatty acid enhancers*: A number of studies have been carried out with fatty and carboxylic acids as enhancers. It is reported that the effectiveness of the carboxylic acid sodium salts for the rectal absorption of several 13-lactam antibiotics is parabolically related to their partition coefficient on a logarithmic basis. Carboxylic acids having metal ion chelating ability enhance the rectal absorption of poorly absorbed drugs including the water-soluble antibiotics. It was suggested that carboxylic acids may serve to make the intercellular space more accessible by temporary removal of calcium ions from the rectal mucosa. With the removal of calcium ions the integrity of the tight junctions is lost. Sodium cefoxitin administered in a triglyceride based suppository form to the small intestine showed promising results. This was supposed to be because of the generation of fatty acids from triglycerides by the action of lipase. In their ionized forms these act as surfactants in the luminal fluid. However, when the same triglyceride suppositories are administered in rectum no increment in absorption was observed, possibly due to the absence of lipase activity in the rectum.

1.9.2.7   Importance of Lymphatic Uptake

As discussed above, drug on rectal administration bypasses the hepatic portal system and makes direct entry to the general circulation. Lymphatic absorption of drugs also account for delivery of drugs directly to the general circulation, especially the water soluble ones. There are reports indicating the uptake of lymphatic uptake of drugs after rectal administration. They concluded that insulin primarily transported via the lymphatic system to the general circulation when administered in the presence of 5-methoxysalicylate. Also, the peak appearance of insulin in the lymph was somewhat earlier than that in the plasma. In the absence of adjuvants rectal administration of insulin failed to produce appreciable concentrations in the plasma and lymph.

1.9.2.8   Transdermal Route

Transdermal delivery has attracted considerable interest as a route for administering peptides and proteins. As early as 1996, Tregear investigated the feasibility of administering proteins and polymers through skin excised from human and animals [176]. More recently, Menasche et al. [177] studied the percutaneous absorption of elastin peptides through rat skin and its subsequent distribution in the body. The small peptides such as thyrotropin releasing hormone (TRH) [178] vasopressin [179, 180] have great difficulty in permeating the skin barrier.

In contemporary therapeutics, oral route is the most preferred route of administration. However, numerous barriers in administering a peptide/protein drug. Secondly, a programed drug delivery, the delivery of peptide and protein, was of utmost importance in order to have a constant and therapeutic level of the drug in the systemic circulation.

This necessitated for the search of an alternative other than oral, vaginal, intravenous or mucoadhesive routes. The site of application/administration that then exploited for this reason was the 'skin.' Skin has been used as the site for topical admin is dermatological drugs to achieve localized pharmacological activity. This mode of drug administration named as transdermal drug delivery since the skin serves as the site for administration of the drug. Until recently, the utility of intact skin as a port for continuous transdermal delivery of drugs was con topical medication. With an insight into the anatomy and physiology of the skin and percutaneous is now evident that through this route benefits of i.v. drug infusion (i.e., direct entry into the systemic and control over drug levels) can be closely duplicated without its inherent hazards. The schematic section of skin is presented in Figure 1.19.

Transdermal drug delivery, serves as the site for administration of systemically active drugs. However, the topically applied drug must be distributed following skin permeation, first into the systemic circulation and then transported to the target tissues, which could be distantly located in the body. Skin which lacks in proteolytic enzymes offers the major advantage for the administration of peptides and proteins. Several transdermal drug delivery systems (TDDS) have recently been developed. These systems can be classified, according to the technological basis involved in their construction, into four categories.

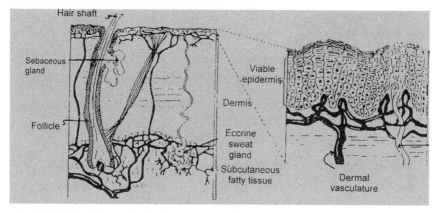

FIGURE 1.19   Schematic cross-section of the skin.

- membrane permeation-controlled;
- adhesive dispersion-type;
- matrix diffusion-controlled;
- microreservoir dissolution-controlled.

Cross-sectional view showing structural components of a TDDS Although these systems have demonstrated usefulness in affording transdermal controlled delivery of pharmaceuticals that are somewhat lipophilic and relatively small in molecular size, they are having certain limitations in the delivery of peptides and protein drugs. Biological molecules greater than 1000 kDa are not favorable candidates for delivery through TDDS as their structural dimensions do not allow them to permeate across the stratum corneum. However, low molecular weight peptides and peptidomimetics may successfully be administered using various TDDS in along with some well identified permeation enhancers. Thus, this route not only has improved patient compliance but also offers the following advantages.

- Elimination of variables that affect G.I. absorption, like pH changes, presence of enzymes, food variations in stomach emptying times, intestinal motilities.
- Elimination of hepatic first-pass phenomenon
- Controlled administration is possible and thereby avoidance of toxic effects. Also drugs with shorter half-life can be administered.
- Administration of drugs with low therapeutic indices are possible.

- Termination of therapy can be achieved by simply removing the topical device.

However, the stratum corneum is expected to deliver drug moieties for some time after removal of delivery system. The disadvantages include:

- A low rate of permeation for most protein drugs due to their large molecular weight and hydrophilicity
- High intra- and inter-patient variability.

Various approaches have been used to improve drug penetration.

1.9.2.8.1 Approaches for Transdermal Delivery

1.9.2.8.1.1 Iontophoresis

Iontophoresis is a method that induces migration of ions or charged molecules when an electric. Allowed to flow through an electrolyte medium. In addition to electrophoresis, it serves as a promising strategy to facilitate the membrane transport of charged molecules depending on their ionic characteristics is that to undergo iontophoresis the peptide and protein molecules must carry a charge. This can be controlling the pH and ionic strength of the solution. Usually, the drug containing in gauze pad is skin under an electrode of the same charge as the drug. The other electrode bearing the opposite charge on the gauze pad soaked in saline at a distal location of the body. A small current corresponding to the patient pain threshold is applied for a sufficient period of time.

Attempts have been made to administer insulin and TRH through transdermal iontophoretic delivery. Delivery of insulin as a highly ionized monomeric form in the presence of iontophoresis produced promising results. Studies with TRH also indicated improvement in delivery. Figure 1.20 illustrates the mechanism of the transdermal iontophoretic delivery of peptide and protein drugs for systemic administration.

The disadvantages associated with this strategy are manifold. There is always the possibility of burning the skin even at low voltages. Denaturation of the peptides and proteins at the pH required to achieve a suitable charge for iontophoresis may be manifested as loss in quaternary, tertiary and secondary structures. The heat generated during the process of iontophoresis can also lead to denaturation

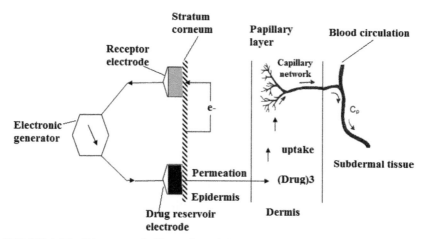

FIGURE 1.20 Diagrammatic illustration of the transdermal iontophoretic delivery of peptide and protein drugs across the skin.

It is use of electric current to drive charged drug molecules into skin by placing them under an electrode of like charged.

- *DC Iontophoresis*
  Siddiqui et al. [181] used the phoresor system or recently marketed DC iontophoretic device, as the power source for direct current and were able to deliver insulin transdermally to diabetic hairless rats, with attainment of a reduction in hyperglycemia. Recently a body wearable DC iontophoresis delivery device, called power patch applicator, was developed [182] and applied to diabetic rabbits [183, 184].
- *Pulse DC iontophoresis*
  By delivering a pulse current with a 20% duty cycle (4 μsec), followed by an 80% depolarizing period (16 μsec), a β-blockers was successfully delivered systemically to five human subjects without polarization induce skin irritation. A transdermal periodic iontotherapeutic system (TPIS) was designed [185] and it is capable of delivering the pulsed direct current with variable combinations of waveform, frequency, on/off ratio and current intensity for a specific duration treatment. Proteins or peptides drugs like insulin, TRH, salmon calcitonin, delta sleep inducing peptide, LHRH, vasopressin, leuprolide can be applied through iontophoresis.

1.9.2.8.1.2   Phonophoresis

Another strategy that has been exploited to enhance transdermal delivery is phonophoresis. In this method ultrasound is applied via a coupling-contact agent to the skin. The drug absorption is enhanced via thermal effect of ultrasonic waves and subsequent temporary alterations in the physical structure of the skin. The limitation posed by denaturation of protein following heat generation during the process is disturbing. In this method ultrasound is applied via a cupling contact agent to the skin. Insulin, IFN γ, erythropoietin can be delivered by this method. Penetration enhancers: Penetration enhancers like oleic acid, dimethylsulphoxide. Surfactants and azone have been used for topical delivery of peptide or proteins. Prodrugs; Prodrug with modeled physicochemical characteristic permeated well across the skin. LHRH, TRH, neurotensin can be delivered by this method.

1.9.2.8.1.3   Penetration Enhancers

Use of penetration enhancers holds promise in delivery of proteins and peptides through transdermal route. For this purpose penetration enhancers like oleic acid and azone has been used. These agents fluidize the intercellular lipid lamellae of the stratum corneum. 3% azone improved the absorption of vasopressin across the skin. Fatty acids are reported to disrupt the packed structure of the lipids in the extracellular spaces. However, considerable skin irritation is reported with penetration enhancers and is thus likely to limit its utility.

1.9.2.8.1.4   Pro drugs

Another strategy that has shown promising results especially with small peptides is production of prodrugs. The enzymes present in the skin are exploited to regenerate the active drug. The terminal pyroglutamyl residue of LHRH, TRH and neurotensin can be readily derivatized to produce prodrugs. These prodrugs have enhanced transdermal permeability and are capable of undergoing spontaneous hydrolysis at physiological pH, in order to regenerate the active drug.

Advantages of transdermal route

- Avoids the hepatic first-pass effect and gastrointestinal breakdown.
- Provides controlled and sustained administration, particularly suitable for the treatment of chronic disease.
- Reduces side-effects, often related to the peak concentrations of the circulating agent;
- Enables self-administration and improves patient compliance, due to its convenience and ease of use.
- Permits abrupt termination of drug effect by simply removing the delivery system from the skin surface.

Limitations of transdermal route

- A low rate of permeation for most of protein drugs due to their large molecular weight.
- High intra- and inter-patient variability.
- Because the skin has a relatively low proteolytic activity, the peptide drugs have poor skin permeability.

1.9.2.9 Pulmonary Route

The respiratory tract offers an alternative site for systemic non-invasive delivery of peptides/proteins. Most of these drugs are readily absorbed through the lung, once they entered the deep lung tissues via transcytosis. They provide larger surface area (70 m^2) as compared to the other mucosal sites including nasal, buccal, rectal and vagina (Figure 1.21). A simple diffusion and carrier mediated transport mechanism operate in lungs. Three devices are currently available for the pulmonary delivery of the protein/peptide drugs: metered dose inhaler, nebulizer and powder inhaler (insufflator). Insulin can be delivered by this route by using devices such as aerosol, dry powder or as a administered with penetration enhancers like 1% azone, 1% fusidic acid or 1% glycerol. Calcitonin can be delivered as dry powder by this route.

The lungs offer an alternative site for systemic delivery of peptides and proteins. They provide larger surface area (70 m^2) as compared to other mucosal sites such as nasal, buccal, rectal and vaginal. Some of the macromolecules have been extensively studied for their absorption from lungs. They include leuprolide, insulin and albumin. The aerosolized

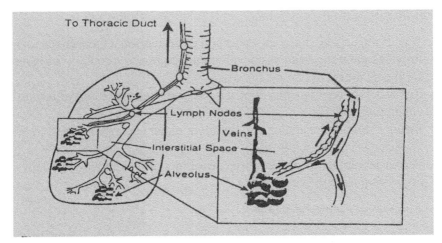

FIGURE 1.21 Schematic representation of macromolecular behavior in the lung interstitial spaces and lymphatics after intrapulmonary administration.

administration of insulin in rabbit, resulted into 40% absorption of the administered dose. This was relatively better than absorption recorded in healthy human subjects (7–16%) receiving aerosolized insulin through nebulizer. A significant increase in absorption with some penetration enhancers such as 1% azone, 1% fusidic acid or 1% glycerol. Albumin in particular was found to be absorbed largely perhaps via pinocytosis process. In order to appreciate the potential of pulmonary route in systemic administration of proteins and peptides it seems necessary to understand the permeability characteristics of lungs, absorption mechanism in the lungs, lung metabolizing capacity, factors controlling dose deposition and safety consideration with regard to permeation. About 90% of the absorptive surface area of lungs is represented by the alveoli which are comprised of heterogeneous population of epithelial cells. Out of these cells, a small group of cells called type III cells are present as free alveolar macrophages. The epithelial cells remain in intimate contact with the vasculature. The distance between air and blood is much less than 1 μm. However, due to its involvement in free exchange of gases it becomes a major barrier to the large molecule, for example, horse-radish peroxidase with a molecular weight of 40,000 deposited on air side fails to reach the interstitium. The principal resistance is offered by alveolar epithelium where the cells are tightly intercalated. The pore size of 6–10 A in alveolar membrane and 40–50 A in pulmonary capillary membrane has been established by Taylor and Gar.

Controlling the release of proteins/peptides via the pulmonary route
The inhalation route is seen as the most promising non-invasive alternative for the delivery of proteins; however, the short duration of activity of drugs delivered via this route brought about by the activities of alveolar macrophages and mucociliary clearance means there is a need to develop controlled release system to prolong the activities of proteins delivered to the lung. Polymeric materials such as (D, L)-poly(lactic glycolic acid) (PLGA), chitosan and poly(ethylene glycol) (PEGs) have been used for controlled release of proteins. Other systems such as liposomes and microcrystallization have also proved effective.

Absorption mechanism through lungs
Similar to the intestine there operates both simple diffusion and carrier mediated transport mechanism in lungs. The diffusion is involved with the absorption of drug molecules ranging up to a molecular size of 75,000. These molecules include neutral molecules such as urea, mannitol, sucrose, ouagine, dihydroougine, cyanocobatarnine and insulin as well as anionic compounds, for example, carboxyfluorescein, heparin, sulfanilic acid, etc. The absorption rates have been recorded to be related to molecular size as the half-life of absorption from the alveolar region ranges from 0.25 minutes for antipyrine to 26.5 minutes for mannitol. The active transport involving carrier in the pulmonary absorption of peptides however is yet to determined, carriers are known to be present in several animal species and they participate in the absorption. 1- aminocyclopentane carboxylic acid, a-methyl-D glucose pyranoside and organic anions such as sodium chromoglycate and phenol red. Carrier mediated transport is usually inferred on the basis of several evidences available as in the case of a-methyl-D-glucose puranoside. The involvement of the carrier is indicated by the sensitivity of the process to inhibition by 0.5 mM phlorizin, 5 mM glucose, 0.5 mM ouabain and Na' depletion. The various factors which play major challenges in pulmonary drug delivery are:

- Reproducibility in dose deposition.
- Site of dose deposition.
- Variation in absorption rates due to variation in epithelial line thickness under physiological conditions.

- Aerodynamics of aerosolized particles.

The safety issue pertaining to the pulmonary route for protein and peptide administration should be considered with regard to immunogenicity. The responses related to later in turn affect the thickness of pulmonary epithelium and so the permeability or the absorption barrier on receiving antigen challenges, a marked response in tight junction as well as bronchial permeability has been frequently recorded. The alteration in permeability of alveolar epithelium may in turn lead to the complication such as lungs edema. Another possibility is change in degree and volume of movement of fluids across the alveolar epithelium. Another safety issue appreciated with pulmonary administration of proteins and peptides is the unlikely event of destabilization of the surfactant film that coats the alveolar surface specially by proteins and peptides. Thus, the full potential of pulmonary route cannot be realized with full justification unless the various cell types in the conduction and non-conducting air ways are characterized for their absorption capacity and mechanism, types and population of Ti proteases and immunological capabilities.

Advantages of pulmonary route

- Provide a direct route to the circulation.
- Reduction in dose requirement upto 50-fold and thus cost effective option for pharmaceutical industries.
- Fast absorption.
- Safe route for drug entry even in patient with lung diseases.
- No triggering of immune function.
- Increase patient compliance with a minimum of discomfort and pain.

Disadvantages of pulmonary route

- Most of the drug is delivered to the upper lung, an area with low systemic absorption.
- Only a small amount of drug can deliver.

1.9.3   PARACELLULAR DELIVERY OF PEPTIDES

Currently, a new rout the paracellular pathway is being explored for delivery of peptides. As opposed to the transcellular pathway, the paracellular pathway is a water-filled pathway, which is amicable to the delivery of polar molecules like peptides and proteins. Another advantage is that by traversing through the area between the two cells the peptide also circumvents the intracellular lysosomal enzymes. A major problem is the low pore size of the paracellular pathway. The pore size can be regulated by various cellular signaling pathways. Paracellular delivery of proteins and peptides may become a clinical reality as it would be possible to design simple and easy to use dosage forms for non-parenteral delivery of proteins and peptides. Polysaccharide hydrogels useful in the development of controlled release formulations for protein drugs. Polysaccharide microspheres; Polysaccharide-conjugated protein drugs; Polysaccharide matrix in protein drug delivery; and Microencapsulation of protein drugs.

1.9.4 LYMPHATIC TRANSPORTATION OF PROTEINS

Proteins and peptides on fatty acylation may form a chylomicron like supramolecular assemblage which as a whole cold passively be taken up by the enterocytes of Peyer's patches and transported to the lymphatic 1) lacteals. Similarly proteins and peptides contained in colloidal dosage form with exterior lipophilic character could be absorbed following oral administration. Some typical examples of lymphotrophic carrier include w/o/w 1 multiple emulsions, fat ultra emulsion, emulsomes and supramolecular biovectors. The absorbed carrier or drug entity typically follows the way as it is established for dietary lipids. Orally ingested lymphotrophs are 2 solubilized or taken up by lecithin/bile salt(s) mixed micelles and passively absorbed into the enterocytes, then at the level of RER (rough endoplasmic reticulum) and golgi bodies the apoprotein(s) as targeting ligands absorbed protein coated lymphotrophs are exocytosed from enterocytes slowly. The latter has been recorded as a rate limiting step. The carrier then joins blood stream/circulation at thoracic duct.

For peptide delivery the lymphatics are assumed to be of prime importance due to the following reasons:

- Lymphatic route being a vital route in some metastatic cancers, lymph nodes lend themselves as a potential target for chemotherapy.

- Circulation antibodies produced by thymus-dependent small lymphocytes, large lymphocytes and macrophages are reported to be implicated in immunological reactions.

Lymphatic system transports slowly large proteins, antibodies and lymphocytes and then they return to the vascular system. The lymphatic capillary network takes up the large particles and molecular complexes that enter the tissue-fluid. After passing through lymph nodes, lymph is transported to the great vein at the base of the neck. The characteristic feature of lymph flow is that it is usually a one-way transport (Figure 1.22). For successful lymphatic delivery of peptidal moieties, following approaches can be utilized:

- For smaller moieties, lympho-selective delivery can be achieved by employing carriers like soluble macromolecules, microparticles, etc. of appropriate sizes.
- Bioavailability of peptides at site of action is to be improved.
- An increment in residence time of peptidal moieties in lymph circulation.

Transport of drugs through lymphatic pathway is to a large extent affected by the site of administration. For instance, dextran after intravenous

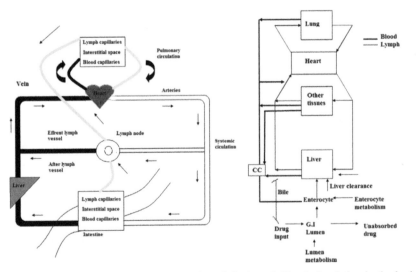

FIGURE 1.22   (a) Schematic representation of the lymph-blood circulation in the body (b) Physiological pharmacokinetic model depicting the relationship of venous and arterial blood flow to lymph flow after oral drug input.

administration shows poor lymph levels in comparison to blood levels. However, better lymph levels were observed after interstitial administration. Still better results have been observed with microparticles injected into tissues like stomach wall and subcutaneous areas. This preferential delivery to lymphatics is attributed to the ability of large molecules to penetrate through intracellular gaps of the lymph capillaries. The G.I. tract is a promising port for peptide administration. The transfer into lymph vessels via absorptive epithelial cells can occur through two pathways.

- Transcellular lipid pathway: Here chylomicrons are formed in the cells and transferred into lymph capillaries.
- Paracellular pathway: This plays a negligible role but is reported to operate by the addition of absorption enhancers. Another important route is transcytosis through Peyer's patches and is believed to be suitable for highly potent moieties like lymphokines and vaccines.

The G.I. epithelial barrier allows easy permeation of lipophilic and small molecules. The limitation that restrains the absorption of peptides and proteins following oral administration has already being discussed. In the case of peptides it has been observed that if the molecule could cross epithelial barrier it may be traced back in systemic circulation. Cyclosporin A is a unique peptide with extra lipophilicity, soluble in lipoidal adjuvants and exhibits good absorption characteristics. When the drug was solubilized in lipoidal micellar solutions, higher lymph levels were obtained following intragastric and rectal administration, since CIA binds to chylomicrons in lymphatic components. Therefore, being related to chylomicron and lipoprotein it is found to be better absorbed following intragastric administration as compared to the rectal. With respect to oral administration endocytotic lymph targeting via non-lymphoid tissue of intestine does not seem to be an approach of potentials. Most of mammalian intestinal cells have capacity to endocytose macromolecules. The process has especially been observed related to mast cells of Peyer's patches so called gut-associated lymph tissues (GALT) consisting of aggregated lymphoid follicles. The Peyer's patches are a standard through luminal-epithelia, lamina propria and lamina submucosa. The size of microparticles may have decisive fate of micro-particles and as a result their peptidal load particles less than 51 μm in diameter localize within mesenteric lymph nodes following their transport through Peyer's patches.

1.9.4.1   Colorectal Transport

The colorectal area as well as small intestine region is predominantly supplied with blood and lymph capillaries and vessels. The colorectal mucosa imposes a barrier that is usually tighter in paracellular transport. However, the route of transport could be improved by addition of permeation enhancers like oleic acid and linoleic acid. Bleomycin (BLM) is a cationic glycopeptide that binds to anionic macromolecules, thus it results into the formation of ion-pair in addition with dextran sulfate. Both BLM and dextran sulfate possess a hydrophilic surface, thus, do not qualify for permeation individually. The macromolecular ion-pair complex being relatively apolar in nature could serve as a lymphotroph leading to lymphoselective delivery.

1.9.4.2   Pulmonary Transport

The lung lymphatics are considered significant in the chemotherapy of cancer of lung origin and in the diseases related to pulmonary lymph nodes. Macromolecules were passed through lung epithelium and were transferred from interstitial space into the lymphatic pathway and subsequently blood circulation (Figure 1.19). FITC labeled dextrans of different molecular weights could successfully investigate size related lymph transport of protein molecules.

Hence it follows that trafficking and targeting through lymphatics could be adopted as a strategy for lymphoselective delivery of peptides where different types of lymphotrophs could be used as carrier units. Furthermore, contribution of aggregated lymphoid tissues such as Peyer's patches, appendix and tonsils is yet to be explored as portal units for successful delivery.

1.9.5   SITE-SPECIFIC PROTEIN MODIFICATION (PROTEIN ENGINEERING)

This approach is being used to improve the stability and specificity of endogenous proteins, in addition, they improve the selectivity and have a prolonged delivery at the active site. One such approach is deletion mutants. The approach has extensively been studied by genetic engineering and cloning of tPA gene and its subsequent expression in eukaryotic cells. An altered pharmacokinetic and thrombolytic property of deletion mutations of human tPA was observed. Using a series of deletion mutants, it has been demonstrated that regions within tPA responsible for liver clearance, fibrin affinity and fibrin specificity are not localized in the same structure. One another approach towards site-specific protein modification is the hybrid proteins. Hybrid proteins have the combination or re-ordered features of one or more proteins as well as their effector functions, protection and recognition properties. These site-specific hybrid proteins can be produced by synthetically linking protein fragments or by using ligated gene fusion processes.

Ligated gene fusion is an approach by which hybrid proteins can be brought about. Gene fusion techniques are used to produce proteins with special but specific properties which vary from the parent protein. This can be well understood by looking into the examples of targeting bacterial and plant toxins to specific cells using hybrids created by ligating toxin and growth factor genes. The approach basically relies upon the deletion of the toxin gene sequence encoding the cell-binding site and allows the hybrid-fusion protein to display the cell specificity of the growth factor. One such example is the CD4-Pseudomonas exotoxin hybrid protein, which shows selective toxicity towards cells expressing the HIV envelope glycoprotein. An increased antiprolifera activity has been recently reported by a hybrid protein of interferon-y and tumor necrosis factor-3 as comp to either interferon-y or tumor necrosis factor-13. Synthetically linked hybrid conjugates constitute another approach by which site-specific prot modification can be brought about. The concept is based on the biological disposition of protein after t chemical linkage to other protein fragments. For example, the toxin gelonin with a circulation half-life of minutes in mice when conjugated with immunoglobulin fragments, increased to days.

1.9.5.1 Enzyme-Peg Conjugates

Polyethylene glycol derivatives of L-asparaginase have been prepared and were observed to have considerable improved properties over the native enzyme. The PEG conjugates were resistant to proteolytic degradation were non-immunogenic in nature. However, its catalytic properties were reduced to 52%. Similar approach h been used to prepare PEG conjugates of other enzymes like uricase, catalase, adenosine deaminase, etc.

1.9.5.2 Protein Glycosylation

The size, surface character and chemical reactivity mainly control pharmacodisposition of proteins and peptides. It has been identified that most of the endocytosis of glycoproteins are through their specific carbohydrate residues which complex with oligosaccharide specific recognition systems contained on plasma surface of target cell. Therefore, glycosylation pattern could be considered as signals which are used by the body for disposition and regaining of its own glycoproteins. This is largely implicated for enzymes, hormones and immune surveillance. One of the classical example describes translocation of lysosomal enzymes to the lysosomes wherein the process of translocation is mediated by phospho-D-mannopyranosyl moiety of lysosomal enzymes and phosphorylation of D-mannose residues. Acyloglycoproteins having biantennary and triantennary oligosaccharides are preferably absorbed by leukocyte lectins whereas triantennary and tetrantennary oligosaccharides are found to rapidly bind to the plasma membrane of hepatocytes in liver. Galactose specific recognition system for D-galactose and N-acetyl-D-galactosamine has been identified in hepatocytes, lymphocytes and on macrophages. Similarly mannose containing oligosaccharides are intercepted by the endothelial cells and Kupffer cells. Apparently, reductive mannosamination may provide an effective strategy for directing their protein to such cells. Carbohydrates as such play no direct role in the biological activity of glycoproteins. Nevertheless, they may influence the stability and conformation of glycoproteins. They may also impart aqueous affinity to the glycoproteins through their hydrophilic character. The well identified biological property (circulation half-life) of glycoprotein may be diversified by altering the distribution of carbohydrates on their surface. Likewise immunogenicity and accessibility to the cellular site of action may be affected. In contrast, to the possible changes in the amino acid composition of a protein through sugar coupling, numerous variations are possible which could be glycoproteins and almost limitless variability and diversity in structure could be imparted. The incorporation of oligosaccharide as affinity handle into macromolecular drug carrier could be utilized effectively for the targeting of small molecules. Various examples have been cited elsewhere in this book. The conjugation of fragment A of diphtheria toxin to galactose containing oligosaccharide such as asialofetuin or asilo or orosomucoid interestingly resulted into targeting of diphtheria components to the hepatocytes. The uptake of such systems has been found to be nonlinear. This is an indicative of ligand mediated endocytosis.

1.9.5.2.1 *Expression Cell Processing*

Protein glycosylation is a rare event in prokaryotic cells such as E. coil. However, some glycosylation have been reported in eukaryotes such as yeast and mammalian cells with different resultant glycosylation patterns. Although recombination technology in bacteria can produce large amount of protein, this is often changed by bacterial proteases and within expression cells. Mammalian cell is adaptively suitable for proteins with complex modification such as y-carboxylation of glutanyl residue and for obtaining homologus human glycosyl pattern with marked effect in biological utilization and therapeutic efficacy of proteins.

1.9.5.3 Modification of Proteases into Peptide Ligases

Peptide ligation of native enzymes tends to high specificity and stereoselectivity. Synthesis of any enzyme capable of catalyzing peptide ligation is advantageous. For example, "subtilisin" a protease has been modified by converting a serine into cysteine or seleno-cysteine which can catalyze peptide ligation a protease (subtilisin) intermediate protein peptide ligase: Chemical modification of a protease by phenylmethyl sulfonyl fluoride to produce a peptide ligase

1.9.5.4 Production of Site Specific Nucleases

The binding properties and DNA recognition properties can be combined using chemical cleavage agent. Cys of E. coli CAP protein has been modified using 5-iodoacetamide-1,10-phenanthroline yielding a DNA cleaving agent, that recognized and cleaved DNA at the center of the recognition site (22 bp) for CAP. This gives restriction enzymes which are capable of recognizing upto 20 bases instead of 6 or 8 bases and are therefore useful in isolating long DNA fragments needed for sequencing and mapping. Specific nucleases can also be produced by fusion of non-specific phosphodiesterases to oligonucleotides defined sequence. This approach can also be used for developing artificial restriction enzymes.

1.9.5.5   Production of Artificial Semisynthetic Oxidoreductase (Flavo-Enzymes)

This can be prepared by incorporating redox-active prosthetic group covalently to the existing sites. In this k 6 method a redox-active group, 10-methyliso-alloxazine derivatives are linked to specific sites of several proteins. The efficiency of these semisynthetic enzymes is comparable with that of naturally occurring flavo-enzymes.

1.9.6 MICROENCAPSULATION OF PROTEIN DRUGS FOR DRUG DELIVERY: STRATEGY, PREPARATION, AND APPLICATIONS

Bio-degradable poly(lactide) (PLA)/poly(lactide–glycolide) (PLGA) and chitosan microspheres (or microcapsules) have important applications in Drug Delivery Systems (DDS) of protein/peptide drugs. By encapsulating protein/peptide drugs in the microspheres, the serum drug concentration can be maintained at a higher constant value for a prolonged time, or injection formulation can be changed to orally or mucosally administered formulation. PLA/PLGA and chitosan are most often used in injection formulation and oral formulation. However, in the preparation and applications of PLA/PLGA and chitosan microspheres containing protein/peptide drugs, the problems of broad size distribution and poor reproducibility of microspheres, and deactivation of protein during the preparation, storage and release, are still big challenges. In this article, the techniques for control of the diameter of microspheres and microcapsules will be introduced at first, and then the strategies about how to maintain the bioactivity of protein drugs during preparation and drug release will be reviewed and developed in our research group. The membrane emulsification techniques including direct membrane emulsification and rapid membrane emulsification processes were developed to prepare uniform-sized microspheres, the diameter of microspheres can be controlled from submicron to 100 μm by these two processes, and the reproducibility of products can be guaranteed. Furthermore, compared with conventional stirring method, the big advantages of membrane emulsification process were that the uniform microspheres with much higher encapsulation efficiency can be obtained, and the release behavior can be adjusted by selecting microsphere size. Mild membrane emulsification condition also can prevent the deactivation of proteins, which frequently occurred under high shear force in mechanical stirring, sonification, and homogenization methods. The strategies for maintaining the bioactivity of protein drug were developed, such as adding additives into protein solution, using solid drug powder instead of protein solution, and employing hydrophilic poly(lactide)–poly(ethylene glycol) (PELA) as a wall material for encapsulation in PLA/PLGA microspheres/microcapsules; developing step-wise crosslinking process, self-solidification process, and adsorbing protein drug into preformed chitosan microsphere with hollow-porous morphology for encapsulation in chitosan microsphere. As a result, animal test demonstrated that PELA microcapsules with uniform size and containing recombinant human growth hormone (rhGH) can maintain higher blood drug concentration for 2 months, and increased animal weight more apparently only by single dose, compared with PLA and PLGA microcapsules; hollow-porous chitosan microsphere loading insulin decreased blood glucose level largely when it was used as a carrier for oral administration

The rapid development of DNA-recombination techniques and other modern biotechnologies promoted more and more protein and peptide drugs to be produced on a large scale, and they are becoming a very important class of therapeutic agents instead of conventional chemical drugs because of their low side effect, specific targeting site, low dosage, and so forth. However, the available protein and peptide drugs are generally characterized by a short biological half-life time, because they are easily hydrolyzed or degraded by enzymes in vivo; thus frequent injection usually is necessary. For example, the half-life time of rh-GH is only 2–3 h by subcutaneous injection, daily injection should be carried out for several years, bringing pains and economic burdens to patients. In addition, other routes such as oral and mucosal administrations are difficult because of their large molecular size and easy degradation character. Microencapsulation techniques have been developed to solve the above problems. Microencapsulation can envelop protein/peptide drugs inside and protect them from degradation by enzymes, and can release them slowly to realize higher constant serum concentration for a prolonged time. Furthermore, the targeting formulation and other formulations such as oral and mucosal administrations may be realized by designing adequate materials, size, surface, and functionalities. Although extensive studies on preparation and applications of encapsulated protein drugs have been performed, only a few products have come to market. Two main difficulties in preparation and applications impeded their development to market:

- The size and size distribution of microspheres/microcapsules are difficult to control, resulting in poor reproducibility in large scale production, as a result, the delivery system is difficult to be approved
- In the case of encapsulation of protein drug, it is also difficult to maintain the bioactivity of protein/peptide drugs during the preparation, storage, and release of the drug, which has rarely been encountered in the encapsulation of chemical drugs.

The protein/peptides will lose their therapeutic effect and even cause the significant side effect if they are deactivated. There are several reasons responsible for protein denaturation.

- Rigorous mechanical shear force will cause the loss of three-dimensional structure of proteins
- the contact of protein with oil/water interface, or crosslinking of protein, will cause the coagulation of proteins
- Hydrophobic interaction between protein and hydrophobic wall material will lead to the denaturation of proteins.

The PLA or PLGA microspheres or microcapsules prepared from poly(dl-polylactide) were the most utilized formulations in Drug Delivery System (DDS) because they are amorphous in the particles, can load drug with higher encapsulation efficiency and can be degraded and absorbed in human body within several months. Furthermore PLA and PLGA have been approved as excipients in injection formulation by the Food and Drug Administration (FDA). Chitosan was considered to be one of the most potential materials for oral or mucosal administration in literatures. We first developed membrane emulsification process to prepare O/W (oil in water) emulsion to obtain uniform PLA microspheres, then we extended membrane emulsification process to W/O/W double emulsion system to prepare uniform PLA microcapsules. Furthermore, we developed it to W/O emulsion system to prepare uniform polysaccharide microspheres and microcapsules. By developing membrane emulsification process to different emulsion types, hydrophilic protein/peptide drugs can be loaded into hydrophobic PLA/PLGA materials or hydrophilic chitosan material. On the other hand, when protein drug is encapsulated in PLA/PLGA microcapsules by W/O/W, protein is easy to contact with water–oil interface to lose its bioactivity as described above. Researchers have developed several methods to reduce bioactivity lose of proteins/peptides, such as adding hydrophilic PEG and copolymers, sugars, carbohydrates, polyols, basic salts of bivalent metal ion, and pre-encapsulating protein in hydrophilic core to shield proteins from contact with interface, thus to reduce the interface/protein ratio. In this article, we developed several strategies to maintain its bioactivity. When protein drug is encapsulated in hydrophilic chitosan, chitosan droplet usually has to be solidified to microsphere by chemical crosslinking, as a result, protein will be cross-linked inside of chitosan and crosslinking between proteins will occur to lose its bioactivity. Researchers have developed several methods to solve this problem, such as employing ionic crosslinking method and utilizing self-assembly method of amphiphilic chitosan after introducing hydrophobic side chain on chitosan. In ionic crosslinking method, for example, it involved the ionic interaction of chitosan/chitosan derivatives and negatively charged macromolecules or a low molecular weight anionic cross linker, such as tripolyphosphate (TPP), sodium sulfate or cyclodextrin (CD) derivatives. Especially, ionic gelation of chitosan with TPP has been extensively used for the preparation of protein and antigen-loaded microspheres.

1.10   TOXICITY PROFILE CHARACTERIZATION

The available literature often lacks data suggesting the possible mechanism(s) of pharmacological and/or 151 toxicological actions of peptide and protein therapeutics in intact mammalian systems. These data are required:

- To determine the maximal tolerated dose by a given route of administration.
- To have an insight into the dose-limiting toxicity and the target organs involved leak.
- To confirm the intrinsic and interactive safety of the proposed formulation to be used.
- To ensure that the manufacturing and formulation processes will provide biologically active and stable material free from contaminating substances that might increase toxicity and/or decrease efficacy.

1.10.1   CLASSES OF TOXICITY OF PROTEINS AND PEPTIDES SU

1.10.1.1   Exuberant Pharmacologic Responses BOH

The dose-related toxicity is the most commonly encountered toxic manifestation of peptides and proteins. The toxicity of this type includes the arteriospasm and hypertension after an overdose of angiotensin II and the profound hypertension and the resultant convulsions after large doses of insulin. To some extent the doses of the various fibrinolytic proteins like tissue-type plasminogen activator, urokinase used in thrombolytic disorders are Cou prone to elicit exuberant pharmacologic response.

1.10.1.2   Generic Toxicity

Generic toxicity is an adverse reaction, which is exhibited by most, if not all, members of a class of proteins. For instance, fever, chill and diaphoresis are observed when monoclonal antibodies are infused that react with circulating cells. Also, fever, fatigue, arthralgias, myalgias, and malaise develop after administration of interferons. Immunogenicity is perhaps the most distinguishing generic toxicity of the peptide and protein drug moieties. This feature has been exploited for therapeutic needs in development of vaccines. Immunogenic behavior of proteins is largely attributed to their molecular weight, extent of structural complexity and hydrophobicity, propensity to aggregate, aromatic amine content and sequence homology to the native protein in the test species. The general pathogenic mechanisms involved in immunologic injury include;

- *Type I reaction*: This mechanism involves the release of active mediators (e.g., bradykinin, histamine, Let serotonin and numerous arachidonic acid metabolites) from mast cells through an antibody-based mechanism. Lip the spectrum of responses observed is wide and includes allergic dermatitis, allergic rhinitis, asthma, urticaria, Ma oesophageal varices, G.I. hypermobility and acute hypotension.
- *Type II reaction*: This mechanism is mediated through the interaction between antibody and antigenic sites at the cell surface. These cells are usually present within the vascular bed but extravascular cells may be targeted as well. Cytotoxicity and cell lysis are the manifestations of this type of reaction. Cytotoxicity mechanisms usually lead to anemias, hemorrhagic conditions, neutropenia and related secondary disorders. Natural activities of a cell may be stimulated or inhibited through receptor binding. They may not be cytotoxic themselves but are capable of leading to disease. For example, long-acting thyroid stimulator (LATS), an antibody to the thyroid cell, brings about increased production of thyroid hormone. Type III reaction: This reaction is mediated through preformed antigen-antibody complexes, usually in antibody excess, which produce damage if they are deposited on the walls of the blood vessel or in the interstitium of vital organs. This mechanism is proposed to be responsible for immunologic injury following protein administration. Type IV reaction: This mechanism of immunologic injury occurs in the absence of antibody and complement. It is mediated by sensitized T lymphocytes. The damage is mediated through direct T-cell cytotoxicity and/or cytokine release, which leads to mobilization and attraction of monocytes, other lymphocytes and neutrophils, leading to augmentation in tissue injury. The examples include cell-mediated lympholysis, allograft rejection, tuberculin skin reaction, skin reactions to poison injections and diphtheria toxoid.

1.10.1.3   Idiopathic Toxicity

This is an unexpected toxicity and is elicited only by some members of a protein therapeutic class. Generally, these toxic reactions are unrelated to the mechanism of action of the protein. Examples, include the 'vascular leak syndrome' associated with long term IL-2 infusion, hypotension observed with TNF-a administration etc.

1.11   PEGYLATION AND ITS DELIVERY ASPECTS

Although various injected peptide and protein therapeutics have been developed successfully over the past 25 years, several pharmacokinetic and immunological challenges are still encountered that can limit the efficacy of both novel and established biotech molecules. PEGylation is a popular technique to address such properties. PEGylated drugs exhibit prolonged half-life, higher stability, water solubility, lower immunogenicity and antigenicity, as well as potential for specific cell targeting. Although PEGylated drug conjugates have been on the market for many years, the technology has steadily developed in respect of site-specific chemistry, chain length, molecular weights and purity of conjugate. These developments have occurred in parallel to improvements in physicochemical methods of characterization. This section describes the achievements in PEGylation processes with an emphasis on novel PEG-drugs constructs, the unrealized potential of PEGylation for non-injected routes of delivery, and also on PEGylated versions of polymeric nanoparticles, including dendrimers and liposomes.

1.12   NOVEL DELIVERY TECHNOLOGIES

The protein and peptide therapeutics have become an important class of drugs due to advancement in molecular biology and recombinant technology. There are more than 100 biopharmaceutical products approved and generating revenue of more than 56 billion US dollars. A safe, effective and patient friendly delivery of these agents is the key to commercial success. Currently, most of therapeutic proteins are administered by the parenteral route which has many drawbacks. Various delivery strategies and specialized companies have evolved over the past few years to improve delivery of proteins and peptides. Polymeric depot and PEGylation technologies have overcome some of the issues associated with parenteral delivery. A considerable research has been focused on non-invasive routes such as pulmonary, per oral and transdermal for delivery of proteins and peptides, in order to increase patient compliance yet their delivery via non-invasive routes remains challenge due to their poor absorption and enzymatic instability. Pulmonary route has shown some success evidenced by recent FDA approval of inhalable insulin. Development of an oral dosage form for protein therapeutics is still the most desirable one but with greater challenge.

1.13 RECENT PROGRESS IN DELIVERY OF PROTEIN AND PEPTIDE BY NONINVASIVE ROUTES

Much progress has been made in the last 5 years toward delivery of protein and peptide drugs by noninvasive routes. The obstacles of instability, poor absorption, rapid metabolism, and nonlinear pharmacokinetics are great challenges for which some solutions are now emerging. Structural modifications of the protein by chemical or recombinant means have improved stability and minimized enzymatic cleavage in some cases. Protection of the protein or peptide drug via liposomes or polymers also offers a means for increasing stability and prolonging half-life. Novel permeation enhancers, which show minimal irritation to mucosal membranes, have become available and show promise for increasing absorption of proteins delivered by a number of noninvasive routes. There are examples in which several of these methods have been used concomitantly to achieve maximum effect; for instance, a bioadhesive microsphere formulation containing a novel permeation enhancer was used to maximize nasal delivery of insulin. Therefore, general methods exist whereby delivery by any noninvasive route may be improved. In some cases, choice of the best route of delivery for a particular drug makes the difference between success and failure. A comparison of the enzyme activity at the various sites of delivery is helpful and, fortuitously, the enkephalins, model peptides whose rate of cleavage and type of degradation products offer information about the type and activity of enzymes present, have been studied extensively.

1.14 COMMERCIAL CHALLENGES OF PROTEIN DRUG DELIVERY

The discovery of insulin in 1922 marked the beginning of research and development to improve the means of delivering protein therapeutics to patients. From that period forward, investigators have contemplated every possible route of delivery. Their research efforts have followed two basic pathways: one path has focused on non-invasive means of delivering proteins to the body; and the second path has been primarily aimed at increasing the biological half-life of the therapeutic molecules. Thus far, the commercial successes of protein delivery by the nasal, oral and pulmonary routes have been more opportunistic rather than the application of platform technologies applicable to every protein or peptide. In several limited cases, sustained delivery of peptides and proteins has employed the use of polymeric carriers. More successes have been achieved by chemical modification using amino acid substitutions, protein pegylation or glycosylation to improve the pharmacodynamic properties of certain macromolecules. Today, commercial successes for protein and peptide delivery systems remain limited. The needle and syringe remain the primary means of protein delivery. Major hurdles remain in order to overcome the combined natural barriers of drug permeability, drug stability, pharmacokinetics and pharmacodynamics of protein therapeutics.

1.15 ALTERNATIVE DELIVERY SYSTEMS FOR PEPTIDES AND PROTEINS AS DRUGS

The emergence of recombinant DNA technology has resulted in the large-scale production of a myriad of genetically engineered proteins and peptides, making many of them available for the first time for potential use as therapeutic entities. In addition, increased knowledge in the area of peptide/polypeptide hormones has resulted in an expansion of research efforts utilizing peptide synthetic chemistry, aimed toward developing novel therapeutic peptide drugs. Proteins and peptides cannot readily be administered by the conventional oral route, and thus alternative delivery methods to circumvent the necessity of frequent injections are being explored.

1.16   POLYMERIC DELIVERY OF PROTEINS AND PLASMID DNA FOR TISSUE ENGINEERING AND GENE THERAPY

In vivo gene expression can be altered by locally delivered DNA and proteins. The ability to deliver bioactive macromolecules, such as proteins and plasmid DNA, over controllable time frames represents a challenging engineering problem. Considerable success has been achieved with polymeric delivery systems that provide the capability to change cell function either acutely or chronically.

1.17   SMART POLYMER FOR PROTEINS AND PEPTIDES [186]

Biodegradable polymeric systems represent promising means for delivering many bioactive agents, including peptide and protein drugs. The development of smart polymer-based injectable drug delivery systems has gained attention over the past few years. The advantages of this delivery system are ease of application, localized delivery for a site-specific action, prolonged delivery periods, decreased body drug dosage with concurrent reduction in possible undesirable effects common to most forms of systemic delivery, the non toxic degradability, and improved patient compliance and comfort.

Smart polymers are macromolecules that display a dramatic physicochemical change in response to small changes in their environment. Smart polymer-based injectable formulations are easy to prepare and form implants at the site of injection upon administration. Smart polymers can be classified according to the external stimulus they respond to (i.e., temperature, pH, solvent, magnetic field, ions, and pressure). Various types of polymers are used, for example

- Temperature sensitive polymers, for example, Poly (ethylene oxide)-poly (propylene oxide)-poly(ethylene oxide) triblock copolymers (PEO-PPO-PEO), poly(n-isopropylacrylamide) (PNIPAAM)
- Phase sensitive polymers, for example, Poly (D, L-lactide), Poly (D, L-lactide-co glycolide)
- pH sensitive polymers, for example, Poly(methacrylic acid g-ethylene glycol) P(MAA-gEG)
- Photo sensitive polymers, for example, PEG, Poly(vinyl alcohol), PEO-PPO.

1.18   HYBRID PROTEIN DELIVERY SYSTEMS [187]

Heterologous hybrid proteins can be designed bearing the combined or reordered features of one or more proteins that display effector functions, protection abilities and recognition properties. Site specific hybrid proteins may be produced by employing ligated gene fusion processes or by chemical linkages of protein fragments.

1.19   LIGATED GENE FUSION HYBRID DELIVERY SYSTEMS

Gene fusion techniques can be applied to blend the diverse properties of parent proteins and there by bioactive protein with improve pharmacological effects can be produced. Hybrid protein between interferon γ and TNF β have been developed. This hybrid preserved the cell specificity of the growth factor, antiviral activity and increase antiproliferative activity.

1.20   SYNTHETICALLY LINKED HYBRID CONJUGATES

Biological disposition and fate of the proteins can be modified by linking them to other proteins. The toxin gelonin has a circulation half-life of 3.5 min (in mice) but on conjugate it to immunoglobulin, the terminal phase blood half-life was increased to days.

1.21   PEPTIDE TARGETING

Natural and synthetic peptides are a class of small ligands that have great potential for such applications. They offer the advantage of providing infinite sequence/structure possibilities that can potentially be designed to bind to any cancer related target. Furthermore, such an approach is expected to yield fewer problems related to immunogenicity. Among potential targets, there are several cell surface receptor systems that have small peptides as ligands that have been shown to be highly expressed in a variety of neoplastic and non-neoplastic cells. Furthermore, receptor-targeting peptides have shown a high level of internalization within tumor cells via receptor-mediated endocytosis. Such a feature of these systems may be of value in facilitating intracellular delivery of the intended payload. The drawbacks related to the use of these compounds are the relatively lower target affinities and the metabolic instability of these compounds that may be extremely sensitive to protease degradation. Improving metabolic stability and pharmacokinetics can be attempted by modifying peptide sequences using specific coded or uncoded amino acids or amino acids with D configurations. Cycling of the N-terminal with the C-terminal or with a side-chain, or the C-terminal with a side-chain and the side-chain with another side-chain, can also be utilized for such purpose. Another advantage is the possibility of designing analogs that can act as antagonists. Cell surface receptor antagonists show the dual advantage of not activating the biological pathways following receptor binding and have also been shown to have higher binding capacities to their agonist counterparts. These attractive physical properties coupled with their smaller size make peptides very appealing candidates for developing new target-specific nanoparticles. Most peptide based targeting ligands are derived from known endogenous proteins capable of binding the target receptor with high affinity. Molecular modeling of new peptide sequences based on the known three-dimensional structure of the target receptor is also a possible strategy for rational design of new compounds, although such an approach requires thorough knowledge of the structure of ligand/receptor interaction. A further possibility for identifying new peptide sequences for recognizing tumor-associated proteins is the use of phage display techniques. Once the binding sequence is identified a number of synthetic strategies have been put in place in order to modify the surface of micelles, liposomes, or nanoparticles in order to display the targeting peptide sequence. One main concern in this part of development is to achieve high coupling efficiency while distancing the bioactive peptide from the nanostructure surface in order to maintain the specific conformation required for high affinity binding to the target. The bioactive peptide may be introduced on the aggregate surface directly during nanostructure preparation by coupling the peptide to an amphiphilic moiety, or introducing the peptide on the surface of the nanostructures after they have been obtained. The first method, usually employed for the obtainment of peptide containing micelles and liposomes, needs a well-purified amphiphilic peptide molecule; it is mixed in appropriate solvents and in the chosen ratio with other amphiphilic molecules and phospholipids; then micelles or liposomes are obtained by evaporating the solvent or using extrusion procedures. The advantage of this approach is that one obtains a well-defined amount of bioactive molecules in the aggregates and there are no impurities. With this approach, however, the bioactive peptide is displayed on the external liposome surface as well as in the inner compartment. For the second approach, peptide coupling after liposome or nanoparticle preparation involves the introduction of suitable activated functional groups onto the external side of liposomes or nanoparticles for covalent or non-covalent peptide binding. To guarantee correct orientation of the targeting ligand, biorthogonal and site-specific surface reactions are necessary. The synthetic strategy should be aimed at optimizing reproducibility and yield of the coupling reaction. Functional groups commonly used are:

- Amine for the amine N-Hydroxysuccinimide coupling method;
- Maleimide for Michael addition;
- Azide for Cu(I)-catalyzed Huisgen cycloaddition (CuAAC); and
- Biotin for non-covalent interaction with avidin or triphosphines for Staudinger ligation, and recently, hydroxylamine for oxime bond.

A considerable number of molecular targets for peptides are either exclusively expressed or overexpressed on both cancer vasculature and cancer cells. They can be classified into three wide categories: integrins; growth factor receptors (GFRs); and G-protein coupled receptors (GPCRs). These receptors offer attractive targets for anticancer therapeutics as they are often implicated in tumor growth and progression. Many nanoparticles and liposomes have been labeled with peptides capable of interacting with these receptors and have been reported in the literature in the last decade. Nanoparticles grafted with the RGD sequence able to bind integrin receptors have been widely evaluated for the treatment of different cancers, such as ovarian cancer, melanoma, and breast carcinoma. Peptides targeting growth factor receptors have been utilized to functionalize liposomes encapsulating chemo-therapeutics. Peptides have also been developed to target the extracellular matrix of the diseased tissues, and this is an important alternative strategy to target unhealthy tissues which can also be incorporated with nanomedicine. This section will focus on delivery systems containing peptides that recognize GPCRs. GPCRs constitute a membrane protein family involved in the recognition and transduction of signals as diverse as light, Ca^{2+}, and small molecule signaling, including peptides, nucleotides, and proteins. The general structural features, obtained by indirect studies as well as X-ray crystallography, indicate the presence of seven transmembrane helices connected by three intracellular and three extracellular loops. The N-terminal domain is directed into the extracellular space and C-terminal points to the intracellular space. Ligand binding to receptor is a crucial event in initiating signals, and the study of how ligands interact with their receptors can reveal the molecular basis for both binding and receptor activation. The ligand binding site for peptides has been found in the N-terminal extradomain or on the portion of the extracellular loops adjacent to the extracellular moiety of the transmembrane helices. Knowledge of the structural details of this interaction could be very useful for designing ligands for targeted delivery. Unfortunately, detailed structural characterization of the ligand-receptor complex for most systems is very difficult to obtain. However several approaches, such as biochemical affinity, photoaffinity labeling, and site-directed mutagenesis have allowed us to determine which amino acid residues are involved in binding. The interest in developing agonist or antagonist peptides against these receptors is based on the biological role these receptor pathways have in specific cancer types. Overexpression of small peptide receptors has been documented for a wide number of cancers. As many as 10^{-5}–10^{-6} receptor molecules per cell or receptor densities in the pmol $\cdot$ mg^{-1} protein range have been reported for a variety of systems, such as somatostatin receptors in neuroendocrine tumors, cholecystokinin (CCK) receptors in medullary thyroid cancer, bombesin receptors in prostate and breast carcinoma, and several others.

1.22 DEVELOPMENT OF DELIVERY SYSTEM FOR PEPTIDE BASED PHARMACEUTICALS

1.22.1 FORMULATION CONSIDERATION

- *Preformulation studies of therapeutic peptides and proteins*
  Preformulation data must be generated to serve as the basis for the formulation development of dosage forms or for the design of delivery

system to achieve optimum physicochemical stability and maximum systemic bioavailability.

• *Surface adsorption behavior of peptide and protein molecules*
   Protein and peptide molecules have a tendency to be adsorbed to a variety of surfaces including glass and plastic.
• *Aggregation behavior of protein and peptide molecule*
   The potential problem is the self aggregation of peptide and protein molecules such as insulin. This has been minimize by the incorporation of additives like urea, dicarboxylic amino acid such as aspartic acid and glutamic acid or other reagents such as glycerols, EDTA, Lysine.

1.22.2   PHARMACOKINETIC CONSIDERATIONS

Due to the very short half-life of protein and peptides, it is more critical to get information about pharmacokinetic of therapeutic peptide and protein. Metabolic degradation by peptidases and proteinases can occur in the vascular endothelium, liver, kidney and non target tissues and even at a site of administration.

1.22.3   ANALYTICAL CONSIDERATION

Bioassay method has been available for detection and potency determination of peptides and proteins, but it is very time consuming, labor intensive. Now a days spectroscopy, chromatography, electrophoretic methods and immunoassays have been available for an analytical determination of protein or peptide. The most commonly used analytical methods include HPLC and RIA. Fast atom bombardment mass spectroscopy is also very useful in peptide and protein analysis.

1.22.4   REGULATORY CONSIDERATION

Biotechnology products regulated under the authority of four federal agencies: the food and drug administration (FDA), the environmental protection agency (EPA), the occupational safety and health administration (OSHA), and U.S. Department of agriculture (USDA).

PolyXen®: PolyXen® is an enabling technology for protein drug delivery. It uses the natural polymer polysialic acid (PSA) to prolong the active life and improve the stability of therapeutic peptides and proteins. It can also be used for small molecule drugs. PSA is a polymer of sialic acid (a sugar). When used for protein and therapeutic peptide drug delivery, polysialic acid provides a protective microenvironment on conjugation. This increases the active life of the therapeutic protein in the circulation and prevents it from being recognized by the immune system. The use of PSA makes PolyXen® a particularly effective form of protein drug delivery. PSA is a naturally occurring polymer, and it is biodegradable, non-immunogenic and non-toxic. This is particularly important where a polymer is to be used to deliver therapeutics chronically or in large dosages.

ImuXen®: ImuXen® is a group of liposomal technologies designed to improve the delivery and effectiveness of DNA, protein and polysaccharide vaccines. ImuXen® technology can help to generate strong protective immune responses, in some cases with a single injected dose. The potential advantages of ImuXen® for DNA, protein and polysaccharide vaccines include:

- Vaccine protection against degradation.
- Efficient delivery of vaccines to the immune system.
- Increased immune responses.
- Protective immunity with a single injection.
- Rapid, simple and scalable manufacture.

The benefits of ImuXen® for DNA, protein and polysaccharide vaccines include:

- Multiple vaccines delivered with a single injection.
- Reduction in the number of doses required.
- Reduction in side effects.
- Potential for oral administration of vaccines.
- Humira is the best-selling new monoclonal antibody in the market.
- Pegasys is the best-selling newcomer to the therapeutic protein, and global pharma market.
- Epogen is the second best-selling therapeutic protein.
- Aranesp, Neulasta and Enbrel are the most successful therapeutic proteins in terms of growth. Aranesp is the fastest-growing therapeutic protein.

1.23   NON-COVALENT PEPTIDE-BASED APPROACH THE DELIVERY OF PROTEINS AND NUCLEIC ACIDS

The recent discovery of new potent therapeutic molecules which do not reach the clinic due to poor delivery and low bioavailability have made of delivery a key stone in therapeutic development. Several technologies have been designed to improve cellular uptake of therapeutic molecules, including cell penetrating peptides (CPPs), which have been successfully applied for in vivo delivery of biomolecules and constitute very promising tools. Distinct families of CPPs have been described; some require chemical linkage between the drug and the carrier for cellular drug internalization while others like Pep-and MPG-families, form stable complexes with drugs depending on their chemical nature. Pep and MPG are short amphipathic peptides, which form stable nanoparticles with proteins and nucleic acids respectively. MPG and Pep based nanoparticles enter cells independently of the endosomal pathway and efficiently deliver cargoes in a fully biologically active form into a large variety of cell lines as well as in animal models. This review will focus on the mechanisms of non-covalent MPG and Pep-1 strategies and their applications in cultured cells and animal models.

Protein and nucleic-acid-based therapeutic molecules such as peptide nucleic acids (PNAs) and their derivatives or siRNA have provided new perspectives for pharmaceutical research [188]. However, their development is limited by their low stability in vivo, their poor cellular uptake and inefficient cellular trafficking. To circumvent these problems, efforts have been harnessed to improve the chemistry of these molecules. Moreover, several delivery systems have been recently developed, including promising tools based on peptide sequences that can cross the cellular barriers, termed protein transduction domains (PTDs) and cell-penetrating peptides (CPPs). Most of the CPPs that have been proposed so far, such as Tat [189], oligo–Arg, Transportan [190], and Penetratin [191], are covalently-linked to their cargoes. These CPPs are internalized by cells together with their cargoes essentially through an endocytotic pathway. Although these covalent PTD-strategies have been successfully used for the delivery of a wide range of cargoes, this technology is limited from the chemical point of view, as it is based on a synthetic disulfide linkage between carrier and cargo, which risks altering the biological activity of the latter. This point was recently pointed out for the delivery of siRNA [192]. In order to offer an alternative to covalent strategies we have proposed a new potent strategy for the delivery of biomolecules into mammalian cells, based on short amphipathic peptide carriers, MPG and Pep-1 [193]. These peptide carriers form stable nanoparticles with cargoes without requiring any cross-linking or chemical modifications. MPG efficiently delivers nucleic acids (plasmid DNA, oligonucleotides, siRNA), whilst Pep-1 improves the delivery of proteins, peptides and protein mimics in a fully biologically active form into a variety of cell lines as well as in animal models [194]. The mechanism through which MPG and Pep-1 deliver active biomolecules does not involve the endosomal pathway and therefore allows the controlled release of the cargo into the appropriate, target subcellular compartment. This review will describe the characteristics of the non-covalent MPG and Pep-1 strategies and their applications for nucleic acid and peptide transduction both in vitro and in vivo. We will focus on the application of these peptide carriers for the delivery of therapeutic nucleic acid-based compounds, specifically siRNAs and PNAs, and discuss the mechanisms involved in the delivery process.

In order to offer an alternative to covalent strategies we have proposed a new potent strategy for the delivery of cargoes into mammalian cells, based on a short amphipathic peptide carriers, which form stable nanoparticles with cargoes without the need for cross-linking or chemical modifications. MPG and Pep-1 technologies have been successfully applied to the delivery of different cargoes (siRNA and peptides) in primary cell lines and in animal model. These peptide-based strategies present several advantages including rapid delivery of cargoes into cells with very high efficiency, stability in physiological buffers, lack of toxicity and of sensitivity to serum. Moreover, the lack of prerequisite for covalent coupling to formation of carrier/macromolecule particles favors the intracellular routing of the cargo and its controlled release in the target subcellular compartment. The final localization of the delivered macromolecule is then determined by its inherent targeting properties, its partners or substrates. A major concern with the cellular uptake of CPPs is to avoid the endosomal pathway or to favor escape of the cargo from early endosomes. MPG and Pep-1 behave significantly differently from other similarly designed cell-penetrating peptides. Although we cannot exclude that MPG or Pep-1 uptake follows several routes, the major cell translocation mechanism is independent of the endosomal pathway and instead involves transient membrane disorganization associated with folding of the carrier into either helical or β-structures within the phospholipid membrane. In conclusion, although, it is clear that MPG and Pep-1 technologies still need to be optimized for systematic in vivo applications, they are already proven powerful tools for basic research and for targeting specific cellular events both in vitro and in vivo, as well as in a therapeutic context for screening potential therapeutic molecules. Both technologies constitute a great alternative to covalent strategies and will no doubt have a major impact on the therapeutic application of siRNA, proteins and peptides.

1.24   PROTEIN TRANSDUCTION TECHNOLOGY

Several proteins can traverse biological membranes through protein transduction. Small sections of these proteins (10–16 residues long) are responsible for this. Linking these domains covalently to compounds, peptides, antisense peptide nucleic acids or 40-nm iron beads, or as in-frame fusions with full-length proteins, lets them enter any cell type in a receptor- and transporter-independent fashion. Moreover, several of these fusions, introduced into mice, were delivered to all tissues, even crossing the blood–brain barrier. These domains thus might let us address new questions and even help in the treatment of human disease. Currently, the ability to ectopically express novel proteins that can either alter the cellular phenotype or provide therapeutic benefit is largely limited to recombinant genetic approaches. The introduction of the transgene and its sustained and regulated expression is often difficult to achieve, however, and can result in undesirable consequences such as immunogenicity, toxicity and an inability to target many cell types. These issues have limited the efficacy of transgenes in vivo. Moreover, the ability to circumvent genetic approaches by the delivery of full-length proteins directly into cells is problematic owing to the bioavailability restriction imposed by the cell membrane. In general, the plasma membrane of eukaryotic cells is impermeable to the vast majority of peptides and proteins. However, this dogma has recently been shown to be untrue with the identification of several protein transduction domains (PTDs) that are capable of transducing cargo across the plasma membrane, allowing the proteins to accumulate within the cell. The three most widely studied PTDs are from the Drosophila homeotic transcription protein antennapedia (Antp) [194], the herpes simplex virus structural protein VP22 [195], and the human immunodeficiency virus 1 (HIV-1) transcriptional activator Tat protein [196]. Transduction across the membrane by these PTDs occurs through a currently unidentified mechanism that is independent of receptors, transporters and endocytosis. Moreover, transduction occurs via a rapid process that at both 37°C and 4°C targets essentially 100% of cells in a concentration-dependent fashion. Significantly, when synthesized as recombinant fusion proteins or covalently cross-linked to full-length proteins, these PTDs are capable of delivering biologically active proteins, such as β-galactosidase, intracellularly. These PTD fusion proteins are found both within the cytoplasm and the nucleus. The identification of short basic peptide sequences from these proteins (Antp, RQIKIWFQNRRMKWKK; Tat(47–57), YGRKKRRQRRR; sequences given in single-letter amino acid code) that confer cellular uptake has led to the recent identification and synthesis of numerous new PTDs [197]. Although different PTDs show similar characteristics for cellular uptake, it is clear that they vary in efficacy in transporting their cargo into the cell. Although there is limited homology between these PTDs, the rate of cellular uptake has been found to strongly correlate to the number of basic residues present, specifically the number of arginine residues. These results indicate the presence of a common mechanism that probably depends on an interaction between the basic charges on the PTD and negative charges on the cell surface. To date, fusions created with the Tat(47–57) PTD show markedly better cellular uptake than similar fusions using the 16 amino acid sequence from Antp; however, recently devised peptide transducers, such as the retro-inverso form of Tat(57–48) or homopolymers of arginine, appear to further increase cellular uptake several fold. Moreover, although the Antp PTD can transduce cells when associated with chemically synthesized peptides [198], the efficiency dramatically decreases with the incorporation of larger proteins. VP22 transduction is somewhat different from that of Tat or Antp. In this system, the DNA encoding the entire VP22 protein is genetically fused to the gene of interest and transfected into cells. The fusion transgene is then transcribed and the translated protein transduces from the primary transfected cells into the surrounding cells to varying levels [199]. Exogenously added VP22 fusion proteins have been reported to be internalized, but little data about the efficiency of this protein delivery mode is available. The direct delivery and efficient cellular uptake of transducing proteins offers several advantages over traditional DNA-based methods for manipulating cellular phenotypes. Consequently, a vast increase in the use of PTD fusion to address biological questions and for the introduction of pharmacologically relevant proteins in vitro and in vivo has now begun.

1.25   PEPTIDE AND PROTEIN DRUGS: DELIVERY PROBLEMS

Many proteins and peptides possess biological activity that makes them potent therapeutics. Enzymes represent an important and, probably, the best investigated group of protein drugs. Their clinical use has already a rather long history. Certain diseases connected with the deficiency of some lysosomal enzymes can be treated only by the administration of exogenous enzymes. In general, therapeutic enzymes include: antitumor enzymes acting by destroying certain amino acids required for tumor growth; enzymes for replacement therapy for the correction of various insufficiencies of the digestive tract; enzymes for the treatment of lysosomal storage diseases; enzymes for thrombolytic therapy; antibacterial and antiviral enzymes; and hydrolytic and anti-inflammatory enzymes. Peptide hormones, first of all insulin, are among the most broadly used drugs. More recently, peptides such as somatostatin analogs become available in the clinic for the treatment of pituitary and gastrointestinal tumors. Peptide inhibitors of angiogenesis including endostatina are currently in different stages of clinical trials and show a great promise for cancer treatment. Research on depsipeptides has also revealed a set of potential anticancer agents. Antibodies against certain cancer-specific ligands can also be considered as protein anticancer drugs. Still, the use of proteins and peptides as therapeutic agents is hampered by the whole set of their intrinsic properties associated with their nature as complex macromolecules, which are, as a rule, foreign to the recipient organism. This leads to low stability of the majority of peptide and especially protein drugs at physiological pH values and temperatures, particularly when these proteins have to be active in the environment different from their normal one. Different processes leading to the inactivation of various biologically active proteins and peptides in vivo include: conformational protein transformation into inactive form due to the effect of temperature, pH, high salt concentration or detergents; the dissociation of subunit proteins into the individual subunits or in case of cofactor-dependent enzymes, enzyme-cofactor complexes and the association of protein or peptide molecules with the formation of inactive associates; non-covalent complexation with ions or low-molecular-weight and high-molecular-weight compounds, affecting the native structure of the protein or peptide; proteolytic degradation under the action of endogenous proteases; chemical modification by different compounds in solution. All these lead to rapid inactivation and rapid elimination of exogenous proteins from the circulation mostly because of renal filtration, enzymatic degradation, uptake by the reticuloendothelial system (RES) and accumulation in non-targeted organs and tissues. Rapid elimination and widespread distribution into non-targeted organs and tissues requires the administration of a drug in large quantities, which is often not economical and sometimes complicated owing to non-specific toxicity. A very important point is also the immune response of the macroorganism to foreign proteins containing different antigenic determinants. There exist also certain problems associated with the biological mechanisms of drug action. Many peptide and protein drugs as well as antibodies exert their action extracellularly, by receptor interaction. Many others, however, have their targets inside the cell. In the latter case, low permeability of cell membranes to macromolecules often represents an additional obstacle for the development of peptide-based and protein-based drug formulations.

1.26 INTRACELLULAR TARGETS AND INTRACELLULAR DRUG DELIVERY

Many pharmaceutical agents, including various large molecules and even drug-loaded pharmaceutical nanocarriers, need to be delivered intracellularly to exert their therapeutic action inside cytoplasm or onto nucleus or other specific organelles, such as lysosomes, mitochondria or endoplasmic reticulum. Intracellular transport of different biologically active molecules is one of the key problems in drug delivery in general. In addition, the introcytoplasmic drug delivery in cancer treatment might overcome such important obstacle in anticancer chemotherapy as multidrug resistance. However, the lipophilic nature of the biological membranes restricts the direct intracellular delivery of such compounds. The cell membrane prevents big molecules such as peptides, proteins and DNA from spontaneously entering cells unless there is an active transport mechanism as in case of some short peptides. Under certain circumstances, these molecules or even small particles can be taken from the extracellular space into cells by the receptor-mediated endocytosis. The problem, however, is that every molecule/particle entering cell via the endocytic pathway becomes entrapped into endosome and eventually ends in lysosome, where active degradation processes under the action of the lysosomal enzymes take place. As a result, only a small fraction of unaffected substance appears in the cell cytoplasm. As a result, many compounds showing a promising potential in vitro, cannot be applied in vivo owing to bioavailability problems. So far, multiple and only partially successful attempts have been made to bring various macromolecular (protein/peptide) drugs and drug-loaded pharmaceutical carriers directly into the cell cytoplasm bypassing the endocytic pathway, to protect drugs from the lysosomal degradation, thus enhancing their efficiency. The methods like microinjection or electroporation used for the delivery of membrane-impermeable molecules in cell experiments are invasive in nature and could damage cellular membrane. Much more efficient are the noninvasive methods, such as the use of pH-sensitive carriers including pH-sensitive liposomes and cell-penetrating molecules. In many cases, to increase the stability of administered drugs, to improve their efficacy and decrease undesired side effects and even to assist in better intracellular delivery various pharmaceutical carriers are used. One can name liposomes among the most popular and well-investigated drug carriers. Liposomes are artificial phospholipid vesicles with the size varying from 50 to 1000 nm, which can be loaded with a variety of drugs. Liposomes are considered as promising drug carriers for well over two decades. They are biologically inert and completely biocompatible; they cause practically no toxic or antigenic reactions; drugs included into liposomes are protected from the destructive action of the external media. Association of drugs with carriers, such as liposomes, results in delayed drug absorption, restricted drug biodistribution, decreased volume of drug biodistribution, delayed drug clearance and retarded drug metabolism. Plain liposomes are rapidly eliminated from the blood and captured by the cells of the reticuloendothelial system (RES), primarily, in liver and spleen, as the result of rapid opsonization of the liposomes. Most liposomes are internalized by phagocytic cells via endocytosis and destined to lysosomes for degradation. The use of targeted liposomes, that is, liposomes selectively accumulating inside the affected organ or tissue, might increase the efficacy of the liposomal drug and decrease the loss of liposomes and their contents in RES. To obtain targeted liposomes, many protocols have been developed to bind corresponding targeting moieties including antibodies to the liposome surface. However, the majority of antibody-modified liposomes still accumulate in the liver, which hinders their significant accumulation in target tissues, particularly those with a diminished blood supply and/or those with a low concentration of a target antigen. Dramatically better accumulation can be achieved if the circulation time of liposomes could be extended leading to the increased total quantity of immunoliposomes passing through the target and increasing their interactions with target antigens. This is why long-circulated liposomes have attracted so much attention over the past decade. It was also demonstrated that unique properties of long circulating and targeted liposomes could be combined in one preparation, where antibodies or other specific binding molecules have been attached to the water-exposed tips of PEG chains.

1.27 HOMING PEPTIDES AS TARGETED DELIVERY VEHICLES

Each normal organ and pathological condition appear to contain organ- or disease-specific molecular tags on its vasculature, which constitute a vascular "zip code" system. In vivo phage display has been exploited to profile this vascular heterogeneity and a number of peptides that home specifically to various normal organs or pathological conditions have been identified. These peptides have been used for targeted delivery of oligonucleotides, drugs, imaging agents, inorganic nanoparticles, liposomes, and viruses. Identification of the receptor molecules for the homing peptides has revealed novel biomarkers for target organs. In tumors many of these receptors seem to play a functional role in tumor angiogenesis. Recently, tumor homing peptides have entered clinical trials. Results from several Phase I and II trials have been reported, and a number of trials are currently ongoing or recruiting patients. In these trials no dose-limiting toxicity has occurred and all combinations of peptide-targeted therapies have been well tolerated. Both tumor growth and its metastatic spread are dependent on sufficient blood supply, which is ensured either by co-opting existing blood vessels or inducing growth of new vessels in a process called angiogenesis. Tumor angiogenesis consists of sprouting of the pre-existing vessels and recruitment of hematopoietic and circulating endothelial precursor cells from the bone marrow. During adulthood, most blood vessels remain quiescent and angiogenesis occurs only in the cycling ovary, in the placenta during pregnancy, and during the wound healing process. However, endothelial cells retain remarkable ability to divide rapidly in response to a physiological stimulus such as hypoxia. Therefore angiogenesis has to be tightly regulated by a balance between pro- and anti-angiogenic molecules. An imbalance in this process contributes to numerous malignant, inflammatory, ischemic, infectious, and immune disorders. Angiogenesis has been implicated in more than 70 disorders so far and the list is ever growing. Cells constituting the vascular endothelium are heterogeneous and the vascular endothelium expresses tissue-specific markers. Each normal organ and pathological condition seems to contain organ- or disease-specific molecular tags on its vasculature, which constitute a vascular "zip code" system. Many, but not all, of these molecules are angiogenesis-related. *In vivo* phage display screening technology has been widely used to identify tissue-specific markers and to survey disease specific differences. Phage display technology was first reported in 1985 when filamentous DNA-containing bacterial viruses (phages) were genetically engineered to express amino acid sequences on their protein coat. Extensive molecular differences in the vasculature have since been discovered and several peptides that target specifically normal and tumor blood vessels, or tumor lymphatic vessels have been isolated. Some of the tumor homing peptides recognize tumor cells in addition to the vascular cells. Furthermore, some peptides home to several different tumors while others recognize features that are specific for certain tumor types. In vivo phage display technology takes advantage of peptide libraries composed of short, random amino acid sequences, which are expressed on the surface of the phage particles. These libraries are injected into the tail vein of a mouse and allowed to circulate for 5–15 min. It is supposed that during this time the homing peptide will bind to its target tissue. The phage can then be rescued from the target organ and amplified, and the whole process is repeated a few times (= biopanning) for the enrichment of a specific phage that displays a homing peptide with an affinity for the target tissue. The phage library consists of billions of polypeptides expressed on the bacteriophage and it is the quality of this library that yields the success of the screening. During in vivo screening a simultaneous negative and positive selection occurs: peptides homing to the target organ will be amplified and peptides homing to all the other organs will be depleted.

1.28   DRUG DELIVERY OF OLIGONUCLEOTIDES BY PEPTIDES

Oligonucleotides are promising tools for in vitro studies where specific down regulation of proteins is required. In addition, antisense oligonucleotides have been studied in vivo and have entered clinical trials as new chemical entities with various therapeutic targets such as antiviral drugs or for tumor treatments. The formulations of these substances were widely studied in the past. Cationically charged peptides from different origins were used, for example, as cellular penetration enhancers or nuclear localization tool. Examples are given for Poly-L-lysine alone or in combination with receptor specific targeting ligands such as asialoglycoprotein, galactose, growth factors or transferrin. Another large group of peptides are those with membrane translocating properties. Fusogenic peptides rich in lysine or arginine are reviewed. They have been used for DNA complexation and condensation to form transport vehicles. Some of them, additionally, have so called nuclear localization properties. Here, DNA sequences, which facilitate intracellular trafficking of macromolecules to the nucleus were explored. Summarizing the present literature, peptides are interesting pharmaceutical excipients and it seems to be feasible to combine the specific properties of peptides to improve drug delivery devices for oligonucleotides in the future.

Antisense oligonucleotides have been used for many years to down regulate specific proteins in cells. The principle of antisense technology, invented in 1978, was the sequence specific binding of an antisense oligonucleotide to target mRNA preventing gene translation. Further antisense methods like triplex forming antisense oligonucleotides targeting DNA, aptamers and small interfering RNA were studied for years. The most described problems in oligonucleotide delivery are degradation by nucleases, insufficient affinity for the target, lack of specificity and most important, the poor bioavailability. To overcome these problems, many chemical modifications were introduced. Another approach is the development of drug delivery systems. Different carrier systems for antisense oligonucleotides have been described in the past years. These systems utilize synthetic polymers like polyethyleneimine, cationic liposomes, dendrimers and various peptides. Such peptides consisting of less than 10 amino acids are called oligo-peptides, if they have more than 10 and up to 100 or more amino acids these are poly-peptides. However, there is no clear definition for the transition of a peptide to a protein. Cationic peptides, for example poly-L-lysine can complex antisense oligonucleotides, protect them against nuclease digestion and enhance the cellular uptake via non-specific endocytosis. To get more specificity in the targeting of cells, various ligands like folic acid, steroids, transferrin, mannose, growth factors were conjugated to poly-L-lysine to improve the uptake of the oligonucleotide peptide complexes or conjugates via receptor-mediated endocytosis. Another functionality of peptides is the endosomolytic potential. Arginine- and histidine-rich peptides mediate an endosomal escape of the antisense oligonucleotides into the cytosol. Furthermore, special amino acid sequence motifs have the ability to transfer oligonucleotides to the cell nucleus, if they include a nuclear localization sequence.

1.29   NANOTHERAPEUTICS FOR NANOBIOCONJUGATION OF PEPTIDE AND PROTEIN

The targeted delivery of therapeutic peptide by nanocarriers systems requires the knowledge of interactions of nanomaterials with the biological environment, peptide release, and stability of therapeutic peptides. Therapeutic application of nanoencapsulated peptides are increasing exponentially and >1000 peptides in nanoencapsulated form are in different clinical/trial phase. Proteinaceous biomolecules such as antibodies, antigens, growth factors, and bioactive peptides are well known for their therapeutic potential in various diseases. These therapeutic proteins and peptides are potential target for development of therapeutic nanoprotein and peptides because of their nanoscale dimension, highly specific therapeutic importance, and numerous other specific enzymatic activities. Encapsulation or bioconjugation of these proteins/peptides on suitable nanocarriers may improve the potential of these therapeutic proteins to create a novel component for therapeutics, disease diagnosis, fluorescent microscopy, imaging, and other life science devices. The protein and peptides nanoparticles are developed by adsorption on the nanocarriers surfaces, encapsulation in nanoparticles, bioconjugation on nanoparticles and by molecular self assembly of small peptides into nanoparticles size. The encapsulated protein on nanoparticles has potential to enhance therapeutic activity of peptides by sustained and targeted delivery, improve stability, better bioavailability, and assemblies of nanocomposites. The properties of protein to interact with each other make them an excellent candidate to serve as a linker to form an ordered nanoparticles structure. These ordered nanoparticles improve the therapeutic potential of encapsulated protein and peptides. Most of the therapeutic proteins and peptides are quite unstable, prone to denaturation compare to conventional therapeutic small molecules. The therapeutic activities of peptides are simply lost due to aggregation, degradation and unfolding. Apart from this, most of the protein lost its activity and structural organization during encapsulation, or bioconjugation on the nanoparticles, for example, *glucose oxidase* encapsulation. The retention of activity, structural identity, and stability of protein and peptides after encapsulation on suitable nanodevices are basic concern in development of protein and peptides nanomedicine. A variety of nanoparticulate systems like polymeric microspheres/nanoparticles, liposomes and solid lipid nanoparticles, etc. are being used for the nanoencapsulation protein and peptides to improve protein drug accumulation inside target cells due to easy and efficient cellular internalization. Further, nanoencapsulation of protein to develop nanomedicines require the complete information about the changes in cell receptors that occur with progression of disease, mechanism and site of action, retention of drugs, multiple administration, molecular mechanisms, stability and therapeutic activity of protein in *vivo* and pathobiology of the disease. The opsonization and clearance of these particles from the body are important point of consideration during the encapsulation. However, these can be minimized by suitable and balanced surface modification of nanocarrier without affecting the drug loading and release mechanisms. Several methods have been developed to mask nanoparticles from the mononuclear phagocytic system. The most preferred method is the adsorption or grafting of poly to the surface of nanoparticles. Addition of poly and PEG-containing copolymers to the surface of nanoparticles results in an increase in the blood circulation half-life of the particles by several orders of magnitude. This method creates a hydrophilic protective layer around the nanoparticles that is able to repel the absorption of opsonin proteins *via* steric repulsion. Advances in protein engineering and materials science have contributed to novel nanoscale targeting approaches that may bring new hope to develop a new therapeutic nanobiomolecules.

1.29.1 METHODOLOGY OF PEPTIDE AND PROTEIN NANOENCAPSULATION

Protein and peptides encapsulation or adsorption on nanocarriers have been achieved by various methods like emulsion polymerization, interfacial polymerization, solvent evaporation, salting out, coacervation, combination of sonication and layer-by-layer technology, solvent displacement/ solvent diffusion, etc. Each methods of protein encapsulation have their own advantage and disadvantage. Since each protein and peptides requires its own specific condition for stability, solubilization, control releases immune elimination. The method of encapsulation is entirely based on the physicochemical activity of protein and its application. A brief description about the each reported protein encapsulated methods along with detail characteristics are provided in subsections.

1.29.1.1 Emulsification–Polymerization

The emulsification–polymerization method is classified in two categories, based on the use of organic or aqueous continuous phase. In the continuous aqueous phase, polymers and protein drugs are dissolved in aqueous solvent without surfactants or emulsifier by using anionic polymerization mechanism with high energy radiation. For example, insulin and cyclosporin A are encapsulated on poly(isobutylcyanoacrylate) nanoparticles of particle size <500 nm and 120 nm respectively. In the continuous organic phase, polymers are dissolved in organic non-solvent by dispersion via surfactants in to solvent. Various enzymes are encapsulated with polyacrylamide nano/microparticles of <1000 nm size nanoparticles.

1.29.1.2   Interfacial Polymerization

In interfacial polymerization, the cyanoacrylate monomer and proteins drug are dissolved in a mixture of an oil and absolute ethanol. Oils have positive influence to reduce the unfolding of protein by ethanol. This mixture is then slowly extruded through a needle into a well stirred aqueous solution, with or without some ethanol containing surfactant. An advantage of interfacial polymerization technique is high efficiency drug encapsulation. Insulin was encapsulated on poly(ethylcyanoacrylate) and poly(isobutylcyanoacrylate) nanoparticles of particle size ~151 nm and 150–300 nm respectively by using this method.

1.29.1.3   Solvent Evaporation

In solvent evaporation method, the polymers along with proteins are dissolved in volatile organic solvent and poured into continuously stirring aqueous phase with or without emulsifier/stabilizer and sonicated. Most of the proteins are likely to denature after the sonication. Thus, a slow and intermittent sonication at low temperature is effective to retain the secondary and tertiary structure of protein drugs. Albumin, and tetanus toxoid are successfully encapsulated on polylactic acid nanoparticles of size 100–120 nm and 150 nm by these methods. Solvent displacement method is similar to solvent evaporation that is based on spontaneous emulsification of the organic internal phase containing partially dissolved polymer along with protein into the aqueous external phase. Insulin on PLA nanoparticles of size ~105–170 nm was nanoencapsulated by these methods.

1.29.1.4   Salting Out

The protein and peptides therapeutic are sensitive to unfolding or inactivation. These problems are minimized by the salting out methods of protein and peptide encapsulation. This procedure is based on the separation of a water miscible solvent from aqueous solution by adding salting out agent like magnesium chloride, calcium chloride, etc. The main advantage of salting out procedure is that it minimizes unfolding stress to protein encapsulates. Various peptides are nanoencapsulated on different nanoparticulate system by using this technique.

1.29.1.5   Coacervation

Coacervation is a process during which a homogeneous solution of charged macromolecules undergoes liquid–liquid phase separation.

Solvent evaporation
Polymer is dissolved in organic solvent, emulsified in aqueous solution. The emulsion is dried under pressure or continues stirring or increasing temperature.

Coacervation
Method: Alcoholic solution of desolvating agent is added to aqueous solution of albumin. Associative phase separation of two polymers in water occurs if there is an electrostatic attraction ion, giving rise to a polymer rich dense phase. This method has been classified into simple and complex processes depending on the number of participating macromolecules. In simple polyelectrolyte coacervation, addition of salt or alcohol normally promotes coacervation. In complex coacervation, two oppositely charged macromolecules can undergo coacervation through associative interactions. The charges on the polyelectrolytes must be sufficiently large to cause significant electrostatic interactions but not so large to cause precipitation. The dilute liquid phase, (usually supernatant), remains in equilibrium with the coacervate phase. These two liquid phases are incompatible and immiscible. BSA has been to lipoic acid-stabilized semiconductor CdSe core/ZnS shell particles. These are some examples of well known nanoencapsulates of protein and peptides on metallic nanoparticles.

1.29.2 *POLYMERIC NANOCARRIERS*

The major advantage of colloidal drug carrier system in protein and peptide therapeutics are the controlled drug targeting, modified body distribution and enhancement of cellular uptake. The polymeric nanocarriers are very promising since they are biodegradable, non-antigenic, relatively easy to prepare and full control on size distribution. A variety of polymeric nanoparticles (natural and synthetic) can be synthesized in laboratory and are also commercially available. Polymeric materials used for the formulation of nanoparticles include synthetic [poly(lactic acids) (PLA), poly(lactic-coglycolic acids) (PLGA), polycaprolactone) (PCL), poly(methyl methacrylates), and poly(alkyl cyanoacrylates)] or natural polymers.

1.29.2.1 Natural Nanocarriers

Chitosan is a modified natural carbohydrate polymer prepared by the partial N-deacetylation of crustacean derived natural biopolymer chitin. It is the second most abundant polysaccharide in nature, and has attracted particular interest as a biodegradable material for mucosal delivery systems. There are at least four methods reported for the preparation of chitosan nanoparticles as ionotropic gelation, microemulsion, emulsification solvent diffusion and polyelectrolyte complex formation. Chitosan nanoparticles have low toxicity and high susceptibility to biodegradation, mucoadhesive properties and has an important capacity to enhance protein drug permeability/absorption at mucosal sites. More importantly, chitosan micro/nanoparticles can be spontaneously formed through ionic gelation using tripolyphosphate as the precipitating agent. This reduces the use of harmful organic solvents during preparation and loading of protein therapeutics. Chitosan solubility is poor above pH 6.0 which is a major drawback of this system. At physiological pH, chitosan is known to lose its capacity to enhance drug permeability and absorption, which can only be achieved in its protonated form in acidic environments. In contrast, quaternized chitosan derivative, N-trimethyl chitosan chloride (TMC) shows perfect solubility in water over a wide range of pH. In addition, these chitosan derivatized nanoparticles have bio-adhesive properties. Thus, it is used for enhancement of permeability and absorption of diverse protein drugs in neutral and basic-pH condition. N-trimethyl chitosan chloride (TMC) nanoparticles to carry proteins were prepared by ionic cross linking of TMC with tripolyphosphate (TPP). The results indicate that different degree of quaternization ($1/\infty$ particle size) of TMC has influenced the physicochemical properties, release profile and degree of loading of different proteins. Thus, the particle size may depend upon the nature and concentration of the loading proteins. Insulin was observed to be directly internalized by enterocytes in contact with intestine and retention of drugs at their absorptive sites by mucoadhesive carriers. Insulin loaded chitosan nanoparticles markedly enhanced intestinal absorption of insulin following oral administration. The hypoglycemia effect and insulinemia levels were significantly higher than that obtained from insulin solution and physical mixture of oral insulin and empty nanoparticles. The mechanism of insulin absorption seems to be a combination of both insulin internalization, probably through vesicular structures in enterocytes and insulin loaded nanoparticles uptake by Payers patches cells. Chitosan nanoparticles are also explored for their efficacy to increase systemic absorption of hydrophobic peptides such as cyclosporin A. The relative bioavailability of cyclosporin A encapsulated chitosan nanoparticles was increased by about 73%. This formulation provides the highest Cmax (2762.8 ng/mL) of Cy-A after 2.17 h. Chitosan nanoparticles administered orally to beagle dogs provide an improved absorption compared to the currently available cyclosporin A microemulsion. It has been shown that ovalbumin loaded chitosan microparticles are taken up by the Peyer's patches of the gut associated lymphoid tissue (GALT). Additionally, after co-administering chitosan with antigens in nasal vaccination studies, a strong enhancement of both mucosal and systemic immune responses was observed. Van der Lubben et al. have demonstrated that large amounts of bovine serum albumin (BSA) or tetanus toxoid (TT) vaccine were easily encapsulated in chitosan nanoparticles. Recently, Alonso's group has developed chitosan nanoparticles as carrier systems for transmucosal delivery. They have shown the enhanced mucosal absorption of chitosan NPs on rats and rabbits. The nanoparticles were in the 350 nm size range, and exhibited a positive electrical charge (+40 mV) and high loading efficiency (50–60%). They have reported important capacity of chitosan nanoparticles for the association of peptides such as insulin, salmon, calcitonin and tetanus toxoid. Their mechanism of interaction with epithelia was investigated using the Caco-2 model cell line. The results showed that chitosan coated systems caused a concentration-dependent reduction in the transepithelial resistance of the cell monolayer.

Gelatin nanoparticles are extensively used in food and medicinal purpose. It is an attractive nanomaterial to exploit in controlled release of peptide therapeutics due to its nontoxic, and biodegradable nature. This polymer works as a polyampholyte consisting both cationic and anionic groups along with hydrophilic functionality. Due to this nature, gelatin molecules are frequently used for encapsulation of both acidic and basic peptides. Gelatin nanoparticles are prepared by desolvation/coacervation or emulsion method. The addition of natural salt or alcohol normally promotes coacervation and the control of turbidity/cross linking that resulted in desired nanoparticles. Gelatin nanoparticles have been used for encapsulation of BSA. These nanoparticles can absorb 51–72% of water, thus the release of BSA from the gelatin nanoparticulate matrix follows a diffusion controlled mechanism. The average diameter of the BSA-containing gelatin nanoparticles is approximately 840 nm. Basic fibroblast growth factor (bFGF) has been successfully loaded on gelatin particles.

1.29.2.2 Synthetic Nanocarriers

Natural polymeric nanoparticles provide a relatively quick drug release. However, synthetic polymers enable extended drug release over periods from days to several weeks. Poly(dl-glycolide-colactide) (PLGA), poly (d, l-lactide) (PLA) and polycaprolactone (PCL) are used to encapsulate proteins drug due to its biodegradable and biocompatible nature. Among these nanoparticles, PLGA (poly-d, 1-lactide-co-glycolide) and PLA nanoparticles are one of the most successfully used biodegradable nanosystem for the development of protein nanomedicines. They undergo hydrolysis in the body to produce the biodegradable metabolite monomers, lactic acid and glycolic acid. Since the body effectively deals with these two monomers, there is very minimal systemic toxicity associated by using PLGA for drug delivery or biomedical applications. Insulin was encapsulated in a blend of poly (fumaric anhydride) (poly(FA) and poly(lactideco-glycolide) (PLGA) at a 50:50 ratio (poly(FA:PLGA)) using the inversion phase method. Animals feeding the poly (FA: PLGA) – encapsulated insulin preparation showed a better ability to regulate glucose load than the controls. This gave an indication that the insulin has crossed the intestinal barrier and was released from the microspheres in a biologically active form. Kawashima et al. evaluated the effectiveness of mucoadhesive polymeric nanospheres in the absorption of calcitonin on chitosan coated elcatonin-loaded PLGA nanospheres. BSA and immuno–globulin (IgG) have been encapsulated on PEO–PLGA nanoparticles. Different amounts of BSA or IgG corresponding to 1%, 2% and 4% theoretical loadings were encapsulated into PEO–PLGA nanoparticles of different composition. The type of used PEO derivative and pH of internal aqueous phase are the most important factors influencing BSA protein encapsulation and release kinetics. Degradation and release characteristics of polyester particles can be improved by the incorporation of polyoxyethylene derivatives with different hydrophilia–lipophilia balance. BSA encapsulated on PEG–PLGA nanoparticles are reported to extend the half-life as 4.5 h than 13.6 min in simple BSA. This extended half-life obviously change the protein bio-distribution in rats compared with that of BSA loaded on PLGA nanoparticles. Similarly, tetanus toxoid is also encapsulated on PLA and PLA–PEG nanoparticles of similar particle size but differed in their hydrphobicity. PLA–PEG nanoparticles led to greater penetration of tetanus toxoid into the blood circulation and in lymph nodes than PLA encapsulated tetanus toxoid. Polyalkylcyanoacrylate (PACA) nanocapsules are used as biodegradable polymeric drug carriers for subcutaneous and oral delivery of octreotide, a long-acting somatostatin analog. It has the ability to reduce secretion of insulin or of prolactin in response to estrogens. Octreotide-loaded nanocapsules reduce prolactin secretion, increased plasma octreotide level, and prolonged therapeutic effect of a somatostatin analog in estrogen-treated rats given orally.

1.30   THERAPEUTIC USE OF PEPTIDE AND PROTEINS

Insulin was the first peptide to be used as therapeutics and since then many therapeutic proteins and peptides have been reported in almost every field of medicine. The protein therapeutics summarizes more than 130 currently used protein therapeutics and 1000 proteins/peptides are in different clinical/trial/approval phase. This therapeutics are currently preferred due to its high specificity and complex set of functions; less interference to normal biological processes; less likely to elicit immune responses; effective replacement treatment in mutated or deleted normal protein and faster clinical development as well as FDA approval time. Today there are more than 300 reports of plant-based production of therapeutic proteins. Various plant isolated cyclic peptides (cyclotides), cysteine proteinases, defensins, plant-made vaccines are well known for its therapeutic importance. These peptides and other therapeutic protein from animal, bacterial, fungal and synthetic are under various stages of research and development for better therapeutic applications. A detail description of FDA approved commercial protein and peptides are reviewed by Leader et al. They have classified the therapeutic protein in four major groups (Group I–IV) based on mechanism and pharmacological actions. Recently, transgenic tobacco cell cultures produced recombinant animal vaccine against Newcastle disease virus was approved by FDA. Insulin, angiotensin I-converting enzyme, inhibitory 1.3 kDa peptide, 15 new antimicrobial peptides form the skin of various amphibians, peptide venom of Bothrops marajoensis snakes, synthetic peptides containing amino acid residue 161–173 of HIV-1 integrase protein are reported to have the great therapeutic importance.

KEYWORDS

- **Drug delivery**
- **Peptides**
- **Protein**
- **Stability**
- **Structure**
- **Synthesis**
- **Targeting**

REFERENCES

1. Lander, E. S., Linton, L. M., Birren, B., Nusbaum, C., Zody, M. C., et al. Initial sequencing and analysis of the human genome. Nature 2001, 409, 860–921.
2. Borgia, J. A., Fields, G. B. Chemical synthesis of proteins. Trends Biotechnol. 2000, 18, 243–251.
3. Mendel, D., Cornish, V. W., Schultz, P. G. Site-directed mutagenesis with an expanded genetic code. Annu Rev Biophys Biomol Struct 1995, 24, 435–62.
4. Fischer, E. Nobel Lectures, Chemistry 1901–1921. Amsterdam: Elsevier; 1966. Syntheses in the purine and sugar group; p. 21–35.
5. Fischer, E., Fourneau, E. Ueber einige Derivate des Glykocolls. Ber. Deutsch. Chem. Ges. 1901, 34, 2868–77.
6. Bergmann, M., Zervas, L. Über ein allgemeines Verfahren der Peptid-Synthese. Ber Deutsch Chem Ges 1932, 65, 1192–201.
7. Du Vigneaud, V., Ressler, C., Swan, J. M., Roberts, C. W., Katsoyannis, P. G., Gordon, S. The synthesis of an octapeptide amide with the hormonal activity of oxytocin. J Am Chem Soc 1953, 75, 4879–80.
8. Merrifield, B. The chemical synthesis of proteins. Protein Sci. 1996, 5, 1947–51.
9. Merrifield, R. B., Stewart, J. M., Jernberg, N. Instrument for automated synthesis of peptides. Anal Chem 1966, 38, 1905–14.
10. Gutte, B., Merrifield, R. B. The synthesis of ribonuclease A. J. Biol. Chem. 1971, 246, 1922–41.
11. Hirschmann, R., Nutt, R. F., Veber, D. F., Vitali, R. A., Varga, S. L., et al. Studies on the total synthesis of an enzyme. V. The preparation of enzymatically active material. J. Am. Chem. Soc. 1969, 91, 507–8.
12. Wlodawer, A., Miller, M., Jaskólski, M., Sathyanarayana, B. K., Baldwin, E, et al. Conserved folding in retroviral proteases: crystal structure of a synthetic HIV-1 protease. Science 1989, 245, 616–21.
13. Sato, T., Aimoto, S. Use of thiosulfonate for the protection of thiol groups in peptide ligation by the thioester method. Tetrahedron Lett 2003, 44, 8085–87.
14. Sievers, A., Beringer, M., Rodnina, M. V., Wolfenden, R. The ribosome as an entropy trap. Proc Natl Acad Sci USA 2004, 101, 7897–901.
15. Page, M. I., Jencks, W. P. Entropic contributions to rate accelerations in enzymic and intramolecular reactions and the chelate effect. Proc Natl Acad Sci USA 1971, 68, 1678–83.
16. Tam, JP., Yu, Q. T., Miao, Z. W. Orthogonal ligation strategies for peptide and protein. Biopolymers 1999, 51, 311–32.
17. Kemp, D. S., Galakatos, N. G., Bowen, B., Tan, K. Peptide synthesis by prior thiol capture. 2 Design of templates for intramolecular O, N-acyl transfer 4, 6-Disubstituted dibenzofurans as optimal spacing elements. J Org Chem 1986, 51, 1829–39.
18. McBride, B. J., Kemp, D. S. Peptide synthesis by prior thiol capture. III. Assessment of levels of racemization during two typical thiol capture coupling reactions. Tetrahedron Lett. 1987, 28, 3435–38.
19. Fotouhi, N., Galaksatos, N. G., Kemp, D. S. Peptide synthesis by prior thiol capture. 6 Rates of the disulfide bond forming capture reaction and demonstration of the overall strategy by synthesis of the C-terminal 29-peptide sequence of BPTI. J. Org. Chem. 1989, 54, 2803–17.

20. Kryukov, G. V., Castellano, S., Novoselov, S. V., Lobanov, A. V., Zehtab, O, et al. Characterization of mammalian selenoproteomes. Science 2003, 300, 1439–43.

21. Kryukov, G. V., Gladyshev, V. N. The prokaryotic selenoproteome. EMBO. Rep. 2004, 5, 538–43.

22. Arner, E. S., Sarioglu, H., Lottspeich, F., Holmgren, A., Bock, A. High-level expression in *Escherichia coli* of selenocysteine-containing rat thioredoxin reductase utilizing gene fusions with engineered bacterial type SECIS elements and co-expression with the selA, selB and selC genes. J Mol Biol 1999, 292, 1003–16.

23. Gieselman, M. D., Xie, L., van der Donk, W. A. Synthesis of a selenocysteine-containing peptide by native chemical ligation. Org. Lett. 2001, 3, 1331–34.

24. Hondal, R. J., Nilsson, B. L., Raines, R. T. Selenocysteine in native chemical ligation and expressed protein ligation. J Am. Chem. Soc. 2001, 123, 5140–41.

25. Quaderer, R., Sewing, A., Hivert, D. Selenocysteine-mediated native chemical ligation. Helv. Chim. Acta. 2001, 84, 1197–206.

26. Hondal, R. J., Nilsson, B. L., Raines, R. T. Selenocysteine in native chemical ligation and expressed protein ligation. J. Am. Chem. Soc. 2001, 123, 5140–41.

27. Hondal R. J., Raines, R. T. Semisynthesis of proteins containing selenocysteine. Methods Enzymol 2002, 347, 70–83.

28. Berry, S. M., Gieselman, M. D., Nilges, M. J., van der Donk, W. A., Lu, Y. An engineered azurin variant containing a selenocysteine copper ligand. J. Am. Chem. Soc. 2002, 124, 2084–85. 29. Besse D, Siedler F, Diercks T, Kessler H, Moroder L. The redox potential of selenocysteine in unconstrained cyclic peptides. Angew Chem Int Ed Engl 1997, 36, 883–85.

30. Roelfes, G., Hilvert, D. Incorporation of selenomethionine into proteins through selenohomocysteine mediated ligation. Angew Chem Int Ed Engl 2003, 42, 2275–77.

31. Huse, M., Muir, T. W., Xu, L., Chen, Y. G., Kuriyan, J., Massague, J. The TGFβ receptor activation process: an inhibitor- to substrate-binding switch. Mol. Cell. 2001, 8, 671–82.

32. Schneider, C. H., de Weck, A. L. Studies on the direct netural penicilloylation of functional groups occurring on proteins. Biochim. Biophys. Acta 1968, 168, 27–35.

33. Friedman, M. Chemistry, biochemistry, nutrition, and microbiology of lysinoalanine, lanthionine, and histidinoalanine in food and other proteins. J. Agric. Food Chem. 1999, 47, 1295–319.

34. Klonowski, W. Non-equilibrium proteins, Comput. Chem. 2001, 25, 349–368.

35. Tüdos, É., A. Simon, F. Á., Dosztányi, Z., Fuxreiter, M., Magyar, C., I. Simon, Non-covalent cross-links in context with other structural and functional elements of proteins, J. Chem. Inf. Comput. Sci. 2004, 44, 347–351.

36. Iyer, P. V., Ananthanarayan L. Enzyme stability and stabilization-aqueous and non-aqueous environment. Process Biochem. 2008, 43, 1019–1032.

37. Villegas V., Viguera A. R., Avilés, F. X., Serrano, L. Stabilization of proteins by rational design of alpha-helix stability using helix/coil transition theory. Fold. Des. 1995, 01, 29–34.

38. Wimmer, R., Olsson, M., Petersen, M. T. N., Hatti-Kaul, R., Petersen, S. B., Müller, N. Towards a molecular level understanding of protein stabilization: the interaction between lysozyme and sorbitol, J. Biotechnol. 1997, 55–85–100.

39. Wang W. Instability, stabilization, and formulation of liquid protein pharmaceuticals. Int. J. Pharm. 1999, 185, 129–188.

40. Sanchez-Ruiz J. M. Protein kinetic stability, Biophys. Chem. 2010, 148, 1–15.

41. Fernández-Lafuente, R. Stabilization of multimeric enzymes: strategies to prevent subunit dissociation, Enzym. Microb. Technol. 2009, 45, 405–418.

42. Maddux, N. R. Joshi, S. B., Volkin, D. B., Ralston, J. P., Middaugh, C. R. Multidimensional methods for the formulation of biopharmaceuticals and vaccines. J. Pharm. Sci. 2011, 100, 4171–4197.

43. Balcão, V. M., Oliveira, T. A., Malcata, F. X. Stability of a commercial lipase from Mucor javanicus: kinetic modeling of pH and temperature dependencies. Biocatal. Biotransf. 1998, 16, 45–66.

44. Serra, I., Serra, C. D., Rocchietti, S., Ubiali, D., Terreni, M. Stabilization of thymidine phosphorylase from Escherichia coli by immobilization and post immobilization techniques. Enzym. Microb. Technol. 2011, 49, 52–58.

45. Arakawa, T., Prestrelski, S. J., Kenney, W. C., Carpenter, J. F. Factors affecting short-term and long-term stabilities of proteins. Adv. Drug Deliv. Rev. 1993, 10, 1–28.

46. Arakawa, T., Prestrelski, S. J., Kenney, W. C., Carpenter, J. F. Factors affecting short-term and long-term stabilities of proteins, Adv. Drug Deliv. Rev. 2001, 46, 307–326.

47. Moutinho, C. G., Matos, C. M., Teixeira, J. A., Balcão, V. M. Nanocarrier possibilities for functional targeting of bioactive peptides and proteins: state-of-the-art. J. Drug Target. 2012, 20, 114–141.

48. Balcão, V. M., Moutinho, C. G. Peptides and proteins: non-invasive delivery, in: J. Swarbrick (Ed.), 4th edition, Encyclopedia of Pharmaceutical Science and Technology, Vol. IV, CRC Press (Taylor & Francis Group LLC), 2013, pp. 2555–2577.

49. Mozhaev, V. V., Martinek, K. Structure-stability relationships in proteins: a guide to approaches to stabilizing enzymes. Adv. Drug Deliv. Rev. 1990, 4, 387–419.

50. Denisov, I. G. Thermal stability of proteins in intermolecular complexes. Biophys. Chem. 1992, 44–71–75.

51. Foit, L., Morgan, G. J., Kern, M. J., Steimer, L. R., von Hacht, A. A., Titchmarsh, J., Warriner, S. L., Radford, S. E., Bardwell, J. C. A. Optimizing protein stability in vivo, Mol. Cell 2009, 36, 861–871.

52. Ragoonanan, V., Aksan, A. Protein stabilization, Transfus. Med. Hemother. 2007, 34, 246–252.

53. Petersen, S. B., Jonson, P. H., Fojan, P., Petersen, E. I., Petersen, M. T. N., Hansen, S., Ishak, R. J., Hough, E. Protein engineering the surface of enzymes, J. Biotechnol. 1998, 66, 11–26.

54. Magyar, C., Tüdös, É., Simon, I. Functionally and structurally relevant residues of enzymes: are they segregated or overlapping? FEBS Lett. 2004, 567, 239–242.

55. Scharnagl, C., Reif, M., Friedrich, J. Stability of proteins: temperature, pressure and the role of the solvent, Biochim. Biophys. Acta 2005, 1749, 187–213.

56. Shenoy, S. R., Jayaram, B. Proteins: sequence to structure and function—current status, Curr. Protein Pept. Sci. 2010, 11, 498–514.

57. Illanes, A., Cauerhff, A., Wilson, L., Castro, G. R. Recent trends in biocatalysis engineering, Bioresour. Technol. 2012, 115, 48–57.

58. Jadhav, S. B. Singhal, R. S. Conjugation of alpha-amylase with dextran for enhanced stability: process details, kinetics and structural analysis. Carbohydr. Polym. 2012, 90, 1811–1817.

59. Berezin, C. F., Glaser, J., Rosenberg, I., Paz, T., Pupko, P., Fariselli, R., Casadio, N., Ben-Tal, ConSeq. The identification of functionally and structurally important residues in protein sequences, Bioinformatics 2004, 20, 1322–1324.

60. Jaenicke R. Protein stability and molecular adaptation to extreme conditions. Eur. J. Biochem. 1991; 202, 715–728.

61. Vieille, C., Zeikus, J. G., Thermoenzymes: identifying molecular determinants of protein structural and functional stability, Trends Biotechnol. 1996, 14, 183–191.

62. Damborsky, J. Quantitative structure–function and structure–stability relationships of purposely modified proteins, Protein Eng. 1998, 11, 21–30.

63. Becktel, W. J., Schellman, J. A. Protein stability curves. Biopolymers 1987, 26, 1859–1877.

64. Somero, G. N. Proteins and temperature, Annu. Rev. Physiol. 1995, 57, 43–68.

65. Lee, B., Vasmatzis, G. Stabilization of protein structures. Curr. Opin. Biotechnol. 1997, 8, 423–428.

66. Torchilin, V. P., Maksimenko, A. V., Smirnov, V. N., Berezin, I. V., Klibanov, A. M., Martinek, K. The principles of enzyme stabilization. IV. Modification of "key" functional groups in the tertiary structure of proteins, Biochim. Biophys. Acta 1979, 567, 1–11.

67. Kristjánsson, M. M., Kinsella, J. E. Protein and enzyme stability: structural, thermodynamic, and experimental aspects, Adv. Food Nutr. Res. 1991, 35, 237–316.

68. Park, H., Lee, S. Prediction of the mutation-induced change in thermodynamic stabilities of membrane proteins from free energy simulations. Biophys. Chem. 2005, 114, 191–197.

69. Shoichet, B. K., Baase, W. A., Kuroki, R., Matthews, B. W. A relationship between protein stability and protein function. Proc. Natl. Acad. Sci. USA 1995, 92, 452–456.

70. Ó'Fágáin. C. Enzyme stabilization—recent experimental progress. Enzym. Microb. Technol. 2003, 33, 137–149.

71. Bhaskara R. M., Srinivasan, N. Stability of domain structures in multi-domain proteins, Sci. Reports 2011, 1, 40–48.

72. Campos, L. A., Garcia-Mira, M. M., Godoy-Ruiz, R., Sanchez-Ruiz J. M., Sancho J. Do proteins always benefit from a stability increase? Relevant and residual stabilization in a three-state protein by charge optimization, J. Mol. Biol. 344 (2004) 223–237.

73. Xie, G., Timasheff, S. N. The thermodynamic mechanism of protein stabilization by trehalose, Biophys. Chem. 1997, 64, 25–43.

74. Jyothi, T. C., Sinha, S., Singh, S. A., Surolia, A., Rao, A. G. A. Napin from Brassica juncea: thermodynamic and structural analysis of stability, Biochim. Biophys. Acta 2007, 1774, 907–919.

75. Foss, T. R., Kelker, M. S., Wiseman, R. L., Wilson, I. A., Kelly, J. W. Kinetic stabilization of the native state by protein engineering: implications for inhibition of transthyretin amyloidogenesis. J. Mol. Biol. 2005, 347, 841–854.

76. Palmer, B., Angus, K., Taylor, L., Warwicker, J., Derrick, J. P. Design of stability at extreme alkaline pH in streptococcal protein G. J. Biotechnol. 2008, 134, 222–230.

77. Castronuovo, G. Proteins in aqueous solutions. Calorimetric studies and thermodynamic characterization, Thermochim. Acta 1991, 193, 363–390.

78. Mateo, C., Pessela, B. C. C., Fuentes, M., Torres, R., Betancor, L., Hidalgo, A., Fernández Lorente, G., Fernández-Lafuente, R., Guisán, J. M. Stabilization of multimeric enzymes via immobilization and further cross-linking with aldehyde-dextran, chapter 12, in: Second edition, J. M. Guisán (Ed.), Methods in Biotechnology—Immobilization of Enzymes and Cells, Vol. 22, Humana Press Inc., Totowa NJ, USA, 2006, pp. 129–141.

79. López-Gallego, F., Betancor, L., Hidalgo, A., Dellamora-Ortiz, G., Mateo, C., Fernández-Lafuente R., Guisán, J. M. Stabilization of different alcohol oxidases via immobilization and post immobilization techniques, Enzym. Microb. Technol. 2007, 40, 278–284.

80. Makhatadze, G. I., Privalov, P. L. Contribution of hydration to protein folding thermodynamics: I. The enthalpy of hydration, J. Mol. Biol. 1993, 232, 639–659.

81. Liao, Y. H., Brown, M. B., Martin, G. P. Investigation of the stabilization of freeze-dried lysozyme and the physical properties of the formulations. Eur. J. Pharm. Biopharm. 2004, 58, 15–24.

82. Miyawaki, O. Hydration state change of proteins upon unfolding in sugar solutions. Biochim. Biophys. Acta 2007, 1774, 928–935.

83. Murphy, K. P. V., Xie B. D., Freire, E. Molecular basis of cooperativity in protein folding. III. Structural identification of cooperative folding units and folding intermediates. J. Mol. Biol. 227, 1992, 293–306.

84. Cheung, J. K., Shah, P., Truskett, T. M. Heteropolymer collapse theory for protein folding in the pressure–temperature plane. Biophys. J. 2006, 91, 2427–2435.

85. Potekhin, S. A., Senin, A. A., Abdurakhmanov, N. N., Tiktopulo, E. I. High pressure stabilization of collagen structure. Biochim. Biophys. Acta 2009, 1794, 1151–1158.

86. Myers, J.K; Trevino, S. R. Increasing globular protein stability, Chem. Today 2012, 30, 30–33.

87. Bizzarri, A. R., Cannistraro, S. Molecular dynamics of water at the protein–solvent interface. J. Phys. Chem. B 2002, 106, 6617–6633.

88. Miyawaki, O. Thermodynamic analysis of protein unfolding in aqueous solutions as a multisite reaction of protein with water and solute molecules, Biophys. Chem. 2009, 144, 46–52.

89. Fenimore, P., Frauenfelder, W. H., McMahon, B. H., Parak, F. G. Slaving: solvent fluctuations dominate protein dynamics and functions, Proc. Natl. Acad. Sci. USA. 2002, 99, 16047–16051.

90. Doster, W. M. Settles, Protein–water displacement distributions, Biochim. Biophys. Acta 2005, 1749–173–186.

91. Luzar, A. Water–hydrogen bond dynamics close to hydrophobic and hydrophilic groups, Faraday Discuss. 1996, 103, 29–40.

92. Kovrigin, E. L., Potekhin, S. A. On the stabilizing action of protein denaturants: acetonitrile effect on stability of lysozyme in aqueous solutions. Biophys. Chem. 2000, 83, 45–59.

93. Wang, W. Protein aggregation and its inhibition in biopharmaceutics, Int. J. Pharm. 289 (2005) 1–30.

94. Ganjalikhany, M. R., Ranjbar, B., Hosseinkhani, S., Khalifeh, K., Hassani, L. Roles of trehalose and magnesium sulfate on structural and functional stability of firefly luciferase, J. Mol. Catal. B Enzym. 2010, 62, 127–132.

95. Kohda, J., Kawanishi, H., Suehara, K. I., Nakano, Y., Yano, T. Stabilization of free and immobilized enzymes using hyperthermophilic chaperonin, J. Biosci. Bioeng. 2006, 101, 131–136.

96. Sakane, I., Ikeda, M., Matsumoto, C., Higurashi, T., Inoue, K., Hongo, K., Mizobata, T., Kawata, Y. Structural stability of oligomeric chaperonin 10: the role of two betastrands at the N and C termini in structural stabilization, J. Mol. Biol. 2004, 344, 1123–1133.

97. Wong, Y. H., Tayyab, S. Protein stabilizing potential simulated honey sugar cocktail under various denaturation conditions, Process Biochem. 2012, 47, 1933–1943.

98. Talbert, J. N., Goddard, J. M. Enzymes on material surfaces. Colloids Surf. B: Biointerfaces 2012, 93, 8–19.

99. Pauling, L., Corey, R. B., Branson, H. R. The structure of proteins: two hydrogen bonded helical configurations of the polypeptide chain. Proc. Natl. Acad. Sci. USA 1951, 37, 205–211.

100. Shah, D., Johnston, T. P., Mitra, A. K. Thermodynamic parameters associated with guanidine HCl- and temperature-induced unfolding of bFGF. Int. J. Pharm. 1998, 169, 1–14.

101. Cléry-Barraud, C., Renault, F., Leva, J., El Bakdouri, N., Masson, P., Rochu, D. Exploring the structural and functional stabilities of different paraoxonase-1 formulations through electrophoretic mobilities and enzyme activity parameters under hydrostatic pressure. Biochim. Biophys. Acta 2009, 1794, 680–688.

102. Khechinashvili, N. N., Volchkov, S. A., Kabanov, A. V., Barone, G. Thermal stability of proteins does not correlate with the energy of intramolecular interactions. Biochim. Biophys. Acta 2008, 1784, 1830–1834.

103. Hinz, H. J., Steif, C., Vogl, T., Meyer, R., Renner, M., Ledermüller, R. Fundamentals of protein stability. Pure Appl. Chem. 1993, 65, 947–952.

104. Arakawa, T., Ejima, D., Kita, Y., Tsumoto, K. Small molecule pharmacological chaperones: from thermodynamic stabilization to pharmaceutical drugs. Biochim. Biophys. Acta 2006, 1764, 1677–1687.

105. Zhou, D., Zhang, G. G. Z., Law, D., Grant, D. J. W., Schmitt, E. A. Physical stability of amorphous pharmaceuticals: importance of configurational thermodynamic quantities and molecular mobility. J. Pharm. Sci. 2002, 91, 1863–1872.

106. Duddu, S. P., Zhang, G., Dal Monte, P. R. The relationship between protein aggregation and molecular mobility below the glass transition temperature of lyophilized formulations containing a monoclonal antibody, Pharm. Res. 1997, 14, 596–600.

107. Makhatadze, G. I., Privalov, P. L. Protein interactions with urea and guadinium chloride. A calorimetric study, J. Mol. Biol. 1992, 226, 491–505.

108. Lavelle, L., Fresco, J. R. Stabilization of nucleic acid triplexes by high concentrations of sodium and ammonium salts follows the Hofmeister series. Biophys. Chem. 2003, 105, 681–699.

109. Tang, X., Pikal, M. J. Measurement of the kinetics of protein unfolding in viscous systems and implications for protein stability in freeze-drying. Pharm. Res. 2005, 22, 1176–1185.

110. Kamiyama, T., Sadahide, Y., Nogusa, Y., Gekko, K. Polyol-induced molten globule of cytochromec: an evidence for stabilization by hydrophobic interaction. Biochim. Biophys. Acta 1999, 1434–44–57.

111. Haque I., Singh, R., Moosavi-Movahedi, A. A., Ahmad, F. Effect of polyol osmolytes on DGD, the Gibbs energy of stabilization of proteins at different pH values, Biophys. Chem. 2005, 117, 1–12.

112. Haque, I., Singh, R., Ahmad, F., Moosavi-Movahedi, A. A. Testing polyols' compatibility with Gibbs energy of stabilization of proteins under conditions in which they behave as compatible osmolytes, FEBS Lett. 2005, 579, 3891–3898.

113. Rodriguez-Larrea, D., Minning, S., Borchert, T. V., Sanchez-Ruiz, J. M. Role of salvation barriers in protein kinetic stability. J. Mol. Biol. 2006, 360, 715–724.

114. Kelch, B. A., Agard, D. A. Mesophile versus thermophile: insights into the structural mechanisms of kinetic stability, J. Mol. Biol. 2007, 370, 784–795.

115. Ahern, T. J., Klibanov, A. M. The mechanism of irreversible enzyme inactivation at 100 °C. Science 1985, 228, 1280–1284.

116. Balcão, V. M., Fernandez-Lafuente, R., Malcata, F. X., Guisán, J. M. Structural and functional stabilization of L-asparaginase from *Escherichia coli* upon immobilization onto highly activated supports: possible biomedical applications using extracorporeal bioreactors. Electron. J. Biotechnol. 2000 (http://www.ejbiotechnology.info/feedback/proceedings/04/poster/p66.html).

117. Balcão, V. M., Mateo, C., Fernández-Lafuente, R., Malcata, F. X., Guisán, J. M. Structural and functional stabilization of L-asparaginase via multi-subunit immobilization onto highly activated supports, Biotechnol. Prog. 2001, 17, 537–542.

118. Balcão, V. M., Mateo, C., Fernández-Lafuente, R., Malcata, F. X., Guisán, J. M. Coimmobilization of L-asparaginase and glutamate dehydrogenase onto highly activated supports. Enzym. Microb. Technol. 2001, 28, 696–704.

119. Jaenicke, R. Glyceraldehyde-3-phosphate dehydrogenase from *Thermotoga maritima*: strategies of protein stabilization. FEMS Microbiol. Rev. 1996, 18, 215–224.

120. Jaenicke, R., Schurig, H. N. Beaucamp, R. Ostendorp, Structure and stability of hyperstable proteins: glycolytic enzymes from hyperthermophilic bacterium Thermotoga maritima, Adv. Protein Chem. 1996, 48, 181–269.

121. Das R., Gerstein, M. The stability of thermophilic proteins: a study based on comprehensive genome comparison, Funct. Integr. Genomics 2000, 1, 76–88.

122. Neri, D. F. M., Balcão, V. M., Carneiro-da-Cunha, M. G., Carvalho Jr., L. B., Teixeira, J. A. Immobilization of β-galactosidase from Kluyveromyces lactis onto a polysiloxane–polyvinyl alcohol magnetic (mPOS–PVA) composite for lactose hydrolysis. Catal. Commun. 2008, 9, 2334–2339.

123. Neri, D. F. M., Balcão, V. M., Costa, R. S., Ferreira, E. C., Torres, D. P. M., Rodrigues, L. R., Carvalho Jr., Teixeira L. B. J. A. β-Galactosidase from Aspergillus oryzae immobilized onto different magnetic supports: a comparative experimental and modeling study of the galacto oligosaccharides production, in: E. C. Ferreira, M. Mota (Eds.), Proceedings of the 10th International Chemical and Biological Engineering Conference—CHEMPOR 2008, Braga, Portugal, September 4–6, 2008, pp. 1036–1041.

124. Neri, D. F. M., Balcão, V. M., Dourado, F. O. Q., Oliveira, J. M. B., Carvalho Jr., L. B., Teixeira, J. A. Galacto oligosaccharides production by β-galactosidase immobilized onto magnetic polysiloxane–polyaniline particles. React. Funct. Polym. 2009, 69, 246–251.

125. Neri, D. F. M., Balcão, V. M., Torres, D. P. M., Rodrigues, L. M., Costa, R., Rocha, I. C. A. P., Ferreira, E. M. F. C., Carvalho Jr., L. B., Teixeira, J. A. Galacto-oligosaccharides production during lactose hydrolysis by free Aspergillus oryzae

β-galactosidase and immobilized on magnetic polysiloxane–polyvinyl alcohol, Food Chem 2009, 115, 92–99.

126. Neri, D. F. M., Balcão, V. M., Dourado, F. O. Q., Oliveira, J. M. B., Carvalho Jr., L. B., Teixeira, J. A. Immobilized beta-galactosidase onto magnetic particles coated with polyaniline: support characterization and galacto oligosaccharides production, J. Mol. Catal. B Enzym. 2011, 70, 74–80.

128. Neri, D. F. M., Balcão, V. M., Cardoso, S. M., Silva, A. M. S., Domingues, M. R. M., Torres, D. P. M. Rodrigues, L. R. M. Carvalho Jr., L. B., Teixeira, J. A. Characterization of galacto oligosaccharides produced by beta-galactosidase immobilized onto magnetized Dacron, Int. Dairy J. 2011, 21, 172–178.

129. Rocha, C., Balcão, V. M., Gonçalves, M. P., Teixeira, J. A. Spent grain as a new carrier for trypsin immobilization, Proceedings of the 34th International Conference of Slovak Society of Chemical Engineering (SSCHE), 2007, pp. 277–284.

130. Arola, S., Tammelin, T., Setälä, H., Tullila, A. M. B. Linder, Immobilization-stabilization of proteins on nanofibrillated cellulose derivatives and their bioactive film formation. Biomacromolecules 2012, 13, 594–603.

131. Bolivar, J. M., Rocha-Martín, J., Mateo, C., Guisán, J. M. Stabilization of a highly active but unstable alcohol dehydrogenase from yeast using immobilization and post immobilization techniques, Process Biochem. 2012, 47, 679–686.

132. Grazu, V., López-Gallego, F., Guisán, J. M. Tailor-made design of penicillin G acylase surface enables its site-directed immobilization and stabilization onto commercial mono-functional epoxy supports. Process Biochem. 2012, 47, 2538–2541.

133. Tang, K. E. S., Dill, K. A. Native protein fluctuations: the conformational-motion temperature and the inverse correlation of protein flexibility with protein stability. J. Biomol. Struct. Dyn. 1998, 16, 397–411.

134. Pessela, B. C., Mateo, C., Filho, M., Carrascosa, A. V., Fernández-Lafuente, R., Guisán, J. M. Stabilization of the quaternary structure of a hexameric alpha-galactosidase from Thermussp. T2 by immobilization and post-immobilization techniques. Process Biochem. 2008, 43, 193–198.

135. Bernal, C., Sierra, L., Mesa, M. Improvement of thermal stability of beta-galactosidase from Bacillus circulans by multipoint covalent immobilization in hierarchical macro-mesoporous silica. J. Mol. Catal. B Enzym. 2012, 84, 166–172.

136. Boscolo, B. Trotta, F., Ghibaudi, E. High catalytic performances of Pseudomonas fluorescens lipase adsorbed on a new type of cyclodextrin-based nanosponges, J. Mol. Catal. B Enzym. 2010, 62, 155–161.

137. Balcão, V. M., Paiva, A. L., Malcata, F. X. Bioreactors with immobilized lipases: state-of the-art, Enzym. Microb. Technol. 1996, 18–392–416.

138. Cowan D. A., Fernández-Lafuente, R. Enhancing the functional properties of thermophilic enzymes by chemical modification and immobilization, Enzym. Microb. Technol. 2011, 49, 326–346.

139. Balcão, V. M., Costa, C. I., Matos, C. M., Moutinho, C. G., Amorim, M., Pintado, M. E., Gomes, A. P., Vila, M. M., Teixeira, J. A. Nanoencapsulation of bovine lactoferrin for food and biopharmaceutical applications. Food Hydrocoll. 2013, 32, 425–431.

140. López-Gallego, F., Betancor, L., Hidalgo, A., Mateo, C., Guisán, J. M., Fernández Lafuente, R. Optimization of an industrial biocatalyst of glutaryl acylase: stabilization of the enzyme by multipoint covalent attachment onto new aminoepoxy Sepabeads. J. Biotechnol. 2004, 111, 219–227.

141. Figueira, J. A., Sato, H. H., Fernandes, P. Establishing the feasibility of using β-glucosidase entrapped in Lentikats and in sol–gel supports for cellobiose hydrolysis. J. Agric. Food Chem. 2013, 61, 626–634.

142. Fuentes, M., Mateo, C., Guisán, J. M., Fernández-Lafuente, R. Preparation of inert magnetic nano-particles for the directed immobilization of antibodies. Biosens. Bioelectron. 2005, 20, 1380–1387.

143. Flickinger, M. C., Schottel, J. L., Bond, D. R., Aksan, A., Scriven, L. E. Painting and printing living bacteria: engineering nanoporous biocatalytic coatings to preserve microbial viability and intensify reactivity, Biotechnol. Prog. 2007, 23, 2–17.

144. Yoshimoto, M., Sakamoto, H., Shirakami, H. Covalent conjugation of tetrameric bovine liver catalase to liposome membranes for stabilization of the enzyme tertiary and quaternary structures, Colloids Surf. B: Biointerfaces 2009, 69, 281–287.

145. Balcão, V. M., Vieira, M. C., Malcata, F. X. Adsorption of protein from several commercial lipase preparations onto a hollow-fiber membrane module, Biotechnol. Prog. 1996, 12, 164–172.

146. Vaidya, B. K., Singhal, R. S. Use of insoluble yeast beta-glucan as a support for immobilization of Candida rugosalipase, Colloids Surf. B: Biointerfaces 2008, 61, 101–105.

147. Serno, T., Geidobler, R., Winter, G. Protein stabilization by cyclodextrins in the liquid and dried state. Adv. Drug Deliv. Rev. 2011, 63, 1086–1106.

148. Fernández-Lafuente, R., Rodriguez, V., Mateo, C., Penzol, G., Hernández-Justiz, O., Irazoqui, G., Villarino, A., Ovsejevi, K., Batista, F., Guisán, J. M. Stabilization of multimeric enzymes via immobilization and post-immobilization techniques. J. Mol. Catal. B Enzym. 1999, 7, 181–189.

149. Abian, O., Wilson, L., Mateo, C., Fernández-Lorente, G., Palomo, J. M., Fernández Lafuente, R. Guisán, J. M., Tam, D. Re, A., Daminatti, M. Preparation of artificial hyperhydrophilic micro-environments (polymeric salts) surrounding enzyme molecules: new enzyme derivatives to be used in any reaction medium. J. Mol. Catal. B Enzym. 2002, 19–20, 295–303.

150. Fernández-Lafuente, R., Rosell, C., Rodriguez, V., Guisán, J. M. Strategies for enzyme stabilization by intramolecular crosslinking with bifunctional reagents. Enzym. Microb. Technol. 1995, 17, 517–523.

151. Bolivar, J. M., Rocha-Martin, J., Mateo, C., Cava, F., Berenguer, J., Fernandéz-Lafuente, R., Guisán, J. M. Coating of soluble and immobilized enzymes with ionic polymers: full stabilization of the quaternary structure of multimeric enzymes. Biomacromolecules 2009, 10, 742–747.

152. Celayeta, J. F., Silva, A. H., Balcão, V. M., Malcata, F. X. Maximization of the yield of final product on substrate in the case of sequential reactions catalyzed by coimmobilized enzymes: a theoretical analysis, Bioprocess Biosyst. Eng. 2001, 24, 143–149.

153. Kotzia, G. A., Labrou, N. E. L-asparaginase from Erwinia chrysanthemi 3937: cloning, expression and characterization, J. Biotechnol. 2007, 127, 657–669.

154. Lumry, R., Eyring, H. Conformation changes of proteins. J. Phys. Chem. 1954, 58, 110–120.

155. Kumar, V., Chari, R., Sharma, V. K., Kalonia, D. S. Modulation of the thermodynamic stability of proteins by polyols: significance of polyol hydrophobicity and impact on the chemical potential of water, Int. J. Pharm. 2011, 413, 19–28.

156. Julca, I., Alaminos, M., González-López, J., Manzanera, M. Xeroprotectants for the stabilization of biomaterials. Biotechnol. Adv. 2012, 30, 1641–1654.
157. Golovina, E. A., Golovin, A. V., Hoekstra, F. A., Faller, R. Water replacement hypothesis in atomic detail—factors determining the structure of dehydrated bilayer stacks, Biophys. J. 2009, 97, 490–499.
158. Kane, R. S., Deschatelets, P., Whitesides, G. M. Kosmotropes form the basis of protein resistant surfaces. Langmuir 2003, 19, 2388–2391.
159. Poddar, N. K., Ansari, Z. A., Singh, R. K. B., Moosavi-Movahedi, A. A., Ahmad, F. Effect of monomeric and oligomeric sugar osmolytes on ΔGD, the Gibbs energy of stabilization of the protein at different pH values: is the sum effect of monosaccharide individually additive in a mixture? Biophys. Chem. 2008, 138, 120–129.
160. Chein, Y. W. Novel drug delivery systems 50, 2nd ed., 637, 676, 679.
161. Chien, Y. W., Chang, S. F. Intranasal drug delivery for systemic medication. Crit. Rev. Ther. Drug carrier Syst. 1987, 4, 67–194.
162. Wieriks J. Resorption of alpha amylase upon buccal application. Arch. Int. Pharmacodyn. Ther. 1964, 151, 126-135.
163. Merkle et al. Self adhesive patches for buccal delivery of peptides. Proc. Int. Symp. Control. Rel. Bio. Mat. 1985, 12, 85.
164. Christie, C. D., Hanzal R. F. J. Clin. Invest., 1931, 10, 787–793.
165. Siddiqui O, Chein Y. W. Non-parenteral administration of peptides and protein drugs. CRC Crit. Rev. Thes. Drug Carrier Syst. 1987, 3, 195–208.
166. Touitou et al. New hydrophilic vehicle enabling rectal and vaginal absorption of insulin, heparin, phenol red, gentamicin. J. Pharm. Pharmacol. 1978, 30, 662–663.
167. Wieriks J. Resorption of alpha amylase upon buccal application. Arch. Int. Pharmacodyn. Ther 1964, 151, 126–135.
168. Aungst et al. Comparison of nasal, rectal, buccal, sublingual and intramuscular insulin efficacy and the effects of a bile salt absorption promoter. J. Pharmacol. Exp. Ther. 1988, 224, 23–27.
169. Ziv et al. Bile salts promotes the absorption of insulin the rat colon. Life Sci. 1981, 29, 803–809.
170. Raz, I., Kidron, M. Rectal administration of insulin. Isr. J. Med. Sci. 1984, 20, 173–175.
171. Saffran, M., Bedra, C., Kumar G. S., Neckers D. C. Vasopressin: A model for the study of effects of additives on the oral and rectal administration of peptide drugs. J. Pharm. Sci. 1988, 77, 33–38.
172. Yoshika, S., Caldwell, C., Higuchi, T., Enhanced rectal bioavailability of polypeptides using sodium 5-methoxysalicylate as an absorption promoter. J. Pharm. Sci. 1982, 71, 593–594.
173. Morimoto et al. Enhanced rectal absorption of [Asul1, 7]-eel calcitonin in rats using polyacrylic acid aqueous gel base. J. Pharm. Sci. 1984, 73, 1366–1368.
174. Morimoto et al. Effect of nonionic surfactants in a polyacrylic acid gel base on the rectal absorption of [Asul1, 7]-eel calcitonin in rats. J. Pharm. Pharmacol. 1985, 37, 759–760.
175. Dalmark M. Plasma radioactivity after rectal instillation of radio-iodine labeled of human albumin in normal subjects and in the patient with ulcerative colitis. Scand. J. Gastroenterol. 1968, 3, 490–496.

176. Tregear R. T. The permeability of skin to albumin, dextrans and polyvinylpyrrolidone. J. Invest. Dermatol. 1996, 46, 2427.

177. Menasche et al. Pharmacological studies on elastin peptides (kappa-elastin). Blood clearance, percutaneous penetration and tissue distribution. Pathol. Biol., 1981, 29, 548–554.

178. Brunette B. R., Marreco D. Comparison between the iontophoretic and passive transport of thyrotropin releasing hormone across excised nude mouse skin. J. Pharm. Sci., 1986, 75, 738–743.

179. Banerjee P. S., Ritschel W. A. Int. J. Pharm. 1989, 49, 189–197.

180. Chein Y. W., Lelawongs P., Siddiqui O., Sun Y., Shi WM. Faciliated tranderdmal delivery of therapeutic peptides/proteins by iontophoretic delivery devices. J. Control. Rel. 1990, 13, 263–278.

181. Siddiqui, O., Sun, Y., Liu, J. C., Chein, Y. W. 1987. Faciliated transdermal transport of insulin. J. Pharm. Sci. 76, 341–345.

183. Sibalis D. Transdermal drug applicator. U. S. Patent 4, 708, 716 (1987).

184. Meyer B. R. Electro-osmotic transdermal drug delivery, in: 1987 Conference Proceedings on the Latest Developments in Drug Delivery Systems, Aster Publishing, Eugene, Oregon, 1987, 40.

185. Chein Y. W., Siddiqui O., Liu J. C. Transdermal iontophoretic delivery of therapeutic peptides/proteins. I. Insulin. Ann. N. Y. Acad. Sci. 1988, 507, 32-51.

186. Alkhaled, T., Singh, J., Recent patent on drug delivery and formulation, 1, 2007, 65–71.

187. Vyas, S. P., Khar, K. R. Targeted and controlled drug delivery, Novel carrier system, CBS publishers and distributors, New Delhi 561.

188. Morris, M. C., Deshayes, S., Simeoni, F., Aldrian-Herrada, G., Heitz, F., Divita, G. A noncovalent peptide-based strategy for peptide and short interfering RNA delivery, in: Ü. Langel (Ed.), Cell-penetrating peptides, CRC press, Boca Raton FL, 2007, pp. 387–408.

189. Fawell, S., Seery, J., Daikh, Y., Moore, C., Chen, L. L., Pepinsky, B., Barsoum, J. Tat-mediated delivery of heterologous proteins into cells, Proc. Natl. Acad. Sci. USA 1994, 91, 664–668.

190. Wender, P. A., Mitchell, D. J., Pattabiraman, K., Pelkey, E. T., Steinman, L., Rothbard, J. B. The design, synthesis and evaluation of molecules that enable or enhance cellular uptake: peptoid molecular transporters, Proc. Natl. Acad. Sci. USA 97 (2000) 13003–13008.

191. Meade, B. R., Dowdy, S. F. Exogenous siRNA delivery using peptide transduction domains/cell penetrating peptides, Adv. Drug Deliv. Rev. 2007, 59, 134–140.

192. Deshayes, S., Heitz, A., Morris, M. C., Charnet, P., Divita, G., Heitz, F. Insight into the mechanism of internalization of the cell-penetrating carrier peptide Pep-1 through conformational analysis, Biochemistry 43, 2004, 1449–1457.

193. Morris, M. C., Depollier, J., Mery, J., Heitz, F., Divita, G. A peptide carrier for the delivery of biologically active proteins into mammalian cells: application to the delivery of antibodies and therapeutic proteins. Handb. Cell Biol. 2006, 4, 13–18.

194. Le Roux I., Joliot, A. H., Bloch-Gallego, E., Prochiantz, A., Volovitch, M. Neurotrophic activity of the antennapedia homeodomain depends on its specific DNA-binding properties. Proc Natl Acad Sci USA 1993, 90:9120–9124.

195. Elliott, G., O'Hare P. Intercellular trafficking and protein delivery by a herpes virus structural protein. Cell 1997, 88, 223–233.
196. Green, M., Loewenstein, P Autonomous functional domains of chemically synthesized human immunodeficiency virus Tat trans-activator protein. Cell 1988, 55, 1179–1188.
197. Ho, A., Schwarze, S. R., Mermelstein, S. J., Waksman, G., Dowdy, S. F. Synthetic protein transduction domains: enhanced transduction potential in vitro and in vivo. Cancer Res 2001, 61:474–477.
198. Chen, Y. N., Sharma, S. K., Ramsey, T. M., Jiang, L., Martin, M. S., Baker, K., Adams, P. D., Bair K. W., Kaelin, W. G. Jr. Selective killing of transformed cells by cyclin/cyclin-dependent kinase 2 antagonists. Proc Natl Acad Sci USA 1999, 96, 4325–4329.
199. Elliott, G., O'Hare, P., Intercellular trafficking of VP22–GFP fusion proteins. Gene Ther 1999, 6, 149–151.
200. Adessi, C., Sotto, C. Converting a peptide into a drug: Strategies to improve stability and bioavailability. Curr Med Chem. 2002, 9, 963–978.

CHAPTER 2

PEPTIDE-MEDIATED NANOPARTICLE DRUG DELIVERY SYSTEM

CONTENTS

ABSTRACT

Innovations in protein research lead to the more development in peptide drug delivery systems. These amino acids linked compounds are originated from either natural or synthetic sources. Here we explored the pharmacokinetic consideration, physico-chemical properties, and methods of preparation, conjugates and various therapeutic applications of different types of peptides. Peptide decorated polymeric nanoconjugates with their mechanism of action and specified targeting strategies and biomedical applications are also discussed well.

2.1 INTRODUCTION

From past decade, researchers are trying to discover novel targeting strategies by drawing the molecular relationship between targeted molecules and their respective binding sites. Several targeting approaches were reported which may overcome the extracellular and intracellular barriers for the selective targeting of therapeutic drug towards the targeted site [1]. The targeted drug delivery systems have some fundamental requirements. Several approaches have been developed to improve the availability of drug at required site more specifically to the diseased cell. Encapsulation and bio-conjugation (antibodies, peptide, folic acid, aptamers, affibodies etc.) are the most common approaches which are used for targeting the therapeutic drug in nanoparticulate drug delivery. This conjugation approach provides cell specificity and selectivity [2].

Nanomedicine is an emerging field from last decade in which many technologies and engineering, for example, nanoparticle, quantum dot, carbon tube, etc. are employed to achieve medical benefits. These technologies have been used for imaging, biosynthesis, repair, defense and improvement of human biological system. Site specific delivery of nanomedicine can be achieved either by changing its phisico-chemical properties or by conjugating it with targeting ligand. Intracellular destinations for nanoparticles can be achieved by cellular internalization via receptor endocytosis, macro pinocytosis and direct interaction [3]. Particle size plays an important role in effecting the efficacy of nanoformulation. Optimum sized (≤ 200 nm) nanoaparticles are proved to be a good candidate for cellular drug delivery as larger particles are easily cleared by reticuloendothelial system. Reduction in particle size imposes various advantages over conventional methods such as improved absorption at desired site, reduces the chances of local inflammation, increase intercellular trafficking and cellular uptake [4–7].

Peptide is a group of several covalently linked amino acids which are originated from either natural or synthetic sources, constituting amine at one terminus and carboxyl at another terminus except cyclic peptide. Based on molecular size it can be dipeptide, tripeptide, oligopeptide or polypeptide for, for example, glutathione is a tripeptide which is used in the form of nano vehicle for targeting gene intracellularly [8]. Peptide can be categories into cell penetrating peptide, cell targeting and organ targeting peptide (Table 2.1).

Cell penetrating peptide can transport drug through membrane translocation whereas cell and organ targeting peptide act by interacting with receptor present on a particular cell [9]. Peptide mediated delivery stood as the most promising targeted delivery to enhance the biological activity with minimal side effects. Peptides can be used as targeting ligands in many nanocarriers due to its unique properties like small size, charge, biocompatibility, easy modification and internalization of cell both actively and passively and cell membrane interaction. Nanoparticulate conjugated peptides solve the problem related to tissue/organ specificity and provide new opportunities for non-selective or highly hydrophobic drugs [10–12]. Peptide can deliver the drug into the cell organelles through receptor-mediated endocytosis [13–16]. Peptide has ubiquitous property that it can mimic the sequence of receptors or surface protein of targeted cell. Cancerous cells are known to be the most attractive sites for peptides because they contains some overexpressed proteins. Overexpression of these proteins or melanoma is selectively treated by peptide therapy in which peptide is act as a core material. Therefore, peptide can be used as targeting ligands or core material which may enhance the cellular uptake of encapsulated material.

Designing and identification of peptide can be performed either from rational design or combinatorial library. Rational design deals with the effect of structure on the function of peptide. This technique suggests the idea to overcome barrier and limitation in activity of previously synthesized peptide. Recent example of this approach is synthesis of CLV3 peptide analog which showed better binding affinity after the addition of proline group [17]. Another method of peptide design is combinatorial library which is faster, less expensive and more powerful. Thousands of compounds can be screened with in limited time [18]. This is well suited for screening of antibodies, peptide, oligonucleotides and cytokines [19]. For the synthesis of combinatorial library members, two supports are required; one is solid support and another is solution support. The reaction

TABLE 2.1   Peptides are Broadly Classified into Cell and Organ Targeting Peptides

Type of peptides	Examples
Arginine rich peptides	Tat, Penetratin, pVEC, HIV-1 Rev, R7W, FHV Coat, BMV Gag, HTLV-II REX, CCMV Gag, P22 N, Polyarginine
Glycine rich peptide	Pg-AMP1, Sarcotoxin IIA, Hymenoptaecin, Attacin, Diptericin, Coleoptericin, Armadillidin, Bactrocerin-1, OdVP1, SK84
Cationic peptide	DVP, Protamine, Human cJun, Human cFos, Yeast GCN4, Penetratin, Islet-1, Fushi-tarazu, Engrailed-2, HoxA-13, Knotted-1, PDX-1, Polyarginine, AIP6, HRSV
Ampipathic peptide	pVEC, M918, ARF, Azurin, hCT, CADY, Pep1, MPG, Transportan, TH, MAP, EB-1, GALA, Erns HIV-1 VPR, Ribotoxin, PreS2-TLM, SG3, BEN-1079
Fusogenic peptides	Kaposi FGF NLS, HA2, VP22
Staple peptide	BID SAHBA, BAD SAHBA, BIM SAHBA, SAH-p53-8, NYAD-1, SAHM1, SAH-gp41, MCL-1 SAHBA, SAH-apoA-I
Signal peptide	Ig(v), BPrPp(1–30), MPrPp(1–28), Kaposi FGF NLS
Proline rich peptide	Cathelicidin, Pyrrhocoricin, Bac7, SAP, Sarcotoxin IIA, Colostrinin
Histidine rich peptides	LAH4
Cystein rich peptide	Defensins, Drosomycin, ToAMP
Antimicrobial peptide	Mastoparan peptide, Chrysophsin-1, Hainanenins, Hb 98–114, Pv-derived peptide, Brevinin, [D4K]B2RP, [E4K]alyteserin-1c, AMPNT-6, Melittin, Magainin 2, Buforin 2, Human, lactoferrin, SynB1, Crotamine, S413-PVrev, Polymyxin E, Nisin A, Buforin II, tachyplesin I, ABP-CM4, polyphemusin, pleurocidin, Hexapeptide WRWYCR, Dermaseptin HNP-1 and HNP-2, PR-39, indolicidin and plant thionins, Histatin-5, hLF(1–11), ABP-CM4, BMAP-28, Cationic a-helical neuropeptides, Non ribosomal peptide, Dermaseptins, Temporins
Synthetic peptide	C105Y, ABP-CM4, L1, L-Bac7(1–35), C15M19, P1, BP100, Penetratin, A3-APO, G1, DS1(1–29)-NH2, Polyarginines, Inv3, EGLA, GALA, JTS, INF

TABLE 2.1 Continued

Type of peptides	Examples
Cell Targeting Peptide	
Synthetic peptide	CTLPHLKMC, RIP (c-terminal), [d]-H6L9, [d]-K6L9, CGNKRTRGC, SMSIARL, VSFLEYR, LTVXPWX, KCCYSL, MCPKHPLGC, Crotamine, PH1 peptide, AWBP, PLAEIDGIELTY, Trp-rich peptides, Pimtide
Antimicrobial peptide	Andropin, Cecropin A, Dermaseptin, Melittin, Tachyplesin, Pr-1, CLIP, Trp rich, Gramicidin, Capreomycin
Metalloproteinase inhibitor	Chlorotoxin, A54 peptide, N-Ac-AA, CTTHWGFTLC, NNF2, Melittin, Tat, FN2, KLD-12 peptide, MMP-2, MMP-9, Buforin II
Glycopeptide	Gliadin, Hep I Peptide, Immunoglobulin G (204–210), Bleomycin, Peptide T
Lipopeptide	Pam3CAG (synthetic), MALP 2, GALA
Cell adhesion peptide	cNGXGXXc, RGD, Vitronectin, cLABL
Hepatocellular carcinoma cell-specific peptide	SP94, A54 peptide
Host defense peptide	Aureins, BMAP-27/BMAP-28, Cecropin A, Citropins, Epinicidin-1, Gaegurin-6/-5, LL-37, Magainins, Melittin, Dolastatin 10, Alloferon-1/-2, PR-39, RGD-tachyplesin, Protegrin-1, Lactoferricin B, Gomesin, Buforin II, Temporin A, Polybia-MPI, Pep27, NK-2, Chlorotoxin, ATWLPPR, ApoPep-1, PIVO-24, RVG, LyP-1, ABP, CTLPHLKMC, PH1 peptide, c(RGDfK), SynB1 – ELP-GRG, [d]-H6L9, [d]-K6L9, EV peptide, Formyl peptide, Polysacchride peptide, Netropsin
Lung specific peptides	NQRGTELRSPSVDLNKPGRH, cNGXGXXc, YIGSR
Islet-binding peptide	RIP1
Collagen-mimetic peptides	GVKGDKGNPGWPGAP
Radiometallated peptide	bombesin (BBN) analogs, a-MSH peptide, guanylin peptides

TABLE 2.1   Continued

Type of peptides	Examples
Neuroglial cell specific peptide	Angiopep, Chlorotoxin, hNgf EE peptide, GL1, TrkB-peptide, SFTYWTNGGSK-Biotin, Lep70–89, trisialoganglioside GT1b, Colostrinin
TNF specific peptide	CLWTVGGGC, P36
Mammmary gland specific	CLHQHNQMC, peptide
Intestinal cell specific	CSKSSDYQC (CSK), GX1 (CGNSNPKSC)
Peptide act on immune system	Polysacchride peptide, Peptide T, Mast cell degranulating peptide, ENF peptide, MSH peptide, Tuftsin FDC-SP (Follicular Dendritic Cell Secreted Protein) related peptide
Protease inhibitor	leupeptin, antipain, chymostatin, pepstatinj
Others	AT3G22820, cyclic RKKH peptide, circular peptide (KGGRAKD), T9,
	Peptide act on immune system
	(Polysacchride peptide, Peptide T, Mast cell degranulating peptide, ENF peptide, MSH peptide, Tuftsin, FDC-SP (Follicular Dendritic Cell Secreted Protein) related peptide)
	Protease inhibitor (leupeptin, antipain, chymostatin, pepstatinj)
Organ Targeting	
Lungs	Vasoactive Intestinal Peptide (VIP) (PHM, PHI, Secretin, rGRF, hGRF, AcTyr' DPbe2, GRF), Gastrin Releasing Peptide (GRP)
Liver	CYP4F2 peptide, YM087, a novel nonpeptide AVP, c-erb A peptide, anti-CYP4F2 peptide, glutathione S-transferase-activating peptide
Pancreases	Neuropeptide Y (NPY), peptide YY (PYY), Insulin, IGF, Proinsulin, C type peptide, Chrmostatin

TABLE 2.1 Continued

Type of peptides	Examples
Kidney	Vasoactive peptide (Urodilatin, Angiotensin, Bradykinin and related kinins, Vasopressin, Endothelin, Neuropeptide Y, ANP, atrial natriuretic peptide), Arterial Natriuretic, Brain Natriuretic, Uroguanylin, Guanylin, Tuberoinfundibular peptide, urotensin II, Endothelins, Kisspeptins, Adrenomedullins, urocortin, orexin-A, calcitonin gene peptide, Glucagon, glucagon-like peptide (GLP- 1), vasoactive intestinal peptide, pituitary adenylate cyclase activating peptide (PACAP), Galanin-like peptide (GALP) (galanin-message-associated peptide (GMAP), M35, M40, Galnon, Galparan, Galmic) gastric inhibitory (GIP), parathyroid hormone-related peptide (PTHrP), Endothelin-1, Arginine Vasopressin Peptide (YM087, [ArgS]vasopressin), Bactenecin, c-erb A peptide)
Cardiac peptide	*Cardiac troponin I, Adenoreceptor peptide* (human β1-adrenoceptor, human β2-adrenoceptor, α-adrenoceptor), *Muscarinic receptor peptide* (M2-muscarinic receptor), *Connexin mimetic peptide* (Cx43Hc, GAP26, GAP27), *RYR2 peptide, Urotensin II-related peptide, glucagon-like peptide, Vasoactive Intestinal Peptide (VIP), N-Terminal Histidine and C-Terminal Isoleucine Amide (PHI), Calcitonin gene-related peptide* (CGRPα, CGRPβ, Amyiin, Calcitonin, Adrenomedullin, vanilloid 1) *Guanylyl cyclases activated peptide* (ANP, Anaritide, Carperitide, BNP, Nesiritide, CNP, enterotoxin, guanylin, uroguanylin, Linaclotide, Vasonatrin peptide, NT-proBNP, MR-proANP, Dendroaspis natriuretic peptide), *Acetylcholine receptor-related peptide* (AChR a179–191), *Opioid peptides* (Calpain, Leupeptin, dynorphin-(1–13), endorphins, enkephalins), *Chromogranin A (CGA)-derived peptides* (Vasostatin-1 (VS-1), Pancreastatin, catestatin, parastatin, leupeptin, antipain, chymostatin, and pepstatin)) *Vasodialating Peptide* (Adrenomedullin (AM), Amylin, CGRP, proadrenomedullin N-terminal, Parathyroid releasing hormone related peptide, Tuberoinfundibular peptide), *Mylin, Ghrelin, Bradykinin-potentiating peptides* (Eribis peptide 94), *Heparin-binding peptide* (RADA16, LRKKLGKA), *Crustacean cardioactive peptide* (hemopressin peptides, Apelin, Milk peptide)
Bone	RYR1 peptide, Parathyroid hormone-related peptide (PTHrP), Collagen peptide, Milk peptide
Skin	Chemerin, collagen peptide, Agouti Related Peptides

TABLE 2.1   Continued

Type of peptides	Examples
Brain	CNP, B type natriuretic, Glucagon like peptide, Insulin related peptide, IGF, Transferrin, (II-A~' Lys') dermorphin analogs (DALDA), Lys 7] dermorphin analogue (K 7DA), Brain derived neurotrophic factor (BDNF) peptide, Peptide nucleic acids (PNA), biotinylated-SS-vasoactive intestinal peptide analog (VIPa), Leptin-derived peptide, Anti-neuroexcitation peptide, CK4-M2GlyR (Tyr-c[D-Cys-Gly-Trp-Cys]-Asp-Phe-NH2), Casoxin C, Omentin, Chemerin, pro-opiomelanocortin related peptide, Thyrotropin-releasing hormone, Ceruletide, Cocaine- and Amphetamine-Regulated Transcript (CART) peptides
Radioimuno therapy	Octreotide, Lanreotide, Octreotate, α-MSH peptide, Glucagon-like peptide-1 (GLP-1), Neuropeptide Vasoactive intestinal peptide, Gastrin releasing peptide, cyclic HWGF, Polysacchride peptide
Intestine	Guanylin, Uroguanylin, T18, gliadin peptide, Casein phosphopeptide, CCK peptide
Reproductive organ	RFamide peptides, Kisspeptin, prolactin-releasing peptide (PrRP), PQRFa peptide, 26RFa/QRFP bombesin-like peptides, neurotensin peptide, oxytocin, Prolactin (PRL)-releasing peptide (PrRP) Gonadotropin-inhibitory hormone (RFamide peptides)
Lung	Cyclic octapeptides, RGD, Laminin receptor binding peptide (YIGSR), gastrin releasing peptide, Vasoactive intestinal peptide (VIP), rVBMDMP
Others	*Cyclic peptide* (RGD, Cyclic citrullinated peptide, Alpha Amantin, Cyanotoxin, Cyclotides, Theta defensin, Aplidine, NR58–3.14.3, Bacterocin, *Cyclo* [EKTOVNOGN] (AFPep), Cyclic lipopeptide, Eptifibatide, Ciclosporin, Urotensin–II, Melanotan II, Crustacean cardioactive peptide, Glycopeptide, Formyl peptide, Chlorotoxin, Viomycin, Phalloidin, Patellamide A)

TABLE 2.1 Continued

Type of peptides	Examples
	Peptide used in nanotechnology
	pH-activated peptides, Material binding peptide, Self assembled peptide, Ultra small peptide, Anuran Antimicrobial Peptides, Bis peptide, dual-affinity peptides, Biosensor peptide, Short lived peptide, Antimicrobial peptides (AMPs), Neuropeptides, VIP, APP (Amyloid Precursor Protein), Tag peptide, Milk peptide, Cell adhesion peptide, Apolipoprotein A-I mimetic peptides, Bolaamphiphilic, Bombesin peptide, Peptide lego, photoresponsive peptide, Chlorotoxin
	Other
	Beefy meaty peptide, WALP peptide, Tomingal like peptide, Phosphopeptide, mammalian liver-expressed antimicrobial peptides (LEAP), Mimitope, Peptide Aptamer, Contryphan

is carried out on resin bed by mixing starting material together. The product bound with the resin and excess product is removed by washing with appropriate solvent. Polymers, protecting group and linker are the three requirements for solid support. Mainly resins are used as a polymer due to its stability in all reaction condition. Linkers are used to attach the compound on the solid support and help in cleavage to get good yield. The commonly used linkers are alkyl-silly, safety catch and photo labile linkers. Protecting group such as Fluorenyl-methyloxycarbonyl (Fmoc) and tert-butyloxycarbonyl (Boc) are used to block and regenerate certain functional group [20].

Purification and elution of compounds are conducted by reverse phase HPLC-C-18, Mass spectrometry and ELISA. The composition of synthesized peptide can be confirmed by amino acid analyzer [21]. The intracellular movement and localization of peptide nano conjugates were studied using a combination of fluorescence spectrophotometry, fluorescence microscopy, and transmission electron microscopy.

Nanoparticle and peptide conjugation showed best combination in targeted drug delivery as it have all properties related to an ideal target drug delivery system. Peptide conjugates on the surface of nanoparticle and carries the nanomedicine to its site of action either cell or tissue. Smaller size, stability, bipodregerable, biocompatible nature and less immunogenicity make them suitable for drug as a ligand in diagnosis, imaging and treatment of a particular disease. Whereas ease in synthesis and modification provides a more economical way to improve the efficiency of whole formulation. The positive facets of excellent biocompatibility and admirable biodegradability with ecological safety and low toxicity with versatile biological activities such non immunogenicity has provided ample opportunities for further development.

2.2 PHARMACOKINETIC CONSIDERATION FOR PEPTIDE-MEDIATED DRUG DELIVERY

The pharmacokinetic information plays an important in the success of a new drug. This information provides a quantitative framework for evaluation and consideration of factors associated with safety and efficacy.

Clinical pharmacokinetic data reveal the relationship between dose and therapeutic response [22]. Peptides are good candidates for treatment of many untreated disease. Research is continuously going on for improvement of delivery of peptide to its target sites. In this, pharmacokinetic and pharmacodynamics are the main issues related to development a new peptide based drug delivery [23]. Pharmacokinetic parameters that have been kept in mind before designing a peptide based drug delivery are: size, charge, plasma protein binding, lipophillicity, route of administration and resistance or susceptibility to the proteolysis degradation [24]. These factors can be modulated via rational design to modulate pharmacokinetic properties of peptide. Pharmacokinetic fate of drug depends on the translocation of peptide in the cell. The translocation of peptide may be carried out via electrostatic interaction, receptor mediated endocytosis and formation of transitory structure. Most of cell penetrating peptide has electrostatic interaction with the membrane of target cell and delivers the drug into that specific cell [25, 26]. Tat (trans activator of transcription) is most common cell penetrating peptide consist of 86–101 amino acids based on their subtype. Protein transduction domain in HIV-tat allows entering the cell by crossing cell membrane and translocating the drug via electrostatic, receptor mediated or through macropinocytosis depending on type of cell and properties of drug to be carried [27]. For optimal drug targeting, elimination of the complete drug–carrier system should be minimal so that it can achieve a required concentration of drug on target site. Drug release at target site should be specific and selective to reduce unwanted effect on healthy tissues. Drug should have enhanced permeation and retention effect to ensure drug accumulation at target site [28]. Enhanced permeation and retention effects of accumulated drug have been observed through passive targeting mechanism. In passive targeting, drug may be accumulated in extracellular matrix and diffuse throughout the body for bioactivity. Passive targeting releases their drug into the milieu rather than specific cell and extracellular drug release may not be effective for macromolecule. Passive targeting is non targeting drug delivery system. It can be achieved either by modification of physiochemical parameter of drug carrier complex (nanoparticles, micelle, liposomes etc.) or by taking the advantages of permeability of diseased tissue such as leaky vasculature, tumor microenvironment and direct application. In active targeting, carriers are attached to the surface of nanoparticles or if passive delivery system attach with a ligand it become active and provide site-specific drug delivery.

This approach can be used to direct the nanoparticles on a specific receptors present in the diseased cell [29]. Active mediate targeting facilitates receptor mediated endocytosis, releasing therapeutic agent into target cell and prevent the drug from reticuloendothelial pathways (Figure 2.1).

Active targeting of drug was found much better approach than passive targeting [30]. Peptides binds to the receptor with high affinity, causes cell internalization and carries drug into the cell [31]. An insect neuropeptide, namely, allatostatin, conjugated it with CdSe-ZnS QDs showed galanin receptor-mediated endocytosis by blocking the cells with a galanin antagonist and have intracellular drug delivery [32].

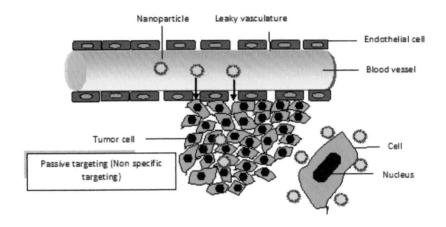

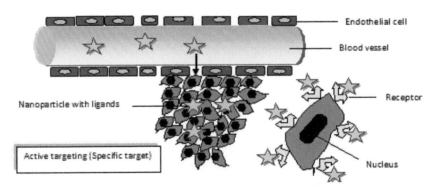

FIGURE 2.1   Nanoparticle mediated specific and non specific targeting of therapeutic agent.

2.3 PEPTIDE-BASED DRUG DELIVERY: METHOD OF PREPARATION

2.3.1 DESIGN AND SYNTHESIS OF PEPTIDE

Designing of new peptide/peptide ligands starts from understanding of conformational and sterochemical requirements of their biologically relevant receptors. In traditional approach, a motif (additional amino acid) was attached to improve biological activity directly or by ligand receptors interaction. Two approaches are used in designing of peptide. One is rational design in which chemical modifications are performed in native peptide and other is combinatorial library (Figure 2.2).

Chemical modification involves changes in physicochemical properties or conformation, introduction to peptide residue and other amino acid, cyclization and PEGylation whereas combinatorial library use the high throughput screening principle [33]. These approaches increase the cell selectivity and reduce its undesirable effect. However, knowledge of structure function relationship of peptide provides a design principle for development of

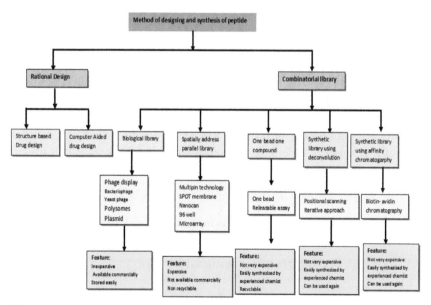

FIGURE 2.2 Illustration of various methods and approaches that are utilized in designing and synthesis of peptides.

peptide based ligands with desirable biological activity profile. The conformational and dynamic properties of amino acid and its analog can be examined by the use of biophysical studies such as X-ray crystallography, nuclear magnetic resonance spectroscopy, and circular dichroism. These techniques help in understanding the process of receptor peptide binding, additionally to find out best working model for interaction. The design and construction of synthetic peptides involves following steps [34] (Figure 2.3).

The conformation stability of synthesized peptide can be increased by inserting a disulfide linkage between cyclic peptide, introduction of amino acid residues with specific coordinate metal ions or association in oilgomeric state [35]. There are a number of characteristics which keep in mind while designing a potent and site specific peptide, for example, peptide size, composition, charge, conformation, lipophillicity, site of application, in vivo potency and route of administration [36, 37].

2.3.2 RATIONAL DESIGN

Rational design of peptide is the strategy of creating new peptide with certain functionality, based upon ability to predict how molecule's structure will affect its behavior through physical model. This can be done either from scratch or by making modification on unknown structure. This technique relays on known target that have a specific biological activity.

2.3.3 STRUCTURE FUNCTION RELATIONSHIP

Peptides are a group of amino acids that are involved in many physiological and biochemical functions. The correlation between the structure and

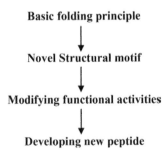

FIGURE 2.3 Design and construction of synthetic peptide.

function not only provide a strong base for understanding of their processes, but also help in improvement or designing of new peptide. For example, Structure function relationship of antimicrobial peptide revealed that cationic charge and amphipathic conformation are required for antimicrobial activity of peptide [38]. The cationic N-terminal form a structure that binds to the target cell surface while C-terminus penetrates into the hydrophobic core of cell membrane that result in a membrane leakage and cell death [39].

Many constraints are used in development of biologically active molecule such as carbogenic fusion of side chain of amino acid, conformation, monomer formation and topographical modulations [40, 41]. Conformation and topographical constraints are found to be more important in designing of highly selective targeting ligands [34]. Peptides are the molecule with flexible conformation and multiple active site of interaction. The main issue in designing of peptide is to find out biologically relevant conformation. Conformational constraint approach has been utilized to design peptide with specific conformational and topographical properties that are important for biological activity [42]. The introduction of conformational constraint improves favorable binding entropy and affinity of ligands. In backbone cyclization, conformational constraints are conferred on a peptide by linking ω-substituted alkylidene chains and replacing N^{α} or C^{α} hydrogen's in a peptidic backbone. This conformation constraint provides stable and biologically active analogs [43]. Lactam is commonly used conformational constraint in peptide designing. Introduction of lactam in peptide enhanced the biological potency of a peptide due to fixation of trans peptide bond and introduction of non covalent interaction [44]. Gamma lactam has been used as conformational constraint in designing of rennin inhibitory peptide [45].

2.3.4 COMPUTER-AIDED DRUG DESIGNING

In-silico method is also used to design the peptide for receptor or target site when there is no idea about the ligand or target. If there is sufficient knowledge about the target or ligands than other rational approaches can be used [46]. The theoretical design approach is a powerful alternative to phage peptide library for protein mimics. Such mini-peptide is more amenable to synthetic chemistry and thus may be useful starting points for the design of small organic mimics [47]. Angiotensin I converting enzyme (ACE) plays an important physiological role in the regulation of hypertension. Virtual screening was used to discover a novel angiotensin I converting enzyme inhibitory peptides from milk casein. Docking studies was used to understand the binding mode between the enzyme and peptide hit [48]. Designing of Peptide for lipid-binding motives of plasma apolipoproteins used molecular modeling approach. This was studied by designing peptide analogs to the helical repeats of the apolipoproteins with variable degrees of salt bridge formation between adjacent peptide chains. The most stable conformation for pairs of synthetic peptides was calculated by energy minimization together with the energy of interaction between peptides. Serrano et al. designed β-sheets peptide successfully using an algorithm PERLA (protein engineering rotamer library algorithm [49–51]. PepDesign procedure calculate new peptides fulfilling the hypothesis, tests the conformational space of these peptides in interaction with the target by angular dynamics and goes up to the selection of the best peptide based on the analysis of complex structure properties. This is the combination of structure function relationship and combinatorial chemistry [52].

2.3.5   COMBINATORIAL LIBRARY DESIGN

Combinatorial library originated with the invention of multipin technology by Gaysen in early 1980. The combinatorial chemistry allows the screening and synthesis of compound rapidly at low cost as compared to traditional method of synthesis. This method has been applicable on peptide, oligonucleotide, proteins, small molecule and oligosaccharide [53]. Common steps used in the process of combinatorial library are: (i) synthesis of library, (ii) screening of library, and (iii) identify lead structure. This approach can find out a hit with infinite number of variation on basic template. A large number of information can be gained with minimum experiment. Solid support for library member is essential for success of combinatorial library because it reduces the need of purification of peptide [54].

There are different approach for synthesize of combinatorial library: (i) biological peptide library using filamentous phage, plasmid or polysomes, (ii) synthetic peptide library that require deconvolution such as an iterative approach, positional scanning, orthogonal approach, (iii) affinity column selection method, (iv) split and mix synthesis, and (v) spatially addressed parallel library [55, 56]. A new compound for specific target can be generated for specific target with in limited time and cost. The choice of library preparation method depends upon the product to be prepared. Protein preparation can be performed from biological approach whereas small molecule such as oligonucleotide, oligosaccharide and oligomers are prepared by synthetic approach [57].

2.3.6 BIOLOGICAL PEPTIDE LIBRARY

Biological library is important tool in designing of peptide for a specific target. It contains a number of microorganisms with specific encoding of DNA or RNA sequence for a certain peptide. Each microorganism represent only single clone. For development of new peptide, the usage of synthetic oligonucleotide, cDNA, genomic DNA fragment and mutagenized specific gene fragment are method of choice for insertion of encoding DNA sequence. This nucleotide sequence can be cloned and expressed as peptide [58].

2.3.7 PEPTIDE PHAGE LIBRARY

Phage library is a method that uses less complex bacteriophage for development of peptide. In 1985, smith discovered gene III of filamentous phage for generation of peptide library and peptide of interest. Filamentous bacteriophage is a type of virus of bacteria defined by long and thin filament that are secreted from host cell without killing them. It contain genome of single stranded DNA and infect gram negative bacteria including Escherichia, Salmonella, Pseudomonas, Xanthomonas, Vibrio, Thermus and Neisseria [59]. Filamentous minor and major coat protein is used for ligation of DNA encoding on the M13 filamentous phage. A phage which shows activity for a specific target will remain on the display while other are washed out to produce more phage. Bounded phages are removed by elution method. Cycle is repeated until relevant phage is not obtained. The repeated cycle of these steps is referred as 'Panning.' F1, fd, T4, T7 and ⋋ are the bacteriophage which are also used in the bacteriophage technique. Plasmid and polysomes were developed to design biological peptide library [60]. This is a powerful method for the discovery of peptide ligands that are used for analytical tools, drug discovery, and target validations. Phage display technology can produce a huge number of peptides and generate novel peptide ligands. Recently, phage display technology has successfully managed to create peptide ligands that bind to pharmaceutically difficult targets such as the erythropoietin receptor. Phase display is a very cost effective and time saving technology as a vast number of peptides can be produced using one phage per peptide. The conformational design of peptides in library is important for selecting high-affinity ligands that bind to every target from a phage peptide library. The biological approach has several problem such as conformational constraint, modification into natural amino acid and limited number of eukaryotic amino acid [61] which can solved by designing scaffold of peptide for phage display libraries.

2.3.8 SYNTHETIC PEPTIDE LIBRARY REQUIRING DECONVOLUTION

The deconvolution technique screens the biological activity of a molecule through iterative process. The active compound obtains from the biological assay and structure can be determined from mass spectrometric analysis. Positional scanning is another library approach in deconvolution method. Positional scanning technique requires eight libraries of four compounds containing two hit mixture one for each active monomer. The advantage with this technique is that it identifies most potent library member rather than weak binder. Other technique used for this library is recursive devolution approach, orthogonal approach and dual recursive approach. Disadvantage with this technique is that it needs sublibraries which are time consuming process and sublibrary with the highest bioactivity determined does not necessarily contain the most bioactive compound (synergistic effect of multiple compounds) [62].

2.3.9 ONE BEAD ONE COMPOUND LIBRARY

One bead one compound is a routine process for screening of peptide. In this technique, each solid phase particle (bead) displayed only one type of peptide with 10^{13} copy of same compound on a single 100 μm bead. When active compound has been identified, peptide is splitted from bead and its sequence is determined. Library of 18 unique peptides can be generated within 5 min by introduction of high-pressure ammonia gas in one bead one compound library. This gas causes the cleavage of peptide from resin bed easily [63]. A million of compound can be screened in a very short time and already assayed library can be re-used. The limitation with this method is that free substance shows different bioactivity as compared to bounded substance.

2.3.10 SPATIALLY ADDRESSED PARALLEL LIBRARY

This technique is based on high throughput screening on solid or solution phase synthesis. Parallel analysis helps to study interaction between large set of oligonucleotides. Main technique used in spatial analysis is multi pin technology, 96 well plates, Spot membrane, nanocan and chemical microarray. Multipin technology was first demonstrated by Geysen et al. in 1984. It allows the processing and synthesis of thousands of discrete peptide simultaneously. The vessel for this technique is microtiter plate. This microtiter plates are usually made up of polyethylene. Each well serves as a separate vessel for amino acid reaction. ELISA assay on solid phase is used to screen peptide of interest. The cleavage of peptide was performed by diketopiperazine under acidic condition. Limitation of this technique is that it can generate libraries of single compound in spatially different manner [64]. About 96 well plates methods are inefficient unless fully automated. In Spot membrane synthesis, cellulose membrane has been used instead of polyethylene rods. Different peptides are synthesized on a single sheet of membrane. Nanocan utilized the synthesis of peptide on resin sealed inside the polyethylene bag. Bags are placed in activated solution of aminoacid for coupling on the resin surface. Each bag contains specific peptide. Microarray utilized the glass, silicone and plastic slide. This technique is used for high through put evaluation of complex protein and peptide. The peptide microarray can be divided into two cateogies: In situ synthesis and chemical ligation. In situ synthesis involves the combination of solid phase synthesis and photolithography. The combination of these techniques develops a new method, for example, light-directed spatially addressable parallel chemical synthesis. It allows the synthesis of thousands of diverse peptide on small glass surface. Chemical ligation method involves the linkage between small molecules on the solid phase or glass surface. Low density and limited chemistry of synthesis are major drawbacks of this approach [65].

2.3.11 SYNTHETIC LIBRARY METHOD USING AFFINITY CHROMATOGRAPHY SELECTION

Affinity chromatography method is used for the screening and synthesis of peptide library. The synthesis of peptide is performed on solid phase resin by coupling the amino acid which is capped by biotin. The peptides are cleaved from the bead and loaded into affinity chromatography on an avidin-agarose column. The specific interaction between biotin and avidin has been utilized in this method. The desired peptide sequence is eluted and undesired retained in the avidin column. The mixture was eluted with the help of reverse phase HPLC and Mass spectroscopy [66]. Cascone et al. purified the monoclonal antibody by affinity chromatography with peptide ligands derived from combinatorial library. High yield is obtained in this process however selectivity is the key issue which remains unsolved till now [67].

Two major approaches used in the production of peptide are enzymatic and chemical approach. In enzymatic approach, protease is most commonly used enzyme for peptide synthesis but their specificity and selectivity limited to small peptide ligands. In case of large peptide construction, unwanted hydrolytic reaction will occur over the formed product and substrates. These reactions can be overcome by the genetic and protein engineering of reaction medium, biocatalyst and the substrate. Low productivity, low yield and high cost are problems involved with enzymatic approach [68]. Another approach, which is commonly adopted for peptide synthesis is chemical approach. Major chemical strategies adopted for the synthesis of peptide on solid support are: (i) Stepwise (ii) Convergent (iii) Chemical ligation. Stepwise construction of peptide involves the attachment of subsequent amino acid in stepwise manner until the desired sequence is not obtained. In convergent synthesis, fragment of peptide are synthesized, purified, characterized and then assembled into complied into desired peptide. Third approach involves link between unprotected peptide. Chemical ligation contained partially protected thioesters bond as a building block and no need of protective group. This approach can be applied for the synthesis of large peptide also [69]. Combination of these approaches can be used for peptide synthesis. Solid-phase peptide synthesis is most commonly used chemical approach for peptide synthesis. It consist a solid matrix on which elongation of peptide chain takes place. The amide bond is formed between incoming carboxyl group of amino acid and amino group of amino acid in solid matrix. This process is repeated until desired sequence and length has been achieved. Synthesis protocol (FMOC and BOC), nature of solid support, coupling reagent and method of cleavage from solid carrier are some variable which affect yield, kinetic and purities of peptide [70, 71].

2.4 PEPTIDE/POLYMER NANOPARTICLE

The rapid advancement of nanotechnology has raised the possibility of using engineered nanoparticles that interact within biological environments for treatment of diseases. Nanoparticles interacting with cells and the extracellular environment can trigger a sequence of biological effects. These effects largely depend on the dynamic physicochemical characteristics of nanoparticles, which determine the biocompatibility and efficacy of the intended outcomes. [72]. There are a number of methods available to encapsulate drug into nanoparticles. Selection of any of the methods should take into consideration factors such as particle size requirement, thermal and chemical stability of the active agents, reproducibility of the release kinetic profiles, stability of the final product, residual toxicity associated with the final products, the nature of the active molecule as well as the type of the delivery device [73]. Encapsulation contains core material and a wall which prevent the material from degradation, undesired leakage and protecting the material from environmental conditions such as light, moisture and oxygen thereby increase the stability of drug and promoting the controlled release of core material [74]. An ideal nanoparticle should have following properties: (1) scalable and cost-effective, (2) biocompatible/biodegradable, (3) non-toxic, (4) non-immunogenic, (5) >100 nm in diameter, and (6) amenable to robust surface modification. For systemic delivery, nanoparticles should also: (1) be stable in blood, (2) avoid the RES, and (3) have prolonged circulation times [75].

When preparing nanoparticle for surface modification with peptide requires specific condition of encapsulation such as solubility, stability, pH etc.

2.4.1 IONIC GELATION METHOD

This is the method used for preparation of nanoparticle based on polysaccharide. Polysaccharides (chitosan, gellan, alginate, xyloglucan and pectin) is dissolved in aqueous or weak acidic medium. The solutions are added dropwise in the solution of counterion under constant magnetic at room temperature for 30 min. The counterion present in solution forms complex, polysaccharide precipitate out to form uniform spherical particle [76]. After this, these nanoparticles are harvested by ultracentrifugation. Supernatant is discarded and washed with distilled water. Nanoparticle prepared from this method display certain advantage such as simple, use mild condition and thermally stable. Physical crosslinking such as electrostatic interaction are less toxic as compared to chemical cross linking agent. Sodium tripolyphosphate (TPP), a polyanion is most commonly used cross linking agent due to its non toxic. Ratio of cross linking agent (TPP/Chitosan molar ratio), stirring speed control the size and yield of nanoparticle, since these parameters affect drug release property [77].

Ionic group present in TPP interact with the groups present in chitosan and cause cross linkage [78–80]. Tat peptide conjugated chitosan nanoparticles were prepared from this method [81]. Polymers used for this method are nonimmunogenic, biocompatible and thermally stable. This method is also applicable for highly hydrophilic drug [82].

2.4.2 EMULSION SOLVENT EVAPORATION TECHNIQUE

Emulsion solvent evaporation comes into picture in early 1970 and found to be a successful approach in the preparation of microsphere and nanoparticles. This method is used for encapsulating hydrophilic as well as hydrophobic drug. A solution of polymer in an organic solvent like dichloromethane, chloroform, methylene chloride or ethyl acetate is mixed with surfactant solution such as poly vinyl alcohol. The mixture is homogenized by vortex and then sonicated to produce oil in water emulsion. Organic phase is evaporated using rota evaporator under vacuum. Nanoparticles are recovered by ultracentrifugation and supernatant is discarded and washed with water to remove extra solvent. The purified nanoparticles are freeze dried [83]. Surfactants are used to stabilize the emulsion in order to prevent the aggregation of polymers. Nevertheless, surfactant removal is an issue in this method.

Surfactant molecules are sometimes harmful in biomedical applications. This method is commonly used with biodegradable polymer such as poly(lactic acid) (PLA), poly(glycolic acid) (PGA), PLGA, poly(e-caprolactone) (PCL), poly(hydroxybutyrate) (PHB) and their copolymers [84, 85]. Diclofenac, curcumin, paclitaxel and ketopofen are the drug which can be encapsulated into nanoparticle by using this method [86]. Drug solubility in dispersion medium is important parameters in controlling encapsulation efficiency in preparation of solid lipid nanoparticle [87].

2.4.3 LAYER-BY-LAYER ASSEMBLY

Layer-by-layer assembly is the deposition of thin film on a core material. The biofilm is formed by sequential layering of oppositely charged material such as one layer of polyanion and other is the layer of polycation to form the polyelectrolyte complex. Layering of two oppositely charged polyelectrolytes occur via electrostatic interaction to form a thin film [88]. A key advantage with technique is the size of nanofilm can be tailored according to the requirement. This technique not only depends on electrostatic interaction but also on hydrogen bonding, covalent bonding, complementary base pairing and hydrophobic interaction. This technique is found applicable in delivery of a large number of drug including dexamethasone, paclitaxel and tamoxifen, docetaxel, doxorubicin, etc. [89–91]. Polysaccharide such as acidic Chitosan, alginate, gelatin, carbomethyl cellulose, poly (lactic acid) and poly ethyleneimine are the polymer used in layer-by-layer assembly [92–94]. Layer-by-layer assembly used as coating on many inorganic nanoparticle. For example, curcumin is a hydrophobic drug showed less bioavailability in vivo. Magnetic nanoparticle when coated with oppositely charged poly(vinyl)pyrrolidone and hyaluronic acid showed enhanced solubility and better absorption via receptor mediated endocytosis [95]. Vaccines are also fabricated by the deposition of oppositely charged polypeptide on calcium carbonate core. LbL nanoparticles induce the immune response by internalizing the dendritic cells without triggering inflammatory cytokines [88].

2.4.4 NANOLITHOGRAPHY

Nanolithography is a high throughput nanofabrication technique which has control over size and shape of nanoparticle. This is one step process which is harvested directly into aqueous buffers using biocompatible substrate. Silicon wafers is used to fabricate peptide and biomolecule nanoparticle less than 50 nm [96]. Enzymatic nanolithography helps in fabrication of FRET peptide in a buffered solution STAPHYLOCOCCAL serine V8 protease and AFM [97]. Some conventional nanotechnique has been used in conjunction with lithography technique. Aromatic peptide bionanosphere on the basis of beta-amyloid peptide was fabricated by using nanolithography with etching process for Alzheimer's disease [98]. Peptide nanolithography yielded arrays of monodisperse gold nanoparticles on the peptide lines on substrates. The width pattern of peptide was determined by number of particle on substrate [99]. Dip pen nanolithography is another technique in lithography which allows direct deposition of a number of materials with the use of AFM (atomic force microscopy). Peptide are small in size so, it can be easily arrayed using this technique [100, 101]. For example, tat peptide was patterned on gold and silica substrate. The peptide nanostructure can be controlled by the feeding time and concentration of ring opening polymerization (tryptophan-N-carboxyanhydride) [102–104]. Colloidal gold particle (5–70 nm) can be prepared by using polymer immobilized substrate. There are three steps for fabrication of gold nanoparticle: (i) generation of oxide/hydroxyl group on a substrate, (ii) polymerization of bifunctional organosilanes and last step involve immersion of substrate into a solution of colloidal Au particles, where surface assembly spontaneously occurs [105, 106]. Gold nanoparticles are easy to prepare via conventional nanotechnique along with additional benefit of size and shape [107]. This technique have wide application in fabrication of gold nanoparticles including probes for sensing/imaging various analytes/targets, including ions and molecules, and excellent drug delivery systems for cancer treatment [108].

2.4.5 STOBER METHOD

Silica nanoparticles are employed as cellular markers and traced their fluorescent property. Lack of heavy metal, water solubility, biocompatibility make it good candidate for biomedical applications. Silica nanoparticle has extensive application in cell targeting, particle tracking, drug carrier, and as imaging agent. The problem related to destruction of dye in polymeric nanoparticle can be overcome by using dye encapsulated silica nanoparticle as contrasting agent [109]. Stober method found to be much better approach than other traditional method for silica nanoparticle synthesis. Stober method was introduced by Werner Stober in 1968 [110]. Peptide based functionalization on silica nanoparticle can be performed by this technique [111]. In an example, RGD peptide is a cell penetrating, bind specifically on integrin receptors in cancer cell when conjugated with fluorescent silica nanoparticle have shown high binding affinity on cancerous cell, was synthesized by Stober method. Tetraethyl ortho silicate is added in water/oil microemulsion containing 3-aminopropyl triethoxy silane, dye and (3-trihydroxyl silyl) propyl methyl phosphonate. (3-trihydroxyl silyl) propyl methyl phosphonate increase the potential so that peptide can easily attach on the surface of nanoparticle [112]. Dimethylamine is used as a catalyst for proper coating. Tetraethyl ortho silicate concentrations have a significant effect on size of nanoparticle [113].

2.4.6 NANOPRECIPITATION (SOLVENT DIFFUSION, SOLVENT DISPLACEMENT)

This method can be used for hydrophobic as well as hydrophilic drug encapsulation. Briefly, polymer and drug dissolved in a polar solvent such as acetone, ethanol or methanol. The solution was poured in a controlled manner through a syringe pump in an aqueous solution of surfactant solution. The solvent was evaporated under reduce pressure. Finally, nanoparticles were washed using distilled water. Nanoparticles are formed instantaneously by this method. Poly (ethylene-maleic anhydride) was used as a surfactant in the preparation of PLGA nanoparticle conjugated with RGD peptide [114]. Polymer concentration, nature of solvent and surfactant play an important role in formation of nanoparticle of desired size and shape. LTVSPWY peptide-modified magnetic nanoparticles for tumor imaging were prepared by solvent diffusion method. Magnetic nanoparticles were used as MRI contrast agents with efficient targeting ability and cellular internalization ability, which make it possible to offer higher contrast and information-rich images for detection of disease [115].

2.4.7   CHEMICAL COPRECIPITATION

Super paramagnetic nanoparticles of magnetite were prepared by a chemical coprecipitation method of ferric and ferrous ions in alkali solution. The reaction steps in our process are as follows: A molar ratio of ferric and ferrous was dissolved in water with sonication. To the solution, alkali was added while stirring at room temperature to supersaturate for the precipitation of the oxide. The rate of reaction was controlled by allowing one drop of alkali per second to react with this solution until a pH of 10, to get a thick dark precipitate. Five grams of citric acid crystals dissolved in 10 ml water was added to this wet precipitate and allowed for further reaction at an elevated temperature of 80°C while stirring for another 90 min. This sample was then washed with distilled water several times for the removal of water soluble byproducts. This is then suspended in distilled water by ultrasound treatment. The obtained fluid was kept for gravity settling of any bare nanoparticles and was then centrifuged at a rotation speed of 3500 rpm to remove any particles that may sediment. Iron oxide nanoparticle conjugated with different peptide such as WSG peptide and arginine rich peptide can be prepared with this method [116–118].

2.4.8   EMULSION DIFFUSION

Emulsion diffusion is a technique used to prepare nanoparticle encapsulating hydrophilic drug. This method was used to encapsulate insulin which is hydrophilic peptide. Preparation of solid lipid nanoparticles from W/O/W emulsions: preliminary studies on insulin encapsulation. Insulin was chosen as hydrophilic peptide drug to be dissolved in the acidic inner aqueous phase (W1) of multiple emulsions and to be consequently carried in SLN. Several partially water-miscible solvents with low toxicity were screened in order to optimize emulsions and SLN composition, after assessing that insulin did not undergo any chemical modification in the presence of the different solvents and under the production process conditions. Emulsion diffusion/Multiple emulsion is used to encapsulate hydrophilic drug in lipid matrix with high encapsulation efficiency. Gallarate M et al. prepared a multiple emulsion with insulin, water soluble peptide dissolved in aqueous phase, glyceryl monostearate as lipid matrix and soy lecithin and Pluronic F68 as surfactants. Insulin was found stable in this microencapsulation technique while another studies showed that insulin formed aggregates in water-organic solvent (o/w) interface during microencapsulation process in the presence of PLGA [119].

2.4.9 COACERVATION TECHNIQUE

Same research group developed coacervation technique to encapsulate hydrophilic as well as hydrophobic drug. Briefly, when the pH of a fatty acid alkaline salt micellar solution in the presence of an appropriate polymeric stabilizer, is lowered by acidification, the fatty acid precipitates owing to proton exchange between the acid solution and the soap; this process was defined as "coacervation." This innovative technique allows to load lipophilic drugs by dissolution in the micellar solution before acidification. Coacervation is a solvent free technique which is used to prepare to solid lipid nanoparticle of peptide using stearic acid. It has targeted as well as sustains the drug release. Hydrophobic ion pairing with anionic surfactants was used to encapsulate Insulin and leuprolide in nanoparticle. Heating process is necessary in the coacervation but it has no effect on peptide stability [120].

2.4.10 SALTING OUT

This method is based on reduction of solubility of certain molecule in a highly concentrated electrolyte solution. It is also considered as modification of emulsification and solvent diffusion. In this method, Polymer drug solution is prepared by dissolving it in water miscible organic solvent such as acetone and tetrahydofuran [121]. This organic phase is added in the aqueous solution of salting out agent and a colloidal stabilizer. Magnesium chloride, calcium chloride, magnesium acetate and sucrose are commonly used salt forming agent whereas polyvinyl pyrrolidone and hydroxyl propyl methyl cellulose can be used as colloidal stabilizer. This technique can be used to encapsulate hydrophilic, hydrophobic or thermolabile drug and can be easily scaled up. The purification of nanoparticle can be performed with of use of cross flow cytometry or ultracentrifugation technique [122].

2.4.11 INTERFACIAL POLYMERIZATION

In interfacial polymerization, the cyanoacrylate monomer and proteins drug are dissolved in a mixture of an oil and absolute ethanol. Oils have positive influence to reduce the unfolding of protein by ethanol. This mixture is then slowly extruded through a needle into a well stirred aqueous solution, with or without some ethanol containing surfactant. An advantage of interfacial polymerization technique is high efficiency drug encapsulation. Insulin was encapsulated on poly(ethyl cyanoacrylate) and poly(isobutyl cyanoacrylate) nanoparticles of particle size ~151 nm and 150–300 nm, respectively, by using this method.

2.4.12 SOLVENT EVAPORATION

In solvent evaporation method, the polymers along with proteins are dissolved in volatile organic solvent (DCM, acetone, CHCl3, EtAc, etc.) and poured into continuously stirring aqueous phase with or without emulsifier/stabilizer and sonicated. Most of the proteins are likely to denature after the sonication. Thus, a slow and intermittent sonication at low temperature is effective to retain the secondary and tertiary structure of protein drugs. Albumin, and tetanus toxoid are successfully encapsulated on polylactic acid (PLA) nanoparticles of size 100–120 nm and 150 nm by these methods. Solvent displacement method is similar to solvent evaporation that is based on spontaneous emulsification of the organic internal phase containing partially dissolved polymer along with protein into the aqueous external phase. Insulin on PLA nanoparticles of size ~105–170 nm was nanoencapsulated by these methods.

2.4.13 SURFACE FUNCTIONALIZATION OF GOLD NPS

Bayraktar et al. demonstrated the ability to disrupt protein–protein interactions using surface-functionalized gold nanoparticles. They have designed nanoparticles that bound selectively to cytochrome c (Cyt c) or cytochrome c peroxidase (CCP), thereby inhibiting enzyme turnover. Surface-functionalized nanoparticles with gold cores (2 nm) are prepared using thiolates with oligo(ethylene glycol) groups terminated in carboxylate

(AuTCOOH) and trimethylamine (Au-TTMA) functionalities. Au-TTMA nanoparticles bind selectively to a negative patch on CCP surface, whereas Au-TCOOH nanoparticles bind selectively to basic amino acid rich surface of Cyt c. Tseng et al. fabricated amino-acid-functionalized gold nanoparticles that modulate the catalytic activity of R-chymotrypsin. They proposed that the amino acid monolayer on the nanoparticles controls both the capture of substrate by the active site and the release of product through electrostatic interactions.

2.4.14 TAYLOR CONE JET METHODS

The secondary structure of adsorbed proteins depends on both the identity of the adsorbed protein and the mode of presentation of chemical functionalities. Taylor Cone Jet Mode is a new technique for protein encapsulation on nanoparticles to protect their structure and function. In this technique, protein solution is dispersed in organic phase (with DCM) dissolved PLGA by controlled sonication processes. Further, electrospray with well defined potential difference is created between the nozzle and the ring by applying high voltage on the nozzle and a lower high voltage in the ring. The emulsion solution was pumped through nozzle to form liquid cone. This creates a thin jet from the apex to break up into monodispersed droplets.

2.5 PEPTIDE POLYMER CONJUGATE

2.5.1 CRITERIA FOR DESIGNING PEPTIDE POLYMER CONJUGATE

- The polymer backbone needs to be biocompatible and biodegradable. Partition coefficient plays an important role in drug transport. Mostly, targeted peptides have log p less than zero and are hydrophilic in nature. So, targeted peptide follow paracellular route whereas lipophilic peptide have high tendency to follow transcellular diffusive route through lipidic membrane.
- It should not have high electrostatic charge.
- Ionization and dissociation constant also play an important role in drug targeting.

- It should have a size above the renal threshold limit to increase its circulation time.
- Chemically, it should provide functional sites to allow covalent attachment to create polymer-drug conjugate.
- High molecular weight of polymer increases the size of conjugate and slows the drug clearance by kidney. The concentration of drug can be maintained above therapeutic level for prolonged time [123, 124].

2.5.2 STRATEGIES FOR COUPLING OF PEPTIDE ON THE SURFACE OF NANOPARTICLE

Targeting with ligand functionalized nanoparticle enhance the cell selectivity and specificity. Ligands size, chemistry of ligands, encapsulated material, coating polymer and nature of linkage can effect the conjugation of ligands on the surface of nanoparticle [125]. Functionalization of nanoparticle with peptide has a wide number of applications in biomedical field. For conjugation of peptide with nanoparticle, it should not lose its functionality and do not interfere with nanoparticle activity. The strategies which are commonly used for conjugation of peptide on the surface of nanoparticles are: (i) Electrostatic interaction (ii) Direct interaction (iii) Covalent conjugation (iv) Non covalent conjugation (Figure 2.4).

2.5.2.1 Electrostatic Interaction

Electrostatic interactions are strong interaction that occurs between cation and anion charges of functional groups in polymer and ligands. The ability of the peptide to interact with the charged gold nanoparticles is directly related to its helical structure and was not found for a random coil peptide with the same net charge. Interestingly, the interaction with nanoparticles seems to induce a fibrillation of the coiled coil peptide. Electrostatic interaction use opposite charge on the surface of nanoparticle and peptide and form charge based nanoparticle conjugation [126]. The nanoparticles are stabilized by a citrate ligand layer through electrostatic interactions. Due to the only electrostatic effects of citrate adsorption binding of proteins and other biomolecules on the NPs surface is possible. Gold nanoparticles reduced and stabilized by citrate have a negatively charged surface

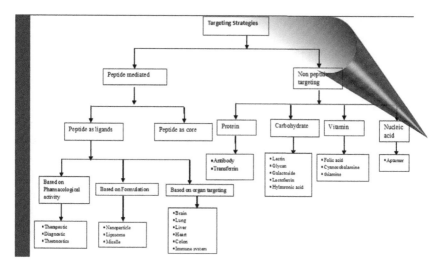

FIGURE 2.4 Illustration of peptide and nonpeptide mediated targeting strategies that are widely used in drug delivery system.

and preferentially bind to molecules including thiol, amine, cyanide or diphenylphosphine functional groups. They also observed that the reactivity of the amino group of amino acids was pH-dependent. Binding of amino groups is preferential at especially low pH, while the role of this group is negligible at neutral and high pH. The thiol moiety of cysteine is a very effective site to interact with gold. Due to this interaction cysteine-capped gold nanoparticles are formed [127, 128]. Peptide containing positive charge binds on the surface of nanoparticle containing negative charge. RGD peptide was conjugated with glycol chitosan nanoparticle through interaction with carboxylic group of peptide and amino group of chitosan backbone [129]. Shape and charge distribution on peptide surface affects the strength of electrostatic interaction [130]. This method is efficient and simple but non specific. Peptide containing positive charge bound on the surface of negatively charged surface of dextran coating iron oxide nanoparticle via electrostatic interaction [131]. Electrostatic interaction plays an important role in the modification of citrate stabilized gold nanoparticle. Amyloid peptide can attach to the surface of gold nanoparticle via polar and N-donors amino acid side chain [132]. Electrostatic interaction between synthetic non-arginine (SR9) peptides and carboxyl functionalized quantum dot can be achieved by using non covalent linkage. This non covalent linkage increases its rate and efficiency of cellular uptake in living cells [133].

2.5.2.2   Direct Interaction

Some ligands bind directly on the surface of nanoparticle after some modification namely guanidination, imidazolinylation, dimethylation and nicotinylation. ε-amino groups of lysine residues in the peptides are modified by guanidination and imidazolinylation whereas dimethylation and nicotinylation targets both N-termini and ε-amino groups of lysine residues [134]. Tat, a commonly used cell penetrating peptide was conjugated directly with the chitosan nanoparticle via disulfide linkage. This linkage results in higher biological activities. Size and chemical properties of cargo can effect the conjugation of peptide on surface of nanoparticle [135]. Cysteine terminated peptide was mixed in a solution of Citrate stabilized Gold nanoparticle, yielded a densely functionalized gold nanoparticle [107]. An alternative strategy, for example, Affinity binding strategy has been used for biconjugation of biomolecule. The conjugation occurs between the affinity of biomolecule and ligands. An advantage with this strategy is that it needs not any functionalization and affinity tag can be introduced naturally. His tag is commonly used affinity tag for binding of divalent substrate. Metal affinity biconjugation occur with a number of metal but CdSe-ZnS have been exploited for the biconjugation of biomolecule on the surface of quantum dot [136]. A cell penetrating peptide (Tat) containing polyhistidine motif which allow the peptide to self assemble on the surface of CdSe-ZnS quantum dots via metal affinity bioconjugation. This conjugation has generated a complex of peptide which performs cellular uptake, endosomal escape and organelle targeting [137]. N-(S-1-carboxyethyl) glycine displayed on the fibers can control the high affinity metal binding along the fiber axis [138]. Another example of metal affinity binding is functionalization of quantum dots with oligohistidine tagged cell penetrating peptide derived from HIV Tat. The association occurs between imidazole groups of terminal hexahistidine residues of peptides and proteins and the ZnS shell of quantum dot. These biconjugation nanoparticles have wide application in imaging, biosensing and theronostic, etc. [139].

2.5.2.3   Non-Covalent Coupling of Peptide and Nanoparticles

Non-covalent coupling of biotin and streptavidin linkage is used to overcome the problem related with covalent coupling. Streptavidin fused HIV tat peptide was attached to biotinylated DNA NPs and reduce aggregation biotinylated DNA NPs [140]. Avidin is a biotin binding protein obtained from egg white of avians having molecular weight of 68 kda whereas streptavidin with equal binding affinity and similar binding site for biotin can be obtained from streptomyces bacteria commercially. Biotin streptavidin are irreversible non covalent bond with high strength and specificity of interaction (K_d: 4×10^{-14}). Streptavidin is a tetramer which is capable of binding four biotin molecules [141]. Thiol groups are introduced on the surface of nanoparticle either by aldehyde quenching with cysteine residue or reaction with 2-iminothiolane. Avidin is attached on the surface of nanoparticles via reaction of amino group of avidin and sulfhydryl group of nanoparticle. Avidin attach on the surface of nanoparticle. Biotinylated peptides bind covalently on the site in Avidin in nanoparticle. This reaction is rapid, stable and highly efficient for bifunctionalization of ligands on nanoparticles [142]. This technique has been explored for conjugation of a number of ligands on the surface of nanoparticles including peptide, protein, fluorescent dye, nucleic acid, ferritin, etc. [143–146]. One major advantage with this technique is that both streptavidin and biotin remain active after dissociation on 70°C, therefore reused for another assay method [147]. Only drawback with this technique is large size of complex, immunogenicity and rapid clearance by reticuloendothelial system.

2.5.2.4   Covalent Coupling of Peptide and Nanoparticles

Covalent coupling is most commonly used method for bifunctionalization of peptide on nanoparticle (Figure 2.5). This biconjugation can be divided into four categories based on type of linkage containing: (i) thioether, (ii) amide bond, and (iii) cycloaddition. Thiolated group of cysteine residue in protein and peptide react with the maleimide functionality to form thioether bond with nanoparticle. This method provides high degree of specificity and faster rate with thiol as compared to the amine at neutral pH. Addition of maleimide-derived coupling reagent; sulfosuccinimidyl-4-(maleimidomethyl)

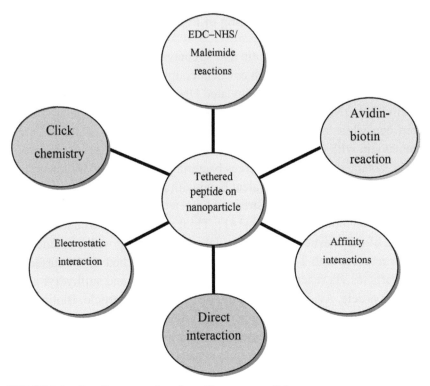

FIGURE 2.5   Coupling strategies of peptide on nanoparticle.

cyclohexane-1-carboxylate (sulfo-SMCC) has been used with the biomolecule as compared to nanoparticle because nanoparticle require several steps of purification. Maleimide functionalized PLGA nanoparticle was used for attachment of cRGD peptide on the surface. Interfacial Activity Assisted Surface Functionalization (IAASF) was used to introduce new functional group on the surface of nanoparticle and utilize partition coefficient between ligand and polymer [148]. Cysteine group containing A 371 peptide coupled on the surface of liposome nanoparticle bind on the neurotrophin receptor tropomyosin-related kinase (Trk) receptor tyrosine kinase, especially TrkB in spiral ganglion cells via maleimide coupling agent [149]. Next strategy for covalent coupling is amide bond formation. Carbodiimide 1-ethyl-3-(3-dimethylaminopropyl) carbodiimide (EDC) is coupling agent which can be used in the formation of covalent link between carboxylic group of polymer of nanoparticle and a free amine group on peptide. N-hydroxysulfo-succinimide (NHS) has been used for conjugation of peptide and protein on the surface of CdTe quantum dot [150]. The benefits related with this method are the aqueous solubility, do not require spacer for conjugation and low toxicity but formation of by product and non specific reactions are major drawbacks [151]. Staudinger and chemical ligation are another approach, used for amide bond formation on peptide. Staudinger was first described by a German scientist in 1919 and coined it Staudinger ligation. This reaction occurs between azide and phosphine to form amide bond [152]. This ligation is biorthogonal, biocompatible and chemoselective.

Chemical ligation is an attractive method for conjugation of thioester modified ligand to the nanoparticle surface. This provides site specificity but have low conjugation yield [153]. RGD peptide was attached on the surface of PDA [2-(pyridyldithio)-ethylamine] nanoparticle by Michael addition [154]. Peptide containing cysteine at carboxy terminus was conjugated with vinyl sulphone derivatized polymer via free thiol through Michael addition [155]. Click chemistry was first introduced by sharpless in 2001. This reaction joins the small unit together to form a substance. Diel-Alder and Huisgen 1,3-dipolar cycloaddition in click chemistry are the most useful reaction for biconjugation. Click chemistry have high degree of stability and selectivity. Huisgen cycloaddition involve copper catalyst for conjugation. This is the major drawback of Huisgen cycloaddition whereas Diel Alder do not need addition of any metal. RGD conjugated with azide present on the surface of nanoparticle via click chemistry and provide a stable, chemoselective with high yield biconjugation [156–158].

2.6 BIOCONJUGATION OF PEPTIDE AND PROTEIN ON NANOPARTICLES

Most of the therapeutic proteins and peptides are encapsulated by adsorption on the surface of nanocarriers. However, various biochemical conjugation reactions are advantageous for the efficient and controlled encapsulation of therapeutic proteins. Furthermore, the activity of the protein polymer/nanoparticles conjugate is closely regulated by altering the response stimulus of the synthetic polymer and the site of attachment to the protein and peptides. These chimeric systems may thus be considered as true molecular-scale devices for the development of delivery systems.

The nanoparticles and specific ligands are activated by incorporation of functional group (–NH2, –SH, –COOH,), and blocking of specific functional group (–NH2, –SH, –CHO, –COOH,). These functionalized nanoparticles and its complementary functionalized ligands are bioconjugated by functional group specific bioconjugation reaction. A variety of bioconjugation reaction are used like maleimide reaction, EDC/NHS chemistry, affinity interactions, click chemistry, streptavidin biotin reaction, etc. for the covalent conjugation of specific ligands on functionalized nanoparticles.

2.6.1   MALEIMIDE REACTIONS

The maleimide reaction is well known for the bioconjugation of protein containing free thiol group on nanoparticles. Thiol groups of proteins or functionalized nanoparticles (GNPs, etc.) react with maleimide functionality to form thioether bond. In addition, thiol-maleimide based couplings are known to proceed efficiently in solution, with site specific attachment of maleimide groups on Fc region of antibodies. It is more beneficial to incorporate this functionality on the biomacromolecule than the nanoparticles, because nanoparticles require several steps of purification in aqueous solution by dialysis over a longer period of time. Maleimide functionalized bovine serum albumin (BSA) as a model biomacromolecule was bioconjugated with thiol-terminated poly(ethylene glycol) (PEG) chains on shell cross-linked knedel-like (SCK) nanoparticles. The particle size was increased from 20 nm to 30 nm after encapsulation of BSA on SCK nanoparticles with >50–60% encapsulation efficiency. Pioneering work has been carried out by Hoffman, Stayton and co-workers for the bioconjugation of peptide with nanoparticles. They engineered single cysteine viasite-directed mutagenesis a mutant of cytochrome b5. This is accessible for reaction with maleimide end-functionalized poly-N-isopropylacrylamide (PNIPAm). Since the native cytochrome b5 does not contain any cysteine residues this substitution provided a unique attachment point for the polymer. The resultant polymer–protein conjugate displayed low critical solution temperature (LCST) behavior and could be reversibly precipitated from solution by variation in temperature. This approach has proved to be very versatile and a large number of nanopolymer protein conjugates incorporating biological components like antibodies, protein A, streptavidin, proteases and hydrolases have now been prepared. Small changes in environmental conditions can then cause large changes in the polymer conformation, leading to reversible "blocking" or "unblocking" of the protein's active site; such changes also can lead to triggered release of a bound ligand from the protein binding site. The bio-logical functions or activities of these conjugate systems are fully similar to their native counterparts, but they are switched on or off as a result of thermally induced polymer phase transitions. Lactoferrin-conjugated PEG–PLA nanoparticles synthesized by thiolated reaction containing 55 molecules per nanoparticles showed an improved brain delivery. Lactoferrin engineering significantly increased the brain uptake of drugs associated to nanoparticles following an intravenous administration. Thiol group is an attractive functionality to have on the surface of nanoparticles or protein, since reversibly linked bioconjugates can be prepared by disulfide formation, which may allow reductive release of a targeting or therapeutic component.

2.6.2 EDC–NHS BIOCONJUGATION

EDC/NHS bioconjugation is frequently used for the bioconjugation of peptide on variety of nanoparticles. Carboxyl groups present on the surface of nanoparticles are activated by using 1-ethyl-3-(3-dimethylaminopropyl) carbodiimide hydrochloride)) (N-hydroxysulfosuccinimide) (EDC/NHS) to form reactive sulfo-NHS ester. This ester reacts with amine functionality of proteins via formation of amide bond. When working with proteins and peptides, experience indicates that EDC-mediated amide bond formation effectively occurs between pH 4.5 and 7.5. Beyond this pH range, however, the coupling reaction occurs more slowly with lower yields. The presence of both carboxylates and amines on the molecules to be conjugated with EDC may result in self-polymerization. This is because the substance can react with another molecule of its own kind instead of the desired target. PLGA nanoparticles have been conjugated to Anti-Fas mAb and IgG using EDC/NHS chemistry. Variations in surface carboxyl density permitted up to 48.5 g coupled Ab per mg of NP. The amount of anti-Fas mAb that could be conjugated to the NP using two-step carbodiimide activation was dependent on the ratio of non-end-capped to end capped PLGA types.

In addition, it was observed that anti-Fas conjugation was not affected by drug loading. The model proteins like BSA, streptavidin or goat antirabbit immunoglobulin G (IgG) are also bioconjugated on magnetic nanoparticles with carboxylic or aminopropyltriethoxysilane (APTS) groups using carbodiimide as a zero-length cross-linker. The average hydrodynamic diameters of the nanoparticles were about, 154 nm and 165 nm and the conjugation yield of proteins were very low. 3-mercaptopropyl acid-stabilized CdTe quantum dot nanoparticles are conjugated with peptides or proteins (active) mediated by N-hydroxysulfo-succinimide (NHS) or 1-ethyl-3(3-dimethylaminopropyl) carbodiimides hydrochloride (EDC). The enzyme-chymotrypsin was nanoencapsulated on polyacrylamide-coated Fe_3O_4 nanoparticles synthesized by photochemical in situ polymerization. Mean particle size of the immobilized enzyme was 31 nm and super paramagnetic properties of Fe3O4 were retained after enzyme immobilization. The wheat germ agglutinin (WGA)-conjugated PLGA nanoparticles loaded with paclitaxel and isopropyl myristate has been prepared by two-step carbodiimide method to use it as an anticancer agents. WDA-conjugated PLGA nanoparticles had lower yields and encapsulation efficiency because they require more purification steps.

2.6.3 AFFINITY INTERACTIONS

An alternative approach taking advantage of affinity interactions has been extensively used for bioconjugation of recombinant proteins on Ni-NTA coated nanoparticles. This approach exploits the interaction between divalent metal ions such as nickel, copper, or cobalt and a short sequence of histidine residues (4–10 repeating units) added to the N- or C-terminus of the protein (histidine-tags). These purification methods generally involve immobilizing the metal ion onto the column packing material using a chelating agent such as nitrilotriacetic acid (NTA). In the case of Cu^{2+} and Ni^{2+} which have six coordination sites, NTA forms a strong complex with four of the metal sites, leaving two additional sites for interaction with the His-tag present on the protein. His-tag HIV-1 Gag p24 protein was coupled to DOGSYNTAYNi nanoparticles utilizing the above approach. The advantages of affinity interactions between divalent Ni nanoparticles are effective for enhancing its interaction with His-tag proteins. The stronger affinity of antigen with the Ni-NPs resulted in superior humoral immune responses in vivo compared to protein adjuvant with alum. Following this method various metallic nanoparticles have been used for the attachment of protein molecule by polymeric hydrophilic nanolayer surface modification.

2.6.4 CLICK CHEMISTRY

Click chemistry is one of the most useful reactions for the bioconjugation of proteins on the nanoparticles. This reaction is based on the reaction between alkyne moiety and azide moiety to from triazole ring. Beauty of this reaction is that alkyne moiety attached to the ligands reacts only with azide moiety of nanoparticles avoiding other side products. Cyclic tumor-targeting peptide LyP-1 (CGNKRTRG) containing thiol and amine groups was bio-conjugated on polymer-coated magneto fluorescent nanoparticles by applying click chemistry. Click nanoparticles are able to stably circulate for hours in vivo following intravenous administration (>5 h circulation time), extravagate into tumors, and penetrate the tumor interstitium to specifically bind p32-expressing cells in tumors. An acetylene-functionalized lipase has been attached to azide-functionalized water-soluble gold nanoparticles in a copper-catalyzed 1,2,3-triazole cycloaddition with full enzymatic activity. Seven fully active lipase molecules are attached to each nanoparticle. Staudinger ligation reaction work similarly as click chemistry but the reacting moieties are different. It is used to ligate modified phosphine and an azide to form an amide bond. The basis of this reaction is the hydrolysis of amidophosphonium intermediate in water. The ligand can be readily attached either before assembly of the complexes (precomplexation) or to the assembled nanoparticle (postcomplexation) by using this bioconjugation methods. The PEGylated polyamidoamine DMEDAPEG-DMEDA-(MBA-DMEDA)n+1-PEG-DMEDA was attached with integrin-binding peptide ligand via the Staudinger ligation. Postcomplexation strategy led to small and discrete toroidal nanoparticles whilst the precomplexation particles showed loose complexes.

2.6.5 STREPTAVIDIN–BIOTIN REACTION

Streptavidin is a tetrameric protein which binds very tightly to a small water-soluble molecule biotin. The specificity of biotin binding to streptavidin provides the basis for developing a bioassay system to detect or quantify analytes. More importantly, both streptavidin and biotin are effectively conjugated to other proteins or labeled with various detection reagents without loss of their binding affinity. The coupling reaction with biotin-(poly(ethylene glycol))amine and polymeric PLGA nanoparticles has been used for the conjugation of various protein and peptides. Biotin binding proteins (avidin, streptavidin, or neutravidin) have been used as cross linkers to conjugate proteins on biodegradable nanoparticles (PLGA–PEG–biotin). Avidin gave the highest level of overall protein conjugation, whereas neutravidin is known to minimize non-specific protein binding to the polymer.

2.7 PEPTIDE-MEDIATED TARGETING

Among various targeting ligands, peptides seem to be advantageous due to their small size, non immunogenic, biocompatible, biodegradable, and also can be modified easily. There are several classes of peptide with various mechanism of action directed against the targeting site. These include: (1) organ targeting, (2) cellular targeting, and (3) sub cellular targeting peptide. Organ targeting peptides target to a specific organ for example drugs may be targeted to the liver because its vasculature is normally leaky or fenestrated or "having loose junctions." This is organ targeting. In this case drug is not released in other tissues because their vasculature is not leaky; cellular targeting peptide attaches to specific receptor present on the cell with in a tissue or organ whereas sub cellular can enter the specific cell and leave the drug intracellularly [159]. Peptides are specific and have high affinity for protein. Due to small, it can easily deliver into cell. For an effective drug delivery, peptide can either initiate the active site of interacting protein or depress the target site. Peptides have dual function, either as drug or targeting ligand.

2.7.1 PEPTIDE ENCAPSULATING NANOPARTICLE

Ideal drug have high specificity, high affinity, solubility and safety with proper pharmacokinetic profile. Peptides contain not all but some properties of ideal drug such as high specificity, affinity, prolonged retention and biodegradability. First peptide introduced in 1920 was insulin by Banting and best for diabetes [160]. Till date there is no alternative approach to replace this peptide drug. Peptide act directly against any disease as in case of diabetes. Peptides have been used to elicit the immune response in form of vaccine to boost up immune system of body. According to one report in a magazine, 40 peptides are in market and 300 in clinical trial. Although, peptides are most important and biologically relevant molecule but this is not out of challenge [161].

Main problem with peptide drug is hydrophillicity, bioavailability and biodegradability. One approach to bypass these problems is encapsulation of peptide drug in the form of nanoparticle which protects the drug from enzyme and absorption barrier [162]. There are so many examples of successful drug delivery of peptide nanoparticle. Peptide vaccine contained tumor specific antigen (8–10 aminoacid) which was encapsulated in PLGA nanoparticle, showed 63 times better cellular uptake compared to free peptide by binding with antigen binding cell [163]. In inflammatory Bowel disease, anti-inflammatory tripeptide Lys-Pro-Val (KPV) encapsulated in nanoparticle was used to deliver peptide to colon and assessed its therapeutic activity in colitis. By using NPs, KPV can be delivered at a concentration that is 12,000-fold lower than that of KPV in free solution, but with similar therapeutic efficacy [164]. Evaluation of Anti neuroexcitation peptide loaded chitosan nanoparticle followed absorption mediated pathways for brain targeting. The nanoparticle delivered antineuroexcitation peptide into the cell [165]. Cell-penetrating anti-inflammatory peptide (KAFAK) has the ability to suppress pro-inflammatory cytokines TNF-α and IL-6 when released from poly (N-isopropylacrylamide)—2-acrylamido-2-methyl-1-propanesulfonic acid nanoparticles. When bovine knee cartilage explants treated with KAFAK loaded nanoparticles rapid and highly selective targeting of only damaged tissue occurred, found to be very effective tool to treat osteoarthritis [166]. Attachment of targeting ligand on the surface of nanoparticle facilitated the transport of peptide drug with prolonged action. White germ agglutinins were used as a targeting ligand with liposomes encapsulated with calcitonin. Calcitonin activity enhanced more than 20 fold as compared to unmodified one [167]. Conjugation of drug and peptide in the form of nanoparticle has also been used successfully in the treatment of various diseases such as multi myeloma. This approach showed better accumulation in diseased tissue [168].

Another approach to deliver peptide drug is formation of solid lipid nanoparticle. Solid lipid nanoparticle can carry both hydrophilic and hydrophobic peptide. In addition, it deliver peptide drug locally as well as systemically. Solid lipid nanoparticles successfully target insulin, calcitonin and somatostatin [169]. Nanoparticle solved the problems related to peptide as well as conjugation of ligand made it more specific and selective approach.

2.7.2 NANOCARRIERS OF PEPTIDE AND PROTEIN ENCAPSULANT

Nanocarriers are one of the useful tools for achieving the main objective of protein therapeutics and its targeted delivery. A variety of nanocarriers have received much attention for the delivery of peptides, proteins, and genes due to their ability to protect protein and peptides from degradation in the gastrointestinal track and also in blood circulation. These protein encapsulated carriers have many advantages, such as a potential for selective targeting, controlled release and increase persistence of proteins therapeutics in the body. The protein therapeutic agents have short half-life due to proteolysis, rapid clearance from the blood stream and repeated administration. Encapsulation of protein drugs on nanoparticles provides sustained release and protects the non-released protein from degradation. The therapeutic protein molecules interact with nanocarriers by forming a coat or adsorption on the surface, or by bioconjugation (direct or using cross linkers). Various methods and nanocarriers systems are used to overcome the problems associated with functional native protein encapsulation on nanoparticles. The choices of encapsulation methods and nanocarriers system entirely depends on the application and nature of protein encapsulate. Some proteins, enzymes, and DNA plasmids are encapsulated in biodegradable polymeric nanofibers by electro-spinning technique to retain their bioactivity. The strength and selectivity of protein–protein interactions make the protein therapeutics as excellent candidates to serve as a linker between nanoparticles ordered structures. In addition, the proteins therapeutic molecules are encapsulated on various kinds of nanoparticles like metallic, polymeric (natural and synthetic), or by self assembly to forms nanoparticles. The proteins and peptides encapsulated on these nanocarriers systems are briefly described in following subsections.

2.7.2.1   Metallic Nanocarriers

Semiconductor and metallic NPs encapsulated protein and peptides therapeutics illustrate unique electronic, optical and catalytic properties. Proteins and peptides encapsulated on nanoparticles (NP) or quantum dot (QD) has yielded composite material with new functionality. Gold NP has been used as nanoelectrode to attach proteins and peptides by dithiol bridge due to inactivation of active site on encapsulation. Horse spleen apoferritin was attached on gold nanoparticles by applying this technique. This enzyme forms a covering layer that stabilizes the gold nanoparticles in solution. Cyt c, hemoglobin and myoglobin are also bioconjugated on silver and gold nanoparticles by using this technique.

The spectral features of protein encapsulated on nanoparticles are broadened and red shifted due to surface binding of these proteins. Similarly, CD11b interleukins conjugated gold nanoparticles show selective binding to macrophage cells. This selective destruction of the target cell line is possible by laser irradiation of plasmon resonant frequency of protein encapsulated on gold nanoparticles. Silver nanoparticles interact with the HIV-1 virus via preferential binding to the gp120 glycoprotein knobs. Due to this interaction, silver nanoparticles inhibit the virus from binding to host cells. Some enzymatic proteins molecules lose their proper function and structural organization during encapsulation on nanoparticles. Small perturbation occurs in the native structure of chymotrypsin on conjugation to silver nanocluster. Ni-NTA modified $Fe3O4–NH^+$ nanoparticles have high efficiency and specificity to His-tagged proteins. These modified nanoparticles are used more efficiently for the purification of His-tagged proteins. Gold and silver nanoparticles produced by citrate reduction have been functionalized with immunoglobulin (IgG) molecules at pH slightly above the isoelectric point of the citrate ligand. This allows effective binding between the positively charged amino acid side chains of the protein and the negatively charged citrate groups of the colloids. Other examples of protein coating through electrostatic interactions include the direct adsorption of heme-containing redox enzymes at citrate-stabilized silver nanoparticles and the binding of basic leucine zipper proteins to lipoic acid-stabilized semiconductor CdSe core/ZnS shell particles. These are some examples of well known nanoencapsulates of protein and peptides on metallic nanoparticles.

2.7.2.2   Polymeric Nanocarriers

The major advantage of colloidal drug carrier system in protein and peptide therapeutics are the controlled drug targeting, modified body distribution and enhancement of cellular uptake. The polymeric nanocarriers are very promising since they are biodegradable, non-antigenic, relatively easy to prepare and full control on size distribution. A variety of polymeric nanoparticles (natural and synthetic) can be synthesized in laboratory and are also commercially available. Polymeric materials used for the formulation of nanoparticles include synthetic (poly(lactic acids) (PLA), poly(lactic-co-glycolic acids) (PLGA), poly(caprolactone) (PCL), poly(methyl methacrylates), and poly(alkyl cyanoacrylates)) or natural polymers (albumin, gelatin, alginate, collagen or chitosan).

2.7.2.3   Natural Nanocarriers

Chitosan is a modified natural carbohydrate polymer prepared by the partial N-deacetylation of crustacean derived natural biopolymer chitin. It is the second most abundant polysaccharide in nature, and has attracted particular interest as a biodegradable material for mucosal delivery systems. There are at least four methods reported for the preparation of chitosan nanoparticles as ionotropic gelation, microemulsion, emulsification solvent diffusion and polyelectrolyte complex formation. Chitosan nanoparticles have low toxicity and high susceptibility to biodegradation, mucoadhesive properties and has an important capacity to enhance protein drug permeability/absorption at mucosal sites. More importantly, chitosan micro/nanoparticles can be spontaneously formed through ionic gelation using tripolyphosphate as the precipitating agent. This reduces the use of harmful organic solvents during preparation and loading of protein therapeutics. Chitosan solubility is poor above pH 6.0 which is a major drawback of this system. At physiological pH, chitosan is known to lose its capacity to enhance drug permeability and absorption, which can only be achieved in its protonated form in acidic environments. In contrast, quaternized chitosan derivative, N-trimethyl chitosan chloride (TMC) shows perfect solubility in water over a wide range of pH. In addition, these chitosan derivatized nanoparticles have bio-adhesive properties. Thus, it is used for enhancement of permeability and absorption of diverse protein drugs in neutral and basic-pH condition. N-trimethyl chitosan chloride (TMC) nanoparticles to carry proteins were prepared by ionic cross linking of TMC with tripolyphosphate (TPP). The results indicate that different degree of quaternization ($1/\infty$ particle size) of TMC has influenced the physicochemical properties, release profile and degree of loading of different proteins. Thus, the particle size may depend upon the nature and concentration of the loading proteins. Insulin was observed to be directly internalized by enterocytes in contact with intestine and retention of drugs at their absorptive sites by mucoadhesive carriers. Insulin loaded chitosan nanoparticles markedly enhanced intestinal absorption of insulin following oral administration. The hypoglycemia effect and insulinemia levels were significantly higher than that obtained from insulin solution and physical mixture of oral insulin and empty nanoparticles. The mechanism of insulin absorption seems to be a combination of both insulin internalization, probably through vesicular structures in enterocytes and insulin loaded nanoparticles uptake by Payers patches cells. Chitosan nanoparticles are also explored for their efficacy to increase systemic absorption of hydrophobic peptides such as cyclosporin A. The relative bioavailability of cyclosporin A encapsulated chitosan nanoparticles was increased by about 73%. This formulation provides the highest C_{max} (2762.8 ng/mL) of Cy-A after 2.17 h. Chitosan nanoparticles administered orally to beagle dogs provide an improved absorption compared to the currently available cyclosporin A microemulsion (Neoral ®). It has been shown that ovalbumin loaded chitosan microparticles are taken up by the Peyer's patches of the gut associated lymphoid tissue (GALT). Additionally, after co-administering chitosan with antigens in nasal vaccination studies, a strong enhancement of both mucosal and systemic immune responses was observed. Van der Lubben et al. have demonstrated that large amounts of bovine serum albumin (BSA) or tetanus toxoid (TT) vaccine were easily encapsulated in chitosan nanoparticles. Recently, Alonso's group has developed chitosan nanoparticles as carrier systems for transmucosal delivery. They have shown the enhanced mucosal absorption of chitosan NPs on rats and rabbits. The nanoparticles were in the 350 nm size range, and exhibited a positive electrical charge (+40 mV) and high loading efficiency (50–60%). They have reported important capacity of chitosan nanoparticles for the association of peptides such as insulin, salmon, calcitonin and tetanus toxoid. Their mechanism of interaction with epithelia was investigated using the Caco-2 model cell line. The results showed that chitosan coated systems caused a concentration-dependent reduction in the transepithelial resistance of the cell monolayer. Gelatin nanoparticles are extensively used in food and medicinal purpose. It is an attractive nanomaterial to exploit in controlled release of peptide therapeutics due to its nontoxic, and biodegradable nature. This polymer works as a polyampholyte consisting both cationic and anionic groups along with hydrophilic functionality. Due to this nature, gelatin molecules are frequently used for encapsulation of both acidic and basic peptides. Gelatin nanoparticles are prepared by desolvation/coacervation or emulsion method. The addition of natural salt or alcohol normally promotes coacervation and the control of turbidity/cross linking that resulted in desired nanoparticles. Gelatin nanoparticles have been used for encapsulation of BSA. These nanoparticles can absorb 51–72% of water, thus the release of BSA from the gelatin nanoparticulate matrix follows a diffusion controlled mechanism. The average diameter of the BSA-containing gelatin nanoparticles is approximately 840 nm. Basic fibroblast growth factor (bFGF) has been successfully loaded on gelatin particles. The average diameter of the gelatin particle-PLGA microsphere composite was 5–18 m, and bFGF loading efficiency was up to 80.5%. The bFGF releasing experiment indicated that this new composite system could release bFGF continuously and protect bFGF from denaturation.

2.7.2.4   Synthetic Nanocarriers

Natural polymeric nanoparticles provide a relatively quick drug release. However, synthetic polymers enable extended drug release over periods from days to several weeks. Poly(dl-glycolide-colactide) (PLGA), poly(d, l-lactide) (PLA) and polycaprolactone (PCL) are used to encapsulate proteins drug due to its biodegradable and biocompatible nature (Table 2.1). Among these nanoparticles, PLGA (poly-d, l-lactide-co-glycolide) and PLA nanoparticles are one of the most successfully used biodegradable nanosystem for the development of protein nanomedicines. They undergo hydrolysis in the body to produce the biodegradable metabolite monomers, lactic acid and glycolic acid. Since the body effectively deals with these two monomers, there is very minimal systemic toxicity associated by using PLGA for drug delivery or biomedical applications.

Insulin was encapsulated in a blend of poly (fumaric anhydride) (poly(FA) and poly(lactideco-glycolide) (PLGA) at a 50:50 ratio (poly(FA:PLGA)) using the inversion phase method. Animals feeding the poly (FA:PLGA) – encapsulated insulin preparation showed a better ability to regulate glucose load than the controls. This gave an indication that the insulin has crossed the intestinal barrier and was released from the microspheres in a biologically active form. Kawashima et al. evaluated the effectiveness of mucoadhesive polymeric nanospheres in the absorption of calcitonin on chitosan coated elcatonin-loaded PLGA nanospheres. BSA and immuno–globulin (IgG) have been encapsulated on PEO–PLGA nanoparticles. Different amounts of BSA or IgG corresponding to 1%, 2% and 4% theoretical loadings were encapsulated into PEO–PLGA nanoparticles of different composition. The type of used PEO derivative and pH of internal aqueous phase are the most important factors influencing BSA protein encapsulation and release kinetics. Degradation and release characteristics of polyester particles can be improved by the incorporation of polyoxyethylene derivatives with different hydrophilia–lipophilia balance. BSA encapsulated on PEG–PLGA nanoparticles are reported to extend the half-life as 4.5 h than 13.6 min in simple BSA. This extended half-life obviously change the protein bio-distribution in rats compared with that of BSA loaded on PLGA nanoparticles. Similarly, tetanus toxoid is also encapsulated on PLA and PLA–PEG nanoparticles of similar particle size (137–156 nm) but differed in their hydrphobicity. PLA–PEG nanoparticles led to greater penetration of tetanus toxoid into the blood circulation and in lymph nodes than PLA encapsulated tetanus toxoid. The results confirm that PLA nanoparticles suffered an immediate aggregation upon incubation with lysozyme, whereas the PEG-coated nanoparticles remained totally stable. The antibody levels elicited following in administration of PEG-coated nanoparticles were significantly higher than those corresponding to PLA nanoparticles. Insulin encapsulated into poly(isobutylcyanoacrylate) (PIBCA) nanoparticles by interfacial polymerization methods protect the insulin against proteolytic enzymes and promote absorption by the intestinal mucosa. There is evidence that PIBCA nanoparticles is able to pass from the gut lumen to the blood compartment by means of a paracellular pathway. These insulin-containing nanocapsules induced a significant hypoglycemic effect for several days in fasting and fed diabetic rats but were ineffective in normal rats. Damge et al. developed insulin-loaded poly (isobutylcyanoacrylate) of

60–300 nm nanocapsules that upon oral administration could deliver insulin directly to the blood. Das and Lin developed double-coated (Tween 80 and PEG 20000) poly(butylcyanoacrylate) nanoparticulate delivery systems (PBCA) for oral delivery and brain targeting of dalargin. Even if they did not elucidate the mechanisms of PBCA nanoparticle transport from gastrointestinal tract to brain, they observed a significant dalargin-induced analgesia with double-coated PBCA nanoparticles compared to single-coated PBCA nanoparticles (either Tween 80 or PEG). Earlier studies described similar approaches to protect the orally delivered cyclosporin A, LHRH, vasopressin, INF and peptidomimetics. Polyalkylcyanoacrylate (PACA) nanocapsules are used as biodegradable polymeric drug carriers for subcutaneous and oral delivery of octreotide, a long-acting somatostatin analog. It has the ability to reduce secretion of insulin or of prolactin in response to estrogens. Octreotide-loaded nanocapsules reduce (higher than 72%) prolactin secretion, increased plasma octreotide level, and prolonged therapeutic effect of a somatostatin analog in estrogen-treated rats given orally. Similarly, luteinizing 180 S.C. Yadav et al./Peptides 32 (2011) 173–187 hormone releasing hormone (LHRH), conjugated with hydroxypropyl methacrylamide nanoparticles are effective formulations for enhancing its stability and improve targeting possibilities. LHRH-loaded copolymerized peptide particle systems (mean size of 100 nm) administered orally show a half-life of LHRH in blood of 12 h than normal 2–8 min. Calcitonin, a peptide secreted by the parathyroid gland of the human body has been encapsulated into polyacrylamide nanospheres, PIBCA nanocapsules and chitosan nanoparticles. This encapsulated calcitonin greatly decreases the level of ionized calcium in blood than simple calcitonin on oral administration. Cyclosporine was also encapsulated in PIHCA by interfacial and emulsion polymerization. This therapeutic peptide is a cyclic nonribosomal peptide produced by the fungus Hypocladium inflatum, initially isolated from a Norwegian soil sample. The nanoparticle formulation had a notably increased bioavailability compared with that of the commercial formulation. Poly (N-isopropyl acrylamide) (PNIPAm) undergoes a sharp coil globule transition in water at 32° C. It is being hydrophilic below this temperature and hydrophobic above it. PEG coating on encapsulated therapeutic proteins have been shown to stabilize the protein without losing their biological function on PNIPAm nanoparticles. Amine functionalized polymeric nanoparticles is being used to enhance charge directed targeting of lysozyme nanoparticles conjugates to bacteria. Activity of lysozyme conjugated to positively charged PNIPAm nanoparticles was approximately twice, while lysozyme conjugated to negatively charged PNIPAm nanoparticles showed little detectable activity. When calcitonin is incorporated in nanoparticles, oral absorption is enhanced in rats and consequently calcium concentration in blood decreases compared to oral administration of a calcitonin solution. BSA was loaded on PNIPAm gels fabricated with higher encapsulation efficiency (40%) in the presence of lower cross-linker contents. An incomplete release of encapsulated BSA from the PNIPAm gels was observed in all cases. Enhanced mass transfer was created by oscillating swelling–deswelling in response to temperature cycling across the LCST. This reveals that lowering in vitro release and temperature did not promote BSA release due to strong BSA–PNIPAm gel interactions.

2.7.2.5   Protein Itself as Nanoparticles

Peptides serve as excellent building blocks for bionanotechnology owing to the ease of their synthesis, small size, relative high stability and chemical/ biological modifiability. Understanding the physicochemical determinants that underlie peptide self assembly is a fundamental step in view of the rational design of new nano building blocks for biotechnological applications or new drugs. A large number of nonpathogenic and pathogenic peptide have been shown to form ordered fibrils under particular solvent, temperature and pH conditions. Native folded structures of proteins and peptides have capability to self assembles as protein fibers, for example, coil–coil or amylogenic peptides. Three classes of protein components as planar crystalline arrays, engineered proteins pores and molecular motors are frequently used for development of protein nanodevices. The surface-layer (Slayer) proteins having square or hexagonal symmetry of 3–30 nm unit cells with 5–10 nm thickness are one of the good examples of planar protein crystalline arrays. Several therapeutic applications have been suggested for S-layer as nanoscale immobilization matrices. This includes S-layer streptavidin fusion nanodevice and IgG binding domain based high density adsorbent for extracorporeal blood purification. This is reported to have 20 times higher capacity to remove autoantibodies. The protein nanopores such as-hemolysin (HL) are developed as medical analytes for single molecule level by stochastic sensing. Various molecular motors like linear kinesin, myosin, ATP synthase have strong potential to be used as nanodevices for various cystis fibrosis diseases. Despite the high sequence diversity, many proteins and peptides aggregate into a common cross-sheet structure, and the resulting fibrils show remarkable ultra-structural and biophysical similarity. These assemblies belong naturally to the realm of the nanoscale. The study of their properties and mechanisms of formation might be a source of inspiration for the development of ordered, rationally designed nanostructures with potentially interesting applications in biotechnology and other fields. Normal proteins are engineered to self assemble into nanodevice and thus self assembly at nanoscale is important for the fabrication of novel supramolecular structure, having applications in the field of nanotechnology and nanomedicines. Peptides represent the most favorable building block for the design and synthesis of nanostructures because they offer a great diversity of chemical and physical properties. They can be synthesized in large amounts, can be modified and decorated with functional elements for diverse nanotherapeutic applications. Self assembly of polypeptide in nature such as amyloid fibrils is associated with human medical disorder. Structural element as short as dipeptides can form well ordered assemblies at the nanoscale. Many novel building blocks that are based on either natural or synthetic amino acids have capability to develop linear or cyclic configurations. The diversity of the side chains of amino acids enables the control of both conformation and functionality. A number of investigator used antigen–antibody interactions and biotin–avidin interactions to study the direct assembly of interacted protein.

2.8   PEPTIDE AS LIGAND

Successful targeting of peptide requires that the target tissue binding site determinant be expressed in amount sufficient to achieve therapeutic level of encapsulated drug. These determinants must be different from normal tissue either in qualitative or quantitative terms. Ideal target are over expressed in diseased tissue than normal one. For example, epidermal growth factor receptors are highly expressed in melanoma cell as compared to normal. Cells in normal tissue have low death rate and dead cells cleared immediately but in tumor have a significant amount of dead cell. This is useful in targeting anticancer drug to the tumor tissue [170]. Distinct anatomy and physiology of tumor vasculature distinguish it from normal site such as abundant fenestration in the tumor vessel, resulting in hypoxia and acidosis, which is potential target for many peptides [171]. Like tumor vasculature, intergin $\alpha v\beta 3$ is not expressed in high ratio in normal ocular tissue but in age macular degeneration disease of eye. Cardiac myocytes cell membrane also has an abnormal level of myosin as compared to healthy heart. These structural abnormalities are a good potential target for peptide. Peptide targets abnormal receptors, cell adhesion and endothelial wall in various organs which increase its site selectivity and specificity. The choice of peptide depends upon the disease pathology, site of action and target molecule (Table 2.2).

- Cancer
- Immune system
- Respiratory disease
- Liver and Pancreases
- Neurological disorder
- Skin
- Vasculature
- Ophthalmic disease
- Other

2.8.1   PEPTIDE NANOCONJUGATE TARGETING CANCER

The side effects and problem related to traditional chemotherapy can be overcome by peptide nanoconjugate with more specific drug delivery. The activity of anticancer drug can be enhanced by coupling peptide on the surface of nanoparticle. Cell penetrating peptides transport the drug into the cell via direct penetration, endocytosis mediated translocation and through the formation of transitory structure (Table 2.3).

The first cell penetrating peptide described in 1988 was tat from HIV. HIV Tat is a small basic peptide containing 86 residues and expressed in *E. coli*. Membrane translocation study across the cell membrane was demonstrated that the 86-residue transcriptional activator protein Tat (encoded by HIV-1) could be efficiently internalized by cells when present in the

TABLE 2.2   Peptide Targeting Disease

Targeted organ/Disease		Peptide	Core material	Polymer	Model	Application
Cancer	Breast	c(RGDfK)/ c(GCGNVVRQGC/ p18–4, RGD, NGR, Lytic peptide/Tat/ H7K(R2)2(arginine rich peptide)/TMT peptide/AP peptide/ VIP	Paclitaxel/ doxorubicin/17- Allylamino- 17-demethoxy geldanamycin	poly(ethylene oxide)- b-poly(e-caprolactone/ poly-(ethylene oxide)- block-distearoylphosphatidyl- ethanolamine/1,2- distearoylphosphadidyl ethanoloamine (DSPE)/ modified glycol chitosan/ poly(L-lactic acid) (PLLA) and polyethylene glycol (PEG)	Human breast adenocarcinoma (MCF-7) cells MDA-MB-435/ human umbilical vein endothelial cells (HUVEC)	Anticancer therapy, angiogenesis and diagnosis
	Liver	AGKGTPSLETTP peptide (A54)/ HAIYPRH (T7)/ Gal peptide/RGD	doxorubicin / DNA/chelated gadolinium/N4- octadecyl-1-β-D- arabinofurano sylcytosine (NOAC)	Stearic acid grafted chitosan Poly (ethylene glycol)/polyethylene glycol-modified polyamidoamine (PAMAM) dioleoylphosphatidyle thanolamine (DOPE)	BEL-7402 cell/ HepG2 cells, (B16BL6)/Human hepatoma cell line SMMC-7721	Targeted anticancer therapy of liver/ diagnosis and imaging liver cancer
	Brain	HAIYPRH (T7)/ TGN/Chlorotoxin/ RGD/Angiopep/ NFL-TBS. 40–63/ Angiopep/F3 peptide/ tLyp-1	Gene/doxorubicin/ docetaxel/ Si RNA/Iron oxide/paclitaxel/ Ferrocenyldiphenol tamoxifen	a-Malemidyl- u-N- hydroxy succinimidyl polyethylene glycol/ polyethylene glycol (PEG)-grafted chitosan and iron oxide/PEG/	U87MG glioblastoma cells/C6 cells and bend. 3/human umbilical vein endothelial cells (HUVEC/9L	Anti-glioma activity/brain targeting and imaging

TABLE 2.2 Continued

Targeted organ/Disease	Peptide	Core material	Polymer	Model	Application
		derivative (Fc-diOH)	poly(ethylene glycol)-co-poly(-caprolactone)/DSPE-mPEG 2000-maleimide	gliosarcoma cells/brain capillary endothelial cells (BCEC)	
Colon	PR b peptide/TrkB-peptide	5-Fluorouracil/DNA	polyethylene glycol/chitosan	CT26. WT cell/monocyte/macrophage cells (RAW 264)	Therapeutic activity for colon cancer
Renal carcinoma	ErbB2 and HA epitope peptides/RGD	Pam2C and Pam3C derivatives/Doxorubicin	Mannosylated L-a-phosphatidyl-DL-glycerol/L-a-phosphatidylcholine/distearoylphosphatidylcholine	renal carcinomacells (RenCa)/SN12C renal carcinoma cells	Vaccine for cancer/targeting renal cancer
Lung	EV/HVGGSSV/PH1 peptide (TMGFTAPRFPHY)/cLABL/E-selecting–binding peptide/YIGSR	STAT5 siRNA/Doxorubicin/cisplatin/cross linked iron oxide(Cy5.5)/Estapor	Liposome-protamine-heparin/DSPE-PEG2000-Maleimide Lipid/PLGA	H460 human lung cancer/LLC human lung adenocarcinoma cell line SPC-A1 and human non-small lung carcinoma cell line H1299/A549 lung epithelial cells	Lung cancer targeting/imaging

TABLE 2.2 Continued

Targeted organ/Disease	Peptide	Core material	Polymer	Model	Application
Lymphatic system	Histidine containing peptide/cyclic RGD/Lys3-bombesin-hydrazinon-icotinamide-Gly-Gly-Cys-NH2 (HYNIC-GGC)	Doxorubicin/siRNA/technetium-99m	phosphatidylethanolamine polyethylene glycol-750/distearoylphosphatidyl ethanolamine-polyethylene glycol/gold	murine melanoma cell line B16-F1/PC-3 cells	specific for tumor lymphatics and vasculature/imaging
Skin	cyclic RKKH peptide	calcipotriol	PEG2000-DSPE	human epidermoid carcinoma cell line A 431	Skin cance
Pancrease	RGD	Doxorubicin	Distearoyl phosphatidylcholine (DSPC), cholesterol dioleoylphosphatidyl ethanolamine (DOPE)	R40P murine pancreatic cancer cells	Targeted therapy of pancreases cancer
EGFR	LHRH peptide/GE11 peptides	Camptothecin/photosensitizer Pc 4	poly(ethylene glycol/poly(ethylene glycol)–poly(ε-caprolactone)	A2780 cell line/A431 cells	Anticancer therapy/Cytotoxic phototherapy
CNS disorder	Brain Pep TGN/RVG peptide/Tat peptide/RVG29-Cyspeptide/ApoE peptides/Low-molecular-weight protamine/Angiopep/Tet-1	coumarin 6/pDNA/miRNA/Rhodamine dye/b-galactosidase/Curcumin 4/Si RNA/coumarin-6/amphotericin B/Ciprofloxacin/curcumin	Poly(ethyleneglycol)-poly(lactide-co-glycolic acid)/PEG and low molecular weight polyethylenimine/disulfide linkage in the branched polyethyl	bEnd.3 cells/Neuro2a cell/16HBE14o-cells/Brain capillary endothelial cells (BCECs)/LAG cell line	Neurodegenerative disorder like SpinoCerebral Ataxia, Parkinson's, Alzheimer's drug delivery across the blood brain barrier

TABLE 2.2 Continued

Targeted organ/Disease	Peptide	Core material	Polymer	Model	Application
			eneimine/PEG -gelatin-siloxane/Chitosan and pluronic acid/1,2-stearoyl-sn-glycero-3-phosphoethanolamine-N-[maleimide(poly(ethylene glycol)/PEGylated chitosan/PLGA		
Respiratory disorder Lung	glucagon-like peptide 1(7–36) (GLP-1)	—	PEGylated phospholipid micelles	lipopolysaccharide (LPS)-induced ALI in mice	Anti inflammatory action in acute lung injury (ALI)
Autoimmune disorder	Tumor specific antigen peptide/RGD/cLABL/	Tumor specific antigen peptide/siRNA	PLGA/PLGA –PEG	Dendritic cell/M cell/human umbilical cord vascular endothelial cells	Peptide immunotherapy
Bone disorder Cartilage	KAFAK	KAFAK peptide	N-isopropylacrylamide and 2-acrylamido-2-methyl-1-propanesulfonic acid	Immortalized human monocytes (THP-1 cells)	Osteoarthritis
Heart disease	Stabilin-2-specific peptide (CRTLTVRKC)/cyclicRGD/	Florescence dye/Cy5.5	Hydrophobic glycol chitosan (HGC)/N-hydroxysuccinimide (NHS)— activated	stable cell/rat carotid injury model/bovine aortic endothelial cells/cardiac cell	atherosclerotic plaques and other inflammatory diseases/thrombosis and restenosis/

TABLE 2.2 Continued

Targeted organ/Disease	Peptide	Core material	Polymer	Model	Application
	atherosclerotic plaque-homing peptide/angiotensin II type 1 receptor (AT1) scrambled peptide		polyethylene glycol carboxyester derivative of distearoylphosphatidyl ethanolamine/modified glycol chitosan-cholanic acid (HGC)/1,2-distearoyl-sn-glycero-3-phosphoethanolamine-[carboxy(polyethylene glycol)		imaging/myocardial infection
GIT Disorder	M cell homing peptide		chitosan	M-cell model	Payer's patch

TABLE 2.3 Biconjugation of Cell Penetrating Peptide With Nanoparticle

Peptides	Method of preparation	Host cell type (for gene expression)	Post transitional modification	Polymer/Core material	Nano particle (size)	Application
Tat-derived peptide (ω-Ahex) GCGGGYGR KKRRQRRR (48–57 amino acids)	Direct chemisorption	Caco-2 cell	Fluorescence iso thiocynate labeling of Tat peptide (FITC-tat)	PEG600-b-PGA containing Super paramagnetic iron oxide	Nanoparticle (65 nm)	In MRI and targeting cancer cell

TABLE 2.3 Continued

Peptides	Method of preparation	Host cell type (for gene expression)	Post transitional modification	Polymer/Core material	Nano particle (size)	Application
TATp (48–57)	seed-mediated growth method	Infinity Telomerase Immortalized primary human fibroblasts	Tat peptide derived from HIV-1	Conjugated with magnetic gold nanoparticle	Nanoparticle (50–80 nm)	MRI
TAT with an amino acid sequence of CGGGYG RKKRRQRRR	seed-mediated growth method	Caco-2 cell, B16, HeLa, MCF-7, and MCF-7/ADR cells	thiol-containing TAT	silver nanoparticle containing Doxorubicin	Nanocrystal (8 nm)	Cancer treatment
Tat peptide (KYGRRRQ RRKKRGC)	sol–gel process	Mice brain	Cys containing tat	gelatin—siloxane nanoparticles	Nanoparticle (180–190 nm)	Brain disorder
TAT	Epoxy conjugation method	MDR1 MDCK cell lines	FITC labeling	PLA encapsulating ritonavir	Nanoparticle 300 nm	CNS delivery of anti-HIV
TAT	Seed incubation	A-549 and human dermal fibroblasts cell lines	FITC (fluorescein isothiocyanate)	Dye doped silica nanoparticles with folic acid	Nanoparticle 80 nm size	Bioimaging

TABLE 2.3   Continued

Peptides	Method of preparation	Host cell type (for gene expression)	Post transitional modification	Polymer/Core material	Nano particle (size)	Application
VFGF derivatives K237-(HTMYYH HYQHHL)	N-terminal PEGylation technique	Human umbilical vein endothelial cells	none	paclitaxel-loaded PEG–PLA nanoparticles	Nanoparticle (80–150 nm)	antiangiogenic activity
Penetratin (domain of the antennapedia Homeodomain) (RQIKIW FQNRRMK WKKGGC)	Coupling to c-cysteine	KB cells	none	PEGylated PLGA NP of siRNA	Nanoparticle (150–160 nm)	intracellular delivery
HR9 C-5H-R9–5H-C	Incubation	Human bronchoalveolar carcinoma A549 cells	none	Semiconductor quantum dots	Quantum dots	intracellular delivery
Arginine-rich CPPs peptide R7	Coupling method	MCF-7 cells	none	PLGA–PEG nanoparticle of vincristine sulfate modified with folic acid	Nanoparticle (230–250 nm)	Cell apoptosis Anticancer
Peptide TAT and fusogenic peptide HA2	Incubation	Hela cells, HEK-293 cells	none	gelatin-silica nanoparticles of DNA	Nanoparticle (147–161 nm)	gene delivery

TABLE 2.3 Continued

Peptides	Method of preparation	Host cell type (for gene expression)	Post transitional modification	Polymer/Core material	Nano particle (size)	Application
TAT peptide	Attachment with Sulfosuccincinimidyl-4-(N-maleidimidomethyl)-cyclohexane-1-carboxylate linker	Human fetal liver tissue, human mesenchymal stem Cells (MSC)	Flouresence iso thiocyanate labeling of Tat peptide (FITC-tat)	Polyacrylamide nanosensors modified to contain free amine groups	nanosensor	MSC based therapeutics
HIV-1 Tat peptide (GRKKRR QRRPPQC)	Incubation and coupling with (1-ethyl-3-3(3-dimethylaminopropyl) carbodiimide hydrochloride (EDAC) and N-hydroxys-ulfosuccinimide (NHSS) linker	Human embryonic kidney cells (HEK293)	None	Plasmid DNA nanoparticle	237–250 nm	Synergistic effect in transfection of pDNA into the cells, in gene delivery
TAT peptide	A solid-support (Rink amide MBHA resin) using Fmoc chemistry	HeLa cells	Fluorescein isothiocyanate (FITC)-labeled Tat peptides	Maleimido-tetra(ethylene glycol)-poly(glycerol monoacrylate) (MAL-TEG-PGA) iron oxide nanoparticle	47–67 nm	Gene delivery

TABLE 2.3 Continued

Peptides	Method of preparation	Host cell type (for gene expression)	Post transitional modification	Polymer/Core material	Nano particle (size)	Application
G3R6TAT	9-fluorenylmethoxy-carbonyl (Fmoc) approach using an Apex 396 peptide synthesizer	Rabbit brain cell	FITC	Loaded with Cholesterol and cationic antimicrobial peptide	nanoparticle	C. neoformans induced meningitis

surrounding tissue culture media [172–174]. One study performed on Tat peptide [48–60] observed that positive charge is essential for cell internalization and glycosaminoglycan's on the cell surface negatively affect their uptake. One problem with the tat peptide is lack of specificity [175]. Tat can transport the small, large as well as lipophilic molecule to the nucleus with enhanced concentration and much faster rate [176–178]. This peptide form conjugate with molecule of small size and enter directly into the cells by the electrostatic interaction, hydrogen bonding and macropinocytosis pathways are followed by larger molecules for transduction into the cell [179, 180]. The peptide used for conjugation should have sequence as their binding receptor. Therefore, a specific peptide can bind a specific receptor for example RGD peptide mainly binds on integrin receptors in the tumor [181].

Gene therapy has potential to revolutionize the treatment of cancer. One major challenge of applying this technology for clinical application is the lack of site-specific carriers that can effectively deliver short interfering RNA (siRNA) to cancer cells. Asparagine–glycine–arginine (NGR) was found to effective in delivery of doxorubicin and siRNA nanoparticles to the cell. NGR peptide acted on c-myc oncogene is overexpressed and activated in various human tumor. It promotes cell growth, transformation, and angiogenesis that play important roles in the progression and metastasis of tumor such as acute lymphoblastic and myeloblastic leukemia, malignant lymphoma, bone, breast, ovarian, prostate, bladder, gastric, and bronchogenic carcinoma [182, 183]. GE11 is an active peptide against the epidermal growth factor receptor (EGFR) which is an important target for anti-cancer therapy, applicable in many cancer types. It was shown to bind to EGFR competitively with EGF and mediate gene delivery to cancer cells with high-EGFR expression. GE11 conjugated liposomes accumulated into the tumor site preferentially, and have better targeting and drug delivery capacities [184].

pH sensitive peptide utilize the change in pH between normal tissue and solid tumor. Normal tissues have more acidic pH than solid tumor. H7K(R2)2 is a pH sensitive peptide which is conjugated with micelle containing paclitaxel. pH is a rate controlled parameter in pH sensitive peptide [185]. pH Low Insertion Peptide (pHLIP) was used to target various imaging agents to acidic tumor by nonfunctionalized nanogold particles where nanogold is covalently attached to the N terminus of pHLIP. The pHLIP technology can substantially improve the delivery of gold nanoparticles to tumors by providing specificity of targeting, enhancing local concentration in tumors, and distributing nanoparticles throughout the entire tumor mass where they remain for an extended period (several days), which is beneficial for radiation oncology and imaging [186].

Tumor metastasis targeting (TMT) peptide is the ligand of XPNPEPZ, a subtype of aminopeptidase P. It has been found to specifically bind to a series of highly metastatic tumor but not to the non metastatic cell. TMT peptide tethered liposomes showed minor effect due to cell specificity [187]. LTVSPWY peptide-modified PEGylated chitosan nanoparticle was used as imaging agents for detection of tumors overexpressing HER2 [188].

2.8.2   PEPTIDE TARGETING DISEASE RELATED TO IMMUNE SYSTEM

Immunity related disorders are the third most common category of disease after cancer and heart disease; they affect a large number of populations worldwide [189]. Many approaches have been used to cure these diseases but Peptide immunotherapy is one such approach that is effective, safe and disease modifying. This approach utilizes short synthetic peptides, comprising T cell epitopes of major allergens, was unable to cross-link allergen-specific IgE molecules on basophils in vitro. Treatment of allergic volunteers with allergen peptides resulted in reduced skin, lung and nasal sensitivity to allergen challenge and improved their subjective ability to tolerate allergen exposure. Peptides immunotherapy reduces pro-inflammatory cytokine secretion from peripheral blood cells, whilst increasing the immunosuppressive cytokine IL-10. Furthermore, peptide therapy was associated with the induction of a population of CD^{4+}. T cells with a suppressive functional phenotype. Thus, peptide therapy may be suitable for the antigen-specific treatment of allergic diseases [190]. RGD (arginine-glycine-aspartic acid) contain short chain of amino acid and prove to be a successful ligand for targeting integrin receptor specially beta 1 integrin receptor. Integrin is a part of cell adhesion molecule (CAM) which plays an important role in various disease state such as cancer, rheumatoid arthritis and diabetes. RGD have preferential binding affinity toward integrin. In addition, it has prolonged half-life, passive retention in diseased tissues, increases the cellular uptake of nanoparticle [191]. In a study, PLGA-based nanoparticles containing RGD peptide bind on beta integrin receptor on the cell surface of human M cell. RGD is peptide containing short chain aminoacid which specifically bind to integrin receptor and induce cellular immune response. The immune response depends on the presence of ligands on the surface of nanoparticle and represents an effective delivery system for oral immunization [192]. RGD peptide functionalized poly(lactide-co-glycolytic) acid (PLGA) nanosystem was used to deliver a STAT1 siRNA to joint tissues in a mouse model of rheumatoid arthritis. Nanoparticle protects the gene from serum degradation and RGD peptide bind on the surface of cell and selectively inhibits macrophage and dandritic cell activation and suppresses the immune system and used in the rheumatoid arthritis [193]. Another cyclo PenITDGEATDSGC (cLABL) peptide was used commonly to target nanoparticle to human umbilical cord vascular endothelial cells (HUVEC) monolayers that have upregulated intercellular cell-adhesion molecule-1 (ICAM-1) expression. It showed higher binding affinity for ICAM-1 receptors on the apical surface of the cell. When PLGA–PEG nanoparticle decorated with cLABL peptide, it mediate rapid binding and internalization which increase cell specificity. This delivery system has dual function; one is inhibition of the infiltration of immune cells via blockage and another is internalization of ICAM-1 receptors coupled with the ability to localize drug delivery to regions of ICAM-1 up regulation [194]. To improve the efficiency of orally delivered vaccines, PEGylated PLGA-based nanoparticles loaded with an antigen, ovalbumin grafted with a specific ligand RGD peptide. The ligand at their surface was designed to target β1 integrin's expressed at apical side of human M cells. A novel photografting was performed to graft the peptide on the surface of nanoparticles. When this formulation administered orally, elicit the immune response [195]. PLGA nanoparticles encapsulating antigenic peptide decorated with endoplasmic reticulum targeting peptide was taken by dendritic cells and used as an anticancer vaccine [196].

2.8.3 PEPTIDE TARGETING RESPIRATORY SYSTEM

Lung is an important organ for targeting many acute or chronic diseases such as cancer, asthma, cystic fibrosis, alpha-1-antitrypsin deficiency and respiratory distress syndrome. The targeting of drug can be achieved by conjugation of peptide and nanoparticle. The peptide used for targeting should be specific to particular pulmonary disease cell so that it doesn't have any effect on healthy cell. Inflammation is a common symptom in acute lung injury observed in gram negative sepsis. PEGylated phospholipid micelle conjugated with glucagon like peptide was effectively reduced inflammation in experimented mice [197]. Phosphotidylserine liposome with high affinity 7-amino acid peptide was formulated to target the plasmid in pulmonary epithelial cell's nucleus. On intra-tracheal instillation, this complex accelerates the liposomal binding and cell internalization in lung epithelial cell. This also increased expression of gene in bronchioles and alveoli with no clinical adverse effect [198].

Lamanin receptors are present on the lung and used as target for effective drug delivery. It belongs to extracellular glycoprotein family which is responsible for cell adhesion, differentiation, migration, signaling and metastasis. A correlation exists between the up regulation of this polypeptide in cancer cells and their invasive and metastatic phenotype. This receptor is found higher in colon and lung cancer tissue. Nanoparticle conjugated with laminin receptor binding peptide (YIGSR) targeting to lung melanoma via interaction to the laminin receptor, found to be 2 times more effective for cancer cell and no effect on healthy cell [199, 200].

A radiolabeled PEGylated liposomal nanoparticle (NPs) when functionalized with somatostatin analog tyrosine-3-octreotide was used to target the somatotatin receptors present in lung cancer. It has a variety of diagnostic and therapeutic application [201]. LTP-1 peptide conjugated PAMAM dendimer promoted the pulmonary epithelium transport of macromolecule cargoes. Bioavailability increased up to 31% when experimented on isolated perfused rat lung [202].

Intercellular adhesion of molecule (ICAM-1) plays an important role in metastatic processes. ICAM-1 is a member of immunoglobin subfamily which is highly expressed in inflammation and immune response. ICAM-1 is a ligand for LFA-1 (integrin), a receptor found on leukocytes. When activated, leukocytes bind to endothelial cells via ICAM-1/LFA-1 and then transmigrate into tissues. In addition, ICAM-1 has been detected on the surface of a variety of primary tumors and tumor cell lines [203]. cLABL peptide decorated nanoparticle which bind to ICAM-1 and inhibit the interaction of LFA-1/ICAM-1 and inhibit the immune response in lung cancer [204].

cLABL suppressed the immune system while tumor specific antigen peptide bind on the dendritic cell and stimulate the immune system [205].

2.8.4 PEPTIDE TARGETING LIVER AND PANCREASES

Hepatic macrophages play an important role in regulation of inflammation and fibrosis in liver injury. Therefore, hepatic macrophage can be used as a target for ligand binding in liver cell. RGD peptide specifically bind to the hepatic macrophage and delivered the drug into the liver cell. RGD conjugated gold nanoparticle decreased kupffer cell number; alter macrophage polymerization and cytokines production. Additionally, this conjugation was effective in low concentration [206]. Octa arginine is another cell penetrating peptide which delivered the lipid nanoparticle of Si RNA to the hepatocytes and liver cell with minimum toxicity and immune response (207). Cyclo RGD peptides have specific affinity to hepatic stellate cells that is vital for hepatocellular function and play an important role in liver fibrosis and repair [208]. Cyclo RGD peptide conjugated liposome encapsulating human growth factor/recombinant human interferon-alpha 1b was used to target hepatic stellate cell. Receptor-mediated liposomes displayed good targeting antifibrotic effects on liver fibrosis [209, 210]. Hepatocellular carcinoma was successfully targeted with dioleoyl phosphatidyl ethanolamine (DOPE) conjugated PEGylated metalloproteinase-2 peptide, which was incorporated in galactosylated liposome. Presence of excess galactose reduced the toxicity and enhance uptake of liposome in Hep G2 cells [211]. T7 (HAIYPRH), transferring receptor specific peptide bind to transferring receptor present in the liver cell. This peptide was used as a ligand to deliver the combination of gene encoding human tumor necrosis factor-related apoptosis into the cell and accumulated in tumor tissue of liver for prolonged time [212, 213]. When this peptide conjugated with galadium ion, it can be used in imaging of liver cancer [214]. Another peptide, A54 (AGKGTPSLETTP) showed its binding efficiency in liver cancer cell via endocytosis [215].

Early diagnosis of pancreatic ductal adenocarcinoma (PDAC) can prevent the death from this cancer. Plectin-1 was used a biomarker in pancreatic ductal adenocarcinoma. Plectin-1 protein is over expressed in a number of cancers but particularly in pancreatic cancer. Plectin-1 targeting peptide (PTP) binds to the plectin-1 protein with high affinity and specificity. Magneto fluorescent nanoparticle conjugated with PTP was used in diagnosis and management of pancreatic ductal adenocarcinoma [216].

2.8.5 PEPTIDE TARGETING NEUROLOGICAL DISORDER

Blood brain barrier act very effectively to protect the brain from chemical and biological agent. This represents a great difficulty in delivering important diagnostic and therapeutic agents to specific regions of brain in many neuronal diseases. These have problem in crossing the blood brain barrier. Nanoparticles or conjugation of nanoparticle with peptide may help in the transfer of drug across barrier and play an important role in neuroscience.

Low density lipoprotein receptors were over expressed on blood brain and glioma cell. Angiopep was used to target lipoprotein receptor-related protein-1 (LRP1) which enhances transcytosis capacity and parenchymal accumulation. Angiopep is a commonly used cell penetrating peptide targeting lipoprotein receptor and enhance the transcytosis and parenchymal accumulation [217]. Angiopep-2 peptide and EGFP-EGF 1 protein when conjugated with PEG-PCL nanoparticle, peptide penetrated the blood brain barrier and protein binds to neuroglial cells enhance the uptake of nanoparticles. The mechanism of Angio pep peptide to internalize the brain capillary endothelial cell a clathrin- and caveolae-mediated energy-depending endocytosis, also partly through marcopinocytosis [218–222]. A study performed on carbon nanotube decorated with angiopep showed less toxicity and good compatibility due to binding of peptide on specific receptors. Additionally, carbon nanotube has large surface area and enhanced accumulation in tumor cell [223, 224]. Nanosized liposomes containing phosphatidic acid decorated with peptides derived from apolipoprotein E were used to target amyloid-β peptide which play an important role in Alzheimer's disease [225]. Same author suggested that cellular uptake and membrane permeability of drug depends on the density of peptide [226]. Chlorotoxin (ClTx), a 4-kD peptide isolated from Leiurus quinquestriatus scorpion venom, bind specifically and selectively to brain glioma cell, improving anticancer activity. Chlorotoxin bind to glioma specific chloride ion channel and matrix metalloproteinase-2 endopeptidase. This leads to inhibition of cell invasion in glioma. Chlorotoxin peptide was used as a ligand to deliver therapeutic as well as diagnostic agent into glioma cell [227–229]. HIV TAT peptide conjugated nanoparticles was used to deliver drug including doxorubicin, therapeutic DNA, SiRNA to brain by crossing blood brain barrier and used to treat neurodegenerative disease [230–232]. Rabies virus glycoprotein (RVG) peptide containing 29 amino acids was exploited as targeting ligands for brain. RVG peptide bind to the acetyl choline receptor in neuronal cells and cellular uptake occurred through receptor mediated endocytosis. This peptide provided a safe and effective delivery of gene [233, 234]. RVG peptide-based DNA delivery systems have many advantages including the ease of synthesis, low immunogenicity, biocompatibility, and biodegradability in vivo [235, 236]. Other peptides which have high affinity for neuronal cell and possessed retrograde property including Blood brain barrier peptide (similopioid peptide) [237], Tyrosine kinase receptor (Trk B) binding peptide [238], Tumor homing and penetrating peptides, tLyp-1 peptide [239] and plant derived Tet peptide [240].

2.8.6 VASCULAR TARGETING PEPTIDE

There are two therapies involved in vascular targeting: Antiangiogenic and vascular disrupting agent. Vascular antiangiogenic agent (VTA) interferes with new blood vessel but don't disturb those blood vessel which already fed tumor or organ. Vascular disrupting agent (VDA) targets the established tumor blood vessels, show immediate effect [241]. Vascular targeting agents are equally effective against cancer, cardiovascular and ophthalmic diseases. Solid tumors are dependent on their blood supply in cancer disease. The vascular targeting peptides disturb their blood supply and cause ischemia and tissue necrosis. So, Vasculature targeting is an effective antitumor therapy. Many peptide has been used to target vasculature for example Fibroblast growth factor peptide (tbFGF), attached on the surface of liposome via electrostatic interaction with fibroblast growth factor receptors. These receptors are over expressed on the surface of a variety of tumor cells and on tumor neovasculature in situ, are potential targets for tumor- and vascular-targeting therapy [242]. F3 peptides that specifically bound to nucleolin, which is highly expressed on the surface of both glioma cells and endothelial cells of glioma angiogenic blood vessels, was utilized to target glioma cell and neovaculature [243].

Peptides featuring the LR(S/T) motif were identified that could specifically bind to the C-type lectin-like molecule-1, a protein preferentially expressed on acute myeloid leukemia stem cells. Micellar nanoparticles of daunorubicin were covalently decorated with C-type lectin-like molecule-1-targeting peptides for eradicating leukemia stem cells and improving leukemia therapy [244]. F3 peptide containing 31 amino acids was used to target cisplatin-loaded nanoparticles (F3-Cis-Np) to tumor vessels. F3-Cis-Np bind with high specificity to tumor endothelial cells *AND* showed cytotoxic activity. Therapy targeting human vasculature in vivo with F3-Cis-Np led to near complete loss of all human tumor vessels in a murine model of human tumor vasculature. The study indicated that F3-targeted vascular therapeutics may be an effective treatment modality in human ovarian cancer [245]. Another strategy for targeting drug in the endothelial lining of blood vessel of tumor cell is homing peptide. Homing peptide can bind to surface molecules specific to organ or tumor cells. RGD is an example of homing peptide, present in adhesive extracellular matrix, blood and cell surface proteins and bind to integrins. Integrin-mediated cell attachment influence and regulates cell migration, growth, differentiation, and apoptosis, the RGD peptides and mimics can be used to probe integrin functions in various cancer including breast and brain cancer [246–249]. Fewer studies are performed on vasculature targeting other than solid tumor. Endothelial cells that line the vascular lumen can express cell-surface proteins that are specific to the endothelium of a particular tissue. These proteins can be used as a target for diagnostic and therapeutic agent. Heart homing peptide was discovered by Zhang et al. This was specific for heart endothelium and could serve as targeting ligand [250]. Binding of drug to the cardiac endothelium should be very rapid for effective targeting of drug. Liposomes targeted with CRPPR (heart homing peptide) bound to the endothelium within 100 seconds after intravenous injection and remained stably bound [251]. Same author observed that CRPPR-conjugated particles with a terminal free amide accumulated on murine aortic endothelial cells to a greater extent than particles conjugated with the same sequence. This heart homing peptide (CRPPR) bound to endothelial neuropilin-1 and transport drug into interstitium [252]. Another peptide that is bound on the endothelium is RGD. An arginine–glycine–aspartic acid (RGD) conjugated liposomes targeted the integrin

GPIIb–IIIa on activated platelets. This conjugation provided a effective therapy in the treatment of cardiovascular disease processes such as atherosclerosis, restenosis and inflammation [253].

Choroidal neovascularization leads to loss of vision in age-related macular degeneration. The disease is characterized by the accumulation of membranous debris on both sides of the retinal pigment epithelial basement membrane. Macular degeneration can be treated by inhibiting or destroying neovascularization. Transferrin and arginine–glycine–aspartic acid (RGD) peptide functionalized on nanoparticles was encapsulating a anti-vascular endothelial growth factor intraceptor plasmid (Flt23K-plasmid). GRGDSPK specifically bind to integrin $\alpha_v\beta_3$ which are upregulated in ocular neovasculature. Integrin $\alpha_v\beta_3$ was overexpressed in ocular tissue from age related macular degeneration patient. Cellular uptake of encapsulated gene can be performed with the help of receptor mediated endocytosis (Figure 2.6).

Peptide and transferrin functionalized nanoparticles inhibited the neovascularization by targeting the gene intracellularly [254].

2.8.7 PEPTIDE TARGETING OPHTHALMIC DISEASE

Arg-Gly-Asp (RGD) containing motif is known to be vital for integrin receptor mediated cell attachment, which influences cell migration, growth, and differentiation. The specific binding between RGD-containing peptides and their integrin receptors also plays an important role in

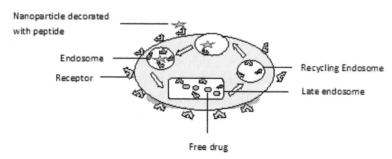

FIGURE 2.6 Illustration of cellular uptake of nanoparticle decorated peptide through receptor mediated endocytosis.

cell targeting. Alkyne-modified KGRGDS peptides were synthesized and coupled to azide functionalized nanoparticles, approximately 400 peptides were bound to each nanoparticle. Coupling RGD containing peptides to nanoparticles may provide a method for targeted drug delivery to the eye after injury [255]. Signal peptide, for example, Serine-threonine-tyrosine peptide sequence was chosen because it is well known to act as a transduction signaling agent within and between retinal pigmented epithelium cells. A nanoformulation of a water-soluble chitosan conjugated with a peptide (serine-threonine-tyrosine) was synthesized and evaluated in retinal cell. Conjugated nano chitosan peptide may promote binding and engulfment. This molecule is an excellent carrier for retinal drug delivery and has the potential to treat age-related macular degeneration [256].

2.8.8 OTHER

Prohibitin is selectively expressed at high levels by the inguinal adipose tissue vasculature. Adipose tissue-specific circular peptide (KGGRAKD) conjugated liposomes were taken up into the isolated primary cultured endothelial cells derived from inguinal adipose tissue via prohibiting-mediated endocytosis [257]. Embryonic stem (ES) cells hold great promise for replacement therapies and studies of developmental biology, due to their ability to differentiate to all lineages of cells while maintaining an undifferentiated state during in vitro culture. APWHLSSQYSRT peptide conjugated nanoparticles have high affinity and specificity towards embryonic cells. Therefore, this peptide can be use in imaging or targeting embryonic cell [258]. Chemotherapy is an important treatment for ovarian cancer. However, conventional chemotherapy has inevitable drawbacks due to side effects from nonspecific biodistribution of the chemotherapeutic drugs. To solve such problem, targeted delivery approaches were developed. The targeted delivery approaches combine drug carriers with the targeting system and can preferentially bring drugs to the targeted sites. Follicle-stimulating hormone receptor (FSHR) is an ovarian cancer–specific receptor. Peptide derived from FSH (amino acids 33–53 of the FSH B chain, named as FSH33), a conjugated nanoparticle, to target FSHR in ovarian cancer was developed. This novel FSH33-NP delivery system showed very high selectivity and efficacy for FSHR-expressing tumor tissues [259].

Insulin loaded trimethyl chitosan chloride nanoparticles (TMC) modified with a CSKSSDYQC (CSK) targeting peptide was used to target goblet cells which have better hypoglycemic activity effect with 1.5 fold than unmodified one [260]. Vincristine sulfate loaded nanoparticles functionalized with Arginine rich cell penetrating peptide R7 and folic acid could be potential vehicle for chemotherapy of breast cancer [261].

2.9 NON-PEPTIDE TARGETING

Chemical nature of targeting ligands plays an important in fate of a drug whether it reaches its target cell in efficient concentration or not. Main problem with peptide are confirmation, size and difficulty in chemical modification. The use of non peptidic ligands over peptide ligands seems to have several advantages: (i) they could be modified and conjugated easily (ii) they have stable confirmation (iii) procedure of production of these ligands are small (iv) they are less expensive. Molecules used for targeting a drug other than peptides are: Lectin, protein, Nucleic acid and vitamin (Figure 2.3).

2.9.1 LECTIN MEDIATED DRUG DELIVERY

Most lectins were isolated from plants mainly seeds and tubers such as cereals, crops and potatoes. Lectin based drug delivery can deliver the drug via two mechanisms: direct and indirect targeting. Direct targeting involves the interaction between external carbohydrates and protein present on the cell surface. The galectin family of carbohydrate receptors represents a promising target for drug delivery in hepatocytes and it is highly expressed in tumor cell which represent a better opportunity for drug targeting cancerous cell [262]. In C type lectin receptors, DC-SIGN binding glycan modified liposomes was used to target antigen on dandritic cell, potentiate tumor-specific CD8+ and CD4+ T-cell response. Liposomes are spherical particles consisting of phospholipid bilayers, which can encapsulate large quantities of hydrophilic or hydrophobic molecules, including proteins and peptides. Consequently, they provide the opportunity to incorporate multiple tumor antigens or peptide mixes as well as different DC activating molecules (e.g., Toll-like receptor (TLR)-agonists).

Due to their composition of naturally derived compounds, liposomes are well tolerated by the body and have low toxicity [263, 264]. Galactose was used as ligands in gene transfection of lipid vesicles on asialoglycoprotein receptors [265]. Hyaluronic acid is a linear polysaccharide nonsulfated glycosaminoglycan distributed widely throughout tissues in body. It is major component of components of the extracellular matrix, hyaluronan contributes significantly to cell proliferation and migration, and may also be involved in the progression of some malignant tumors. Hyaluronic acid receptors are overexpressed in malignant cell with the highest metastatic potential. Hyaluronic acid modified chitosan nanoparticle encapsulating 5-fluorouracil showed enhances binding and internalization with better tissue specificity [266]. Nanoparticle functionalized with lectin bind to sugar complexes (glucose/mannose, N-acetyl glucosamine, n-acetyl galactosamine, L-fucose and sialic acid) attaches to proteins and glycolipids present on the cell. Due to its non immunogenicity, lectin can be used as targeting ligands for delivery of nanoparticle in GIT, nasal mucosa, lung, buccal, ocular and brain, etc. [267].

Wheat germ agglutinins (WGA) obtained from tritium aestivum was covalently attached to carbopol liposomes which retained the biological cell binding of activity of WGA up to 10 fold in case of chlorpromazine and 20 fold in calcitonin loaded liposome [271]. WGA bind on the carbohydrates sites such as sialic acid and N-acetyl glucosamine and transport the drug intracellularly including paclitaxel [272]. WGA conjugated Solid lipid nanoparticle increased oral bioavailability of insulin and prevent the encapsulated drug from degradation [273]. In spite of lectin's interesting biological potential for drug targeting and delivery, a potential disadvantage of natural lectins may be large size molecules that results in immunogenicity and toxicity. In addition, Receptor saturation is the main problem involved in designing of lectin mediated drug delivery [274]. Smaller peptides which can mimic the function of lectins are promising candidates for drug targeting.

2.9.2 PROTEIN AS TARGETING AGENT

Transferrin (80 kDa) is a glycoprotein having highly specific binding sites for ferric ions. The affinity of binding depends on the pH, decrease upon increase in neutrality. Transferrin receptors are highly expressed in tumor cell. Therefore, transferrin binding nanoparticle transport the drug intracellularly via receptor mediated endocytosis [272, 273]. Transferrin's are also used as targeting ligand in delivering the gene into brain due to expression of transferrin receptor into brain [274, 275]. Indomethacin loaded PLA nanoparticles were coupled to Transferrin (TF), showed minimum drug leakage and high binding affinity [276]. In some study transferrin when combined tumor necrosis factor (TNF)-related apoptosis-inducing ligand (TRAIL) encapsulating doxorubicin in human serum albumin nanoparticle showed higher capacity of killing different type of tumor cells in various tissue organs [277]. When transferring functionalized nanoparticles are placed in biological fluid, protein adsorb on the surface of nanoparticles to form protein corona. This protein corona interferes in cell internalization and loses specificity of drug targeting [278]. Additionally, lactoferrin receptors are also overexpressed in brain capillaries and bronchial epithelial cell. Therefore, it can be suggested as a good target for lungs and brain drug delivery [279].

Epidermal growth factor receptors is the cell surface receptors for member of epidermal growth factor family, a subfamily of kinase receptors which are over expressed in majority of human cancer [280]. EGF, a 6 kDa protein containing only 53 amino acids has been demonstrated numerous application in cancer therapy [281, 282]. For example, EGF was attached to single wall carbon nanotube loaded with cisplatin or quantum dots are used for targeted therapy and tumor imaging [283, 284].

Apolipoprotein are protein which binds specifically to lipoproteins. They transport the oil soluble substance through lymph and blood. Apolipoprotein can be classified into different category, in which apolipoprotein B are low density lipoprotein whereas all others are high density lipoprotein. Apolipoprotein E is produced mainly in liver and brain. Thereby, It is suggested a new target for brain and liver drug delivery. A numerous application has been demonstrated it role as targeting ligand in brain drug delivery. For example, apolipo E conjugated nanoparticle bind on lipoprotein receptors and internalize the brain capillary endothelial cell via receptor mediated endocytosis. Apolipoprotein conjugated with covalent linkage on the surface of nanoparticle. Therefore, it does not require any absorptive process [285, 286].

Antibody conjugated nanoparticles was used to initiate antigen-specific immune responses in dendritic cells [287]. Membrane type-1 matrix metalloproteinase (MT1-MMP), which plays an important role in angiogenesis, is expressed on angiogenic endothelium cells as well as tumor cells. Then, the MT1-MMP might be useful as a target molecule for tumor and neovascularity. Fab fragments of antibody against the MT1-MMP were modified at distal end of polyethylene glycol (PEG) of doxorubicin-encapsulating liposomes, DXR-sterically stabilized immunoliposomes. MT1-MMP antibody (Fab) is a potent targeting ligand for the MT1-MMP expressed cells. Antibody conjugated liposomes resulted in acceleration of cellular uptake of lioposomes owing to the incorporated antibody after extravasation from capillaries in tumor [288]. Liposomal nanocarriers loaded with doxorubicin conjugated with dual ligands, for example, folic acid and a monoclonal antibody. These ligands targeted the folate receptor and the epidermal growth factor receptors. Advantage with dual target is that it target only tumor cell and have good selectivity over single ligand approach [289, 290]. Legumain is a member of the asparaginyl endopeptidase family that is over-expressed in response to hypoxic stress on mammary adenocarcinoma, colorectal cancer, proliferating endothelial cells, and tumor-associated macrophages (TAMs). It can serve as a target in ovarian and colon cancer. RR-11a, a synthetic enzyme inhibitor of Legumain coupled with nanoparticle encapsulating doxorubicin improved tumor selectivity and drug sensitivity, leading to complete inhibition of tumor growth and reduction of systemic toxicity [291, 292]. Fibroblast growth factor receptors (bFGF) are highly expressed in cancerous cell and proved to be a good target for binding of anticancer drug. Truncated bFGF fragments chemically linked with cholesterol micelles containing paclitaxel showed potential application in tumor targeting [293].

Nanoparticle containing dual targeting ligands showed higher cellular uptake as compared to single ligands anchored nanoparticle and have specific cancer cell activity [294]. Monoclonal antibodies have monovalent affinity, in that they bind to the same epitope or target cell. These antibodies bind specifically to cancerous cell antigen and induce an immunological response against that cell [295].

Antibody targeting domain of platelet receptor was used as targeting ligands for antithrombotic properties. An antibody (designated C-EL2Ab), which targets the C-terminus of the 2nd extracellular loop (C-EL2) of the thromboxane A2 receptor (TPR), selectively blocks TPR-mediated platelet aggregation [296].

Herceptin is a monoclonal antibody that interferes with the function of HER2+ (Human Epidermal growth factor Receptor 2-positive). Its main use is to treat breast cancer. The HER receptors are proteins that are embedded in the cell membrane and communicate molecular signals from outside the cell (molecules called EGFs) to inside the cell, and turn genes on and off. The HER proteins stimulate cell proliferation. In some cancers, notably certain types of breast cancer, HER2 is over-expressed, and causes cancer cells to reproduce uncontrollably. high concentration of herceptin-functionalized magnetic nanospheres in the cancer cells offers great potential in cancer cell targeting and treatment [297]. Cetuximab conjugated O carboxymethyl chitosan nanoparticles was used to target paclitaxel Epidermal Growth Factor Receptor (EGFR) over-expressing cancer cells and can be used a promising target candidate for the targeted therapy of EGFR over expressing cancers [298].

2.9.3 VITAMIN AS TARGETING LIGANDS

Folic acid (vitamin B9) is a vital nutrient required by the cell for nucleotide synthesis. Folate binds to folic acid receptor, a glycozylphosphatidy-inositol-linked protein which is over expressed in cancer cell. So, Folic acid has been widely used as targeting ligands to deliver anticancer drug to cancer cell due to high binding affinity with folic acid receptors [299, 300]. When comparison was performed between different type of ligands like folate, dextran and galactose anchored poly(propylene imine) dendrimer. Folate showed higher selectivity towards cancerous cell and better uptake compared to other targeting ligands [301]. Vitamin B12 (Cobalamin) is highly hydrophilic large molecule, bind with receptors for Vitamin B12 (transcobalamin) which is expressed on plasma membrane. These receptors are highly expressed on human tumor cell including lung, brain, kidney, ovary and uterus. Due to high specificity and affinity, vitamin B12 is a potential candidate for tumor targeting of diagnostic and chemotherapeutic agent [302]. Tocopherol (vitamin E) is a natural lipophilic antioxidant. It stops the production of reactive oxygen species when fat undergoes oxidation and acts as peroxyl radical scavenger, preventing the propagation of free radical in tissues. Vitamin E analog (Trolox) functionalized gold nanoparticle has been proven to be a high efficient way to enhance the antioxidant activity [303]. Folic acid increased the cellular uptake of nanoparticle (encapsulating therapeutic as well as diagnostic activity) into KB cells [304].

2.9.4   NUCLEIC ACID (APTAMERS) AS LIGAND

Short synthetic sequences of nucleic acid (DNA and RNA) are known as aptamers that have been used as targeting ligand to bind to the target site with high affinity and specificity. Cancer cell specific aptamers can be used to functionalize nanoparticles for more effective drug delivery. Nanoparticles functionalized with aptamers targeted the EGF receptors and have excellent cellular uptake via receptor mediated endocytosis [305]. Bioconjugates of nanoparticle and aptamers could efficiently target the prostate specific antigen protein. These nanoparticles have following desirable characteristics: (i) negative surface charge, which may minimize nonspecific interaction with the negatively charged nucleic acid aptamers; (ii) carboxylic acid groups on the particle surface for potential modification and covalent conjugation to amine-modified aptamers; and (iii) presence of PEG on particle surface, which enhances circulating half-life while contributing to decreased uptake in non targeted cells [306]. In vitro, high affinity aptamers were isolated from random library using systematic evolution of ligands by exponential enrichment (SELEX) techniques. Based on aptamer sequential sequence, highly pure compound can be synthesized at low cost than peptide. Aptamers is a smart targeting agent for therapeutic and contrasting agent. In addition, it is a good electrochemical sensor for medical diagnosis [307]. Aptamer was well suited for targeting of drug encapsulated nanoparticle for specific cancer cell [308]. Silver nanoparticle conjugated with aptamers was used for detection of Adenosine triphosphate. Aptasensor could help in detection of concentration of ATP in micromolar scale [309, 310].

2.10   NANOENCAPSULATION IMPROVES THE THERAPEUTIC IMPORTANCE OF PROTEINS AND PEPTIDES

The foremost aim of the protein and peptide drug delivery systems is the interaction or internalization of protein and peptides drug into target cells.

The encapsulation of these drugs on the suitable nanocarrier is the first step in the development of targeted peptide delivery. The desired feature of these efficient drug delivery systems is to deliver therapeutic peptides to active site at the right time in a therapeutically effective concentration and at highest patient convenience/compliance, with minimal side effects and low production cost. The isolation, and costly downstream process for the production of cost effective therapeutic proteins and peptides are the major obstacle in the development of protein therapeutics. Various proteases present at potential site of administration, may inactivate and degrade these proteins therapeutics. The elimination of these peptide drugs by body defense system is another major reason for less utilization of protein and peptides therapeutics. These problems create the decline of bioavailability and increases potential immune response against the protein therapeutics. The alternative route of administration of the peptides therapeutics like pulmonary, transdermal, and nasal may be beneficial than conventional route, towards the reduction of above mentioned problems and increase in their bioavailability. A variety of methods have been widely proposed to efficient protein delivery in vivo as well as in vitro for the delivery of protein and peptides therapeutics. The most efficient non nanotechnological methods for efficient cellular delivery are transcriptional activator of transcription (TAT) from HIVI, third helix of antennapedia homeodomain and VP22 protein of herpes simplex virus. These methods are used to improve the internalization of peptide drugs due to their unique potential to enter the cells in culture when added exogenously. Though these carriers systems have potential to deliver protein into the cells but, many of them have shown inefficient delivery for the protein therapeutics to their actual target site. In these strategies, a number of other problems, such as complex manipulation, cellular toxicity and immunogenicity are reported. However, the encapsulation of protein and peptides drugs on suitable nanocarriers is one of the emerging techniques with tremendous potential to reduce the problem (stability, degradation, proteolysis etc.) associated with protein and peptide delivery. In addition, encapsulation of proteins and peptides in nanocarriers in physiologically active form provides resistant to organic solvents, moisture, acidic environment, enzymatic degradation, and immunological elimination. This technique is being widely accepted practical approach for the development of protein and peptide nanotherapeutics.

The retention of protein structure and activity on nanoscale support are critical for their therapeutic applications. The other fundamental interest is to understand the effect of size and surface chemistry of nanomaterial on structure, activity, and stability of nanoparticles conjugated proteins. The study of lysozyme and human carbonic anhydrase adsorbed on silica nanoparticles reveals the change in structure, activity and loss of a helical contents depending upon the size of nanoparticles. This finding suggests that the smaller nanoparticles are advantageous for protein stability due to higher surface curvature for the protein nanoencapsulation. The retention of protein structure on nanoparticles is also protein dependent. It is also reported that SWCNTs (single walled carbon nanotubes) stabilize the protein under harsh conditions like high temperature, organic solvents, etc. The surface of nanoparticles also provides the control over the structure and function of adsorbed chymotrypsin. The disruption of protein–protein interaction using surface-functionalized gold nanoparticles is reported with cytochrome and cytochrome c peroxidase. Cellular localization of protein encapsulated on nanoparticles is significantly different from those of small-molecule probes and is extremely sensitive to the surface charge of nanoparticles. The nanoparticles are localized in lysosome, nucleus, endosome in animal cells. Oleylated QDs are localized mainly in the cell membrane. However, coumarin-six loaded chitosan/GMO formulation is localized in nuclear material in MDA-MB-231 cells. The application of carbon-coated magnetic iron nanoparticles in pumpkin by pulverization induces the distribution of these nanoparticles to parenchyma, xylem vessels, cortex and epidermis of the plant cells. This observation provides some highlights in the cellular movement of nanoparticles in vascular plant.

2.11 THE STABILITY OF NANOENCAPSULATED PEPTIDE AND PROTEIN

The activity, structural composition and stability of protein nanomaterial interaction depend on the type of nanomaterial and protein itself. The protein stability may be influenced/improved by nanoparticles interaction. The protein stabilization agents, such as gelatin, BSA and other low-value proteins (for protein-based stabilization), mannitol (to regulate the osmotic pressure of the medium), PVP, PEG, Pluronic F-68, etc. are used for the improvement of stability of nanoencapsulated proteins. It is reported that fabricated amino acid functionalized gold nanoparticles modulates the activity of the chymotrypsin. Electrostatic nanocomplexes (100 nm) consisting of lactoglobulin and pectin showed a very good colloidal stability. The hydrophilic peptide insulin formulated in to biodegradable nanoparticles have been stabilized by the formation of a phospholipid complex. Influenza HA loaded PLGA nanoparticles reveals that the encapsulated proteins are stable inside the nanoparticles and does not degrade during the production process as well as on long term storage. Zhu et al., recently reported that combined physical and chemical immobilization of glucose oxidase in alginate microspheres improves stability of encapsulation and activity. The retention of alcohol dehydrogenases (ADH) activity was dependent on the concentration of ADH in the feed solution. However, the addition of other proteins, such as bovine serum albumin and -lactoglobulin, exhibited an additional improvement of the retention of ADH activity. The inlet and outlet temperature of the drying air was another key factor for the enzyme activity on the spray drying method of protein encapsulation. Biomimetic silica entrapment of chemically derivatized horseradish peroxidase for amperometric sensing applications shows very high stability in comparison to nanoencapsulated native enzymes. Itoh et al., group has reported the similar activity and much higher stability (four times) of catalase upon encapsulation in mesoporous silica synthesized in the pores of an alumina membrane.

2.12 THERAPEUTIC USE OF THE NANOPROTEIN MOLECULES

Proteomic technology to choose appropriate targets is an active area of research. However, till date no clinically effective targets have been identified. The fusion proteins, engineered antibody, dimerization, detection of specific conformation of target receptors together with nanocarriers encapsulation provides novel strategies to cure target cells. The use of peptide as targeting agents results in increased intracellular drug delivery in different murine tumor models. Likewise, several protein and peptide encapsulated nanosystems with different compositions and biological properties have been extensively investigated for drug and gene delivery applications.

The small size of nanoparticles was used to spy at the cellular machinery level with least interference in function. Protein nanotechnology have potential to use in the study of immunocomplex (antigen–antibody) formation, drug and gene delivery, biodetection of pathogen, detection of proteins, probing of DNA structure, tissue engineering, separation and purification of biological molecules and cells, etc. These peptide nanomedicines reduce the frequency of drug administration, and improved patient compliance. This would offer the protection and improved pharmacokinetics of easily degradable, short half-life of protein in vivo and in vitro conditions. Leaky blood vessels and poor lymphatic drainage in tumors provides the permeability of <200 nm nanoparticles in the tumor cells. Targeting cancer with mAb and its feasibility has been clinically demonstrated. Over 17 different mAb are approved by the US Food and Drug Administration (FDA) for cancer treatment. The rituximab (rituxan) and trastuzumab (herceptin binds to ErbB2 receptor) mAb was used for non-Hodgkin's lymphoma and breast cancer treatment respectively. An anti-VEGFmAb as angiogenesis inhibitor for treatment of colorectal cancer Bevacizumab (Herceptin) was approved in 2004. Over 200 delivery systems based on antibodies encapsulated on nanoparticles or their fragments are in preclinical and clinical trials. The tissue rejection could be minimized by using the specific protein coated nanoparticles. This nanoparticle reduces the chances of rejection as well as to stimulate the production of osteoblasts. One of the most exciting applications of protein nanoparticles in tissue engineering and regeneration are the ability of these nanostructures to provide a permissive environment for axonal regeneration in the central nervous system (CNS) after injury. When the peptide scaffold was applied to damage optic tracts in hamsters, regenerated axons could reconnect, and target tissues with sufficient density as to enable the functional return of vision.

2.13   CONCLUSION

Among all drug delivery systems the major challenge for new drug molecules is to improve their bioavailability and cellular uptake. Bio-nanotechnology has opened a wide range of applications for drug delivery systems. So, association of modern bio-nanotechnology and peptide science has become a key stone in therapeutic development and drug delivery systems. To improve drug-targeting properties the major constrain in bionanotechnology is surface functionalization which is usually complicated and rather ineffective. Decoration of well ordered nano structures with therapeutic peptide is excellent way to target the effective dose of drug at specific site with reduced side effects. This review will focus on the broad classification with the therapeutic application of peptide decorated nano conjugates. Various types of peptide nano structures are classified with their different methods of preparations and specified applications in therapeutic world.

KEYWORDS

- **Drug delivery**
- **Nanoparticle**
- **Peptide**
- **Targeting**

REFERENCES

1. Muro, S., Challenges in design and characterization of ligand-targeted drug delivery systems. *Journal of Controlled Release.* 2012, 164(2), 125–37.
2. Wu, H. C., Chang, D. K., Peptide-mediated liposomal drug delivery system targeting tumor blood vessels in anticancer therapy. *Journal of Oncology* 2010, 8, doi, 10.1155/2010/723798.
3. Cheng, R., Feng, F., Meng, F., Deng, C., Feijen, J., Zhong, Z., Glutathione-responsive nano-vehicles as a promising platform for targeted intracellular drug and gene delivery. *Journal of Controlled Release.* 2011, 152, 2–12.
4. Farokhzad, O. C., Cheng, J., Teply, B. A., Sherifi, I., Jon, S., Kantoff, P. W., Richie, J. P., Langer, R., Targeted nanoparticle-aptamer bioconjugates for cancer chemotherapy in vivo. *Proceedings of the National Academy of Sciences of the United States of America.* 2006, 103(16), 6315–20.
5. Yang, Y. Y., Wang, Y., Powell, R., Chan, P., Polymeric core-shell nanoparticles for therapeutics. *Clinical and experimental pharmacology & physiology.* 2006, 33(5–6), 557–62.
6. Moshfeghi, A. A., Peyman, G., A. Micro- and nanoparticulates. *Advanced drug delivery reviews.* 2005, 57(14), 2047–52.

7. Davda, J., Labhasetwar, V., Characterization of nanoparticle uptake by endothelial cells. *International journal of pharmaceutics*. 2002, 233(1–2), 51–9.

8. Delehanty, J. B., Boeneman, K., Bradburne, C. E., Robertson, K., Bongard, J. E., Medintz, I. L., Peptides for specific intracellular delivery and targeting of nanoparticles: implications for developing nanoparticle-mediated drug delivery. *Therapeutic Delivery*. 2010, 1(3), 411–433.

9. Vivès, E., Schmidt, J., Pèlegrin, A., Cell Penetrating and Cell Targeting Peptides in Drug Delivery. *Biochim biophysica Acta*. 2008, 1786(2), 126–38.

10. Harivardhan, R. L., Patrick, C., Nanotechnology for therapy and imaging of liver diseases. *Journal of Hepatology*. 2011, 55, 1461–1466.

11. Bolhassani, A., Potential efficacy of cell-penetrating peptides for nucleic acid and drug delivery in cancer. *Biochimica et Biophysica Acta*. 2011, 1816, 232–246.

12. Zhang, X. X., Eden, H. S., Chen, X., Peptides in cancer nanomedicine: Drug carriers, targeting ligands and protease substrates. *Journal of Controlled Release* 2012, 159, 2–13.

13. Lee, T. Y., Wu HANC., Tseng YUNL., Lin, C. T., A novel peptide specifically binding to nasopharyngeal carcinoma for targeted drug delivery. *Cancer Research*. 2004, 64(21), 8002–8008.

14. Sapra, P., Allen, T. M., Internalizing antibodies are necessary for improved therapeutic efficacy of antibody-targeted liposomal drugs. *Cancer Research*. 2002, 62(24), 7190–7194.

15. Chang DK., Chiu, C. Y., Kuo, S. Y., Antiangiogenic targeting liposomes increase therapeutic efficacy for solid tumors. *Journal of Biological Chemistry*. 2009, 284(19), 12905–12916.

16. Pastorino, D. D., Paolo, D. D., Piccardi, F., Enhanced antitumor efficacy of clinical-grade vasculature-targeted liposomal doxorubicin. *Clinical Cancer Research*. 2008, 14(22), 7320–7329.

17. Kondo, S., Yokomine, K., Aya, N., Sakagami, Y., Plant Analogs of the CLV3 Peptide: Synthesis and Structure–Activity Relationships Focused on Proline. *Cell Physiol*. 2011, 52(1), 30–36.

18. Kay, B. K., Kasanov, J., Yamabhai, M., Screening Phage-Displayed Combinatorial Peptide Libraries. *Methods*. 2001, 24(3), 240–246.

19. Eric, B. T., Dane, W. K., Yeast surface display for screening combinatorial polypeptide libraries. *Nature Biotechnology*. 1997, 15, 553–557.

20. Pandeya, S. N., Thakkar, D., Combinatorial chemistry: A novel method in drug discovery and its application. *Indian Journal of chemistry*. 2005, 44B, 335–348.

21. Edelstein, M., Scott PE., Sherlund, H., Hansen A. L., Hughes, J. T., Design considerations for pilot scale solid phase peptide synthesis reactors. *Chemical Engineering Science*. 1986, 41(4), 617–624.

22. Mark, G., Sneader, W., Pharmacokinetic Considerations in Rational Drug. *Design Clinical Pharmacokinetics* 1994, 26(3), 161–168.

23. Mrsny, R. J., Genentech, A. D., Proteins and Peptides. Pharmacokinetic, Pharmacodynamic, and Metabolic Outcomes. *Drugs and the Pharmaceutical Sciences Informa Healthcare* 2009, 278.

24. Lister, J. J., Moyer, B. R., Dean, R. T., Pharmacokinetic considerations in the development of peptide-based imaging agents. *Journal of nuclear medicine*. 1997, 41(2), 111–8.

25. Muñoz, A., Harries, E., Contreras-Valenzuela, A., Carmona, L., Read, N. D., Marcos, J. F., Two Functional Motifs Define the Interaction, Internalization and Toxicity of the Cell-Penetrating Antifungal Peptide PAF26 on Fungal Cells. *PLoS One.* 2013, 8(1), e54813.

26. Song, Y. C., Sun, G. H., Lee, T. P., Huang, J. C., Yu, C. L., Chen, C. H., Tang, S. J., Su, K. H., Arginines in the CDR of anti-dsDNA autoantibodies facilitate cell internalization via electrostatic interactions. *European journal of immunology.* 2008, 38, 3178–3190.

27. Kim, A., Shin, T. H., Shin, S. M., Pham, C. D., Choi, D. K., Cellular Internalization Mechanism and Intracellular Trafficking of Filamentous M13 Phages Displaying a Cell-Penetrating Transbody and TAT Peptide. *PLoS ONE.* 2012, 7(12), e51813.

28. Mitsuru, H., Fumiyoshi, Y., Pharmacokinetic considerations for targeted drug delivery. *Advanced Drug Delivery Reviews.* 2013, 65, 139–147.

29. Rajni, S., Kim, G. J., Nie, S., Shin, D. M., Nanotechnology in cancer therapeutics: bioconjugated nanoparticles for drug delivery. MOLECULAR CANCER THERAPEUTICS. 2006, 5, 1909–17.

30. Danhier, F., Ucakar, B., Magotteaux, N., Brewster, M. E., Préat, V., Active and passive tumor targeting of a novel poorly soluble cyclin dependent kinase inhibitor. *International journal of pharmaceutics.* 2010, 392(1–2), 20–8.

31. Sate, H., Sugiyama, Y., Tsuji, A., Horikoshi, I., Importance of receptor-mediated endocytosis in peptide delivery and targeting: kinetic aspects. *Advanced Drug Delivery Reviews.* 1996, 19, 445–467.

32. Anas, A., Okuda, T., Kawashima, N., Nakayama, K., Itoh, T., Ishikawa, M., Biju, V., Clathrin-mediated endocytosis of quantum dot-peptide conjugates in living cells. *ACS Nano.* 2009, 3(8), 2419–29.

33. Merzlyak, A., Lee, S. W., Engineering phage materials with desired peptide display: rational design sustained through natural selection. *Bioconjug chem.* 2009, 20(12), 2300–10.

34. Hruby V. J., Sharma, S. D., Designing peptide and protein ligands for biological receptors. *Current Opinion in Biotechnology.* 1991, 2, 599–605.

35. Kevin, H. M., Recent advances in the design and construction of synthetic peptides: for the love of basics or just for the technology of it. *Tibtech.* 2000, 18, 214–217.

36. David, B., Takashi, S., The design of peptide analogs for improved absorption. *Journal of controlled release.* 1994, 29, 283–291.

37. Rana, M., Chatarjee, S., Kochhar, S., Pereira, B. M. J., Antimicrobial peptide: A new dawn for regulating fertility and reproductive tract infection. *Journal of endocrinology and reproduction.* 2006, 10(2), 88–95.

38. Jon-Paul, S. P., Robert, E. W. H., The relationship between peptide structure and antibacterial activity. Peptides 2003, 24, 1681–1691.

39. Nissen-Meyer, J., Rogne, P., Oppegård, C., Haugen, H. S., Kristiansen, P. E., Structure-function relationships of the non-lanthionine-containing peptide (class II) bacteriocins produced by gram-positive bacteria. *Curr Pharm Biotechnol.* 2009, 10(1), 19–37.

40. Brännström, K., Öhman, A., Olofsson, A., Aβ Peptide Fibrillar Architectures Controlled by Conformational Constraints of the Monomer. *PLoS ONE.* 2011, 6(9):e25157.

41. Meyer, F. M., Collins, Borin, Bradow, J., Liras, S., Limberakis, C., Mathiowetz, A. M., Philippe, L., Price, D., Song, K., James, K., Biaryl-Bridged Macrocyclic Peptides: Conformational Constraint via Carbogenic Fusion of Natural Amino Acid Side Chains. Journal of organic chemistry 2012, 77(7), 3099–3114.

42. Hrub, V., Al, O. F., Kazmierski, W., Emerging Approaches, in the Molecular Design of Receptor-Selective Peptide Ligand: Conformational Topographical and Dynamic Considerations. Biochem Journal 1990, 268, 249–262.

43. Chaim, G., David, H., Michael, C., Zvi, S., Gerardo, B., Backbone cyclization: A new method for conferring conformational constraint on peptides. *Biopolymers.* 1991, 31(6), 745–750.

44. Freidinger, R. M., Perlow, D. S., Vaber, D. F., Protected lactam bridged dipeptides for use as conformational constraint in peptide. *Journal of organic chemistry.* 1982, 47, 104–109.

45. Suvit, T., Pals, T. D., Turner, S. R., Kroll, T. L., Conformationally constrained Rennin inhibitory peptides: Gamma lactam bridged dipeptide isosteres as conformational restriction. *Journal of medicinal chemistry.* 1988, 31, 1369–1376.

46. Alberto, B., Joseph, L., Bennett, J. A., Jacobson, H. I., Andersen, T. T., Design and synthesis of biologically active peptides: A 'tail' of amino acids can modulate activity of synthetic cyclic peptides. *Peptides.* 2011, 32, 2504–2510.

47. Feng, J., Li, Y., Shen, B., The design of antagonist peptide of hIL-6 based on the binding epitope of hIL-6 by computer-aided molecular modeling. *Peptides.* 2004, 25, 1123–1131.

48. Wang, Z. L., Zhang, S. S., Wang, W., Feng, F. Q., Shan, W. G., A Novel Angiotensin I Converting Enzyme Inhibitory Peptide from the Milk Casein: Virtual Screening and Docking Studies. *Agricultural Sciences in Chinas.* 2011, 10(3), 463–467.

49. Brasseur, R., Vanloo, B., Deleys, R., Lins, L., Labeur, C., Taveirne, J., Ruysschaert, J. M., Rosseneu, M., Synthetic model peptides for apolipoproteins. Design and properties of synthetic model peptides for the amphipathic helices of the plasma apolipoproteins. *Biochimica et Biophysics Acta.* 1993, 1170, 1–7.

50. Sood, D. V., Baker, D., Recapitulation and Design of Protein Binding Peptide Structures and Sequences. *J. Mol. Biol.* 2006, 357, 917–927.

51. Manuela, L. P., Emmanuel, L., Marina, R. A., Luis, S., Computer-aided Design of b-sheet *Peptides. Mol. Biol.* 2001, 312, 229–246.

52. Decaffmeyer, M., Lins, L., Charloteaux, B., VanEyck, M. H., Thomas, A., Brasseur, R., Rational design of complementary peptides to the hAmyloid 29–42 fusion peptide: An application of Pep Design. *Biochimica et Biophysica Acta.* 2006, 1758, 320–327.

53. Lam, K. S., Leb, M., Krchn, V., The "One-Bead-One-Compound" Combinatorial Library Method. Chem. Rev. 1997, 97, 411–448.

54. Vetter, S. W., Zhong, Y., Combinatorial Chemistry and Peptide Library Methods to Characterize Protein Phosphatases. *Methods in Enzymology.* 2003, 366, 260–283.

55. Lam KS., Lake, D., Salmon SE., Smith, J., Chen, M. L., Wade, S., Abdul, L. F., Knapp, R. J., Leblova, Z., Ferguson RD., Krchnak, V., Sepetov, N. F., Leb, M., A One-Bead One-Peptide Combinatorial Library Method for B-Cell Epitope Mapping. *A Companion to Methods in Enzymology.* 1996, 9, 482–493.

56. Lam, K. S., Michal, L., Synthesis of a One-Bead One-Compound Combinatorial Peptide Library. *Combinatorial Peptide Library Protocols: Methods in Molecular Biology.* 1996, 87, 1–6.

57. Lam, S., Leb, M., Krchn, V., The "One-Bead-One-Compound" Combinatorial Library Method Kit. *Chem. Rev.* 1997, 97, 411–448.
58. Mersich, C., Jungbauer, A., Generation of bioactive peptides by biological libraries. *J Chromatogr B Analyt Technol Biomed Life Sci.* 2008, 861(2), 160–70.
59. Jasna, R., Nicholas, J. B., Julian, S., Dragana, G., Marjorie, R., Filamentous Bacteriophage: Biology, Phage Display and Nanotechnology Applications. *Curr. Issues Mol. Biol.* 13, 51–76.
60. Molek, P., Strukelj, B., Bratkovic, T., Peptide Phage Display as a Tool for Drug Discovery: *Targeting Membrane Receptors. Molecules.* 2011, 16, 857–887.
61. Uchiyama, F., Tanaka, Y., Minari, Y., Tokui, N., Designing Scaffolds of Peptides for Phage Display Libraries. *Journal of Bioscience and Bioengineering.* 2005, 99(5), 448–456.
62. Barnes, C., Balasubramanian, S., Recent developments in the encoding and deconvolution of combinatorial libraries. *Current Opinion in Chemical Biology.* 2000, 4, 346–350.
63. Brown, M. J., Hoffmann, W. D., Alvey, C. M., Wood, A. R., Verbeck Guido, F., Petros Robby, A., One-bead, one-compound peptide library sequencing via high-pressure ammonia cleavage coupled to nanomanipulation/nanoelectrospray ionization mass spectrometry. *Analytical Biochemistry.* 2010, 398(1), 7–14.
64. Valerio, R. M., Benstead, M., Bray, M. A., Campbell, A. R., Maeji, N. J., Synthesis of Peptide Analogues Using the Multipin Peptide Synthesis Method. *Analytical Biochemistry* 1991, 197, 168–177.
65. Resmi, C. P., Xuan, H., Shao, Q. Y., Recent Advances in Peptide-Based Microarray Technologies. Combinatorial Chemistry & High Throughput Screening 2004, 7, 547–556.
66. Quesnel, A., Delmas, A., Trudelle, Y., Purification of synthetic peptide libraries by affinity chromatography using the avidin-biotin system. *Anal Biochem.* 1995, 231(1), 182–7.
67. Liu, R., Enstrom, A. M., Lam, K. S., Combinatorial peptide library methods for immunobiology research. *Experimental Hematology.* 2003, 31, 11–30.
68. Guzmán, F., Barberis, S., Illanes, A., Peptide synthesis: chemical or enzymatic. *Electronic Journal of Biotechnology.* 2007, 10(2), 279–314.
69. Saburo, A., Contemporary method for peptide and protein synthesis. *Current organic chemistry.* 2012, 5(1)45.
70. Nilsson BL., Soellner MB., Raines, R. T., Chemical synthesis of proteins. *Annual Reviews of Biophysics and Biomolecular Structure.* 2005, 34, 91–118.
71. Jeffrey, A. B., Gregg, B. F., Chemical synthesis of proteins. *Tibtech.* 2000, 18, 243–251.
72. Naahidi, S., Jafari, M., Edalat, F., Raymond, K., Khademhosseini, A., Chen, P., Biocompatibility of engineered nanoparticles for drug delivery. *Journal of Controlled Release.* 2013, 166, 182–194.
73. Dasha, M., Chiellini, F., Ottenbrite, R. M., Chiellini, E., Chitosan-A versatile semi-synthetic polymer in biomedical applications. *Progress in Polymer Science.* 2011, 36, 981–1014.
74. Fanga, Z. X., Bhandari, B., Encapsulation of polyphenols: a review. *Trends in Food Science & Technology.* 2010, 21, 510–523.

75. Patel, T., Zhou, J., Piepmeier J. M., Saltzman, W. M., Polymeric nanoparticles for drug delivery to the central nervous system. *Advanced Drug Delivery Reviews* 2012, 64, 701–705.

76. Ştefania, R., Silvia, V., Marcel, P., Cornelia, L., Revue, R., Polysaccharides based on micro- and nanoparticles obtained by ionic gelation and their applications as drug delivery systems. *De chimie.* 2009, 54(9), 709–718.

77. Fàbregas, A., Miñarro, M., García-Montoya, E., Pérez-Lozano, P., Carrillo, C., Sarrate, R., Sánchez, N., Ticó, J. R., Suñé-Negre, J. M., Impact of physical parameters on particle size and reaction yield when using the ionic gelation method to obtain cationic polymeric chitosan-tripolyphosphate nanoparticles. *International journal of pharmaceutics.* 2013, 446(1–2), 199–204.

78. Emmanuel, N., Koukaras, Sofia, A., Papadimitriou, Dimitrios, N., Bikiaris, George, E., Froudakis. Insight on the Formation of Chitosan Nanoparticles through Ionotropic Gelation with Tripolyphosphate. *Mol. Pharmaceutics.* 2012, 9, 2856−2862.

79. Shah, D., Londhe, V., Optimization and characterization of levamisole-loaded chitosan nanoparticles by ionic gelation method using 2(3) factorial design by Minitab 15. *Ther Deliv.* 2011, 2(2), 171–9.

80. Doustgani, A., Farahani, E. V., Imani, M., Doulabi, A. H., Dexamethasone Sodium Phosphate Release from Chitosan Nanoparticles Prepared by Ionic Gelation Method. *Journal of Colloid Science and Biotechnology.* 2012, 1(1), 42–50.

81. Katas, H., Shamiha, N. N., Dzulkefli, N., Sahudin Shariza. Synthesis of a New Potential Conjugated TAT-Peptide-Chitosan Nanoparticles Carrier via Disulfide Linkage. *Journal of Nanomaterials.* 2012, 7.

82. Bhalekar, M. R., Patil, K. P., Kshirsagar, S. J., Mohapatra, S., Formulation optimization and in vivo evaluation of mucoadhesive xyloglucan microspheres. *Pharmaceutical Chemistry Journal.* 2011, 45(8), 503–508.

83. Mainardes, R. M., Evangelista, R. C., PLGA nanoparticles containing praziquantel: effect of formulation variables on size distribution. *International Journal of Pharmaceutics.* 2005, 290, 137–144.

84. Takami, A., Masanori, B., Mitsuru, A., Biodegradable Nanoparticles as Vaccine Adjuvants and Delivery Systems: Regulation of Immune Responses by Nanoparticle-Based Vaccine. *Adv Polym Sci.* 2012, 247, 31–64.

85. Chia, W. L., Wen, J. L., Polymeric nanoparticles conjugate a novel heptapeptide as an epidermal growth factor receptor-active targeting ligand for doxorubicin. *Int J Nanomedicine.* 2012, 7, 4749–4767.

86. Mathew, A., Fukuda, T., Nagaoka, Y., Hasumura, T., Morimoto, H., Curcumin Loaded-PLGA Nanoparticles Conjugated with Tet-1 Peptide for Potential Use in Alzheimer's Disease. *PLoS ONE.* 2012, 7(3):e32616.

87. Dongfei, L., Sunmin, J., Hong, S., Shan, Q., Juanjuan, L., Qing, Z., Rui, L., Qunwei, X., Diclofenac sodium-loaded solid lipid nanoparticles prepared by emulsion/solvent evaporation method. *Journal of Nanoparticle Research.* 2011, 13(6), 2375–2386.

88. Powell, T. J., Palath, N., DeRome ME., Tang, J., Jacobs, A., Boyd, J. G., Synthetic nanoparticle vaccines produced by layer-by-layer assembly of artificial biofilms induce potent protective T-cell and antibody responses in vivo. *Vaccine.* 2011, 29, 558–569.

89. Laboutaa, H. I., Schneidera, M. Tailor-made biofunctionalized nanoparticles using layer-by-layer technology. *Int.J. Pharm.* 2010, 395(1–2), 236–242.

90. Oh, K. S., Lee, H., Kim, J. Y., Koo, E. J., Lee, E. H., Park, J. H., Kim, S. Y., Kim, K., Kwon, I. C., Yuk, S. H., The multilayer nanoparticles formed by layer-by-layer approach for cancer-targeting therapy. *J Control Release.* 2013, 165(1), 9–15.

91. Zhao, Q., Han, B., Wang, Z., Gao, C., Peng, C., Shen. Hollow chitosan-alginate multilayer microcapsules as drug delivery vehicle: doxorubicin loading and in vitro and in vivo studies. *J Nanomedicine.* 2007, (1), 63–74.

92. Prouty, M., Zonghuan, L., Carola, L., Yuri, L., Layer-by-Layer Engineered Nanoparticles for Sustained Release of Phor21-βCG(ala) Anticancer Peptide. *Journal of Biomedical Nanotechnology.* 2007, 3(2), 184–189.

93. Mohanta, V., Madras, G., Patil Layer-by-Layer Assembled Thin Film of Albumin Nanoparticles for Delivery of Doxorubicin. *J. Phys. Chem. C* 2012, 116(9), 5333–5341.

94. Yong-Hua, J., Yan, L., Shu, W., Kai, Z., Ying-Gang, J., Yu, F., Layer-by-Layer Assembly of Poly(lactic acid) Nanoparticles: A Facile Way to Fabricate Films for Model Drug Delivery. *Langmuir.* 2010, 26(11), 8270–8273.

95. Manju, S., Sreenivasan, K., Enhanced Drug Loading on Magnetic Nanoparticles by Layer-by-Layer Assembly Using Drug Conjugates: Blood Compatibility Evaluation and Targeted Drug Delivery in Cancer Cells. *Langmuir* 2011, 27(23), 14489–14496.

96. Glangchai, L. C., Caldorera-Moore, M., Shi, L., Roy, K. J., Nanoimprint lithography based fabrication of shape-specific, enzymatically-triggered smart nanoparticles. *Control Release.* 2008, 125(3), 263–72.

97. Nakamura, C., Miyamoto, C., Obataya, I., Takeda, S., Yabuta, M., Miyake, J., Enzymatic nanolithography of FRET peptide layer using V8 protease-immobilized AFM probe. *Biosensors and Bioelectronics* 2007, 22(9–10), 2308–2314.

98. Han, T. H., OK T, Kim, J., Shin, D. O., Ihee, H., Lee, H. S., Kim, S. O., Bionanosphere lithography via hierarchical peptide self-assembly of aromatic triphenylalanine. Small 2010, 6(8), 945–51.

99. Nurxat, N., Samia, M., Linglu, Y., Hiroshi, M., Biomineralization Nanolithography: Combination of Bottom-Up and Top-Down Fabrications to Grow Arrays of Monodisperse Au Nanoparticles along Peptide Lines on Substrates. *Angew Chem Int Ed Engl.* 2009, 48(14), 2546–2548.

100. Ginger, D. S., Zhang, H., Mirkin, C. A., The evolution of dip-pen nanolithography. *Angew Chem Int Ed* 2004, 43, 30–45.

101. Salaita, K., Wang, Y., Mirkin, C. A., Applications of dip-pen nanolithography. *Nat. Nano.* 2007, 2, 145–155.

102. Cho, Y., Ivanisevic, A., SiOx surfaces with lithographic features composed of a TAT peptide. *J Phys Chem B* 2004, 108, 15223–15228.

103. Cho, Y., Ivanisevic, A., TAT peptide immobilizationon gold surfaces: a comparison study with a thiolated peptide and alkylthiols using, AFM XPS, and FT-IRRAS. *J Phys Chem B,* 2005, 109, 6225–6232.

104. Cho, Y., Ivanisevic, A., Mapping the interaction forces between TAR RNA and TAT peptides on GaAs surfaces using chemical force microscopy. *Langmuir,* 2006, 22, 1768–1774.

105. Grabar, K. C., Freeman, R. G., Hommer, M. B., Natan, M. J., Preparation and Characterization of Au Colloid Monolayers. *Anal. Chem.* 1995, 67, 735–743.

106. Janice Duy, Laurie, B., Connell, Wolfgang Eck, Scott, D., Collins, Rosemary, L., Smith, J., Preparation of surfactant-stabilized gold nanoparticle–peptide nucleic acid conjugates. *Nanopart Res* 2010, 12, 2363–2369.

107. Patel, P. C., Gilijohann, D. A., Seferos, S. D., Mirkin, A. C., Peptide antisense nanoparticles. Proceeding of National academy of science 2008, 105(45), 17222–17226.

108. Tao, L., Xiuxia, H., Zhenxin, W., Application of Peptide Functionalized Gold Nanoparticles In: Functional Nanoparticles for Bioanalysis, Nanomedicine, and Bioelectronic Devices. *The ACS Symposium Series.* 2(4), 55–68.

109. Kotsuchibashi, Y., Zhang, Y., Ahmed, M., Ebara, M., Aoyagi, T., Narain, R. J. Fabrication of FITC-doped silica nanoparticles and study of their cellular uptake in the presence of lectins. *J Biomed Mater Res A.* 2013, 101(7), 2090–2096.

110. Stöber, W., Fink, A., Bohn, E., Controlled growth of monodisperse silica spheres in the micron size range. *Journal of Colloid and Interface Science* 1968, 26(1), 62–69.

111. Mrinmoy, D., Ghosh, S. P., Rotello, V. M., Applications of Nanoparticles in Biology. *Adv. Mater.* 2008, 20, 4225–4241.

112. Mascharak, S., Synthesis of fluprescent silica nanoparticle conjugated with RGD peptide for detection of invasive human breast cancer cells. *Young Scientist Journal,* 2010, 8, 37–41.

113. Kobayashi, Y., Katakami, H., Mine, E., Nagao, D., Konno, M., Liz-Marzán, L. M., Silica coating of silver nanoparticles using a modified Stober method. J Colloid Interface Sci. 2005, 283(2), 392–6.

114. Zhang, N., Chittapuso, C., Ampassavate, C., Siahaan, J. T., Berkland, C., PLGA Nanoparticle-Peptide Conjugate Effectively Targets Intercellular Cell-Adhesion Molecule-1. *Bioconjug Chem.* 2008, 19(1), 145–152.

115. Li-Yong, J., Li-Li, C., Le-Jian, W., Xiao-Ying, Y., Ri-Sheng, Y., Min-Ming, Z., Yong, Z. D., Actively-targeted LTVSPWY peptide-modified magnetic nanoparticles for tumor imaging. *International Journal of Nanomedicine.* 2012, 7, 3981–3989.

116. Mary, A. P. R., Narayanan, T. N., Sunny, V., Sakthikumar, D., Yoshida, Y., Joy, P. A., Anantharaman, M. R., Synthesis of Bio-Compatible SPION-based Aqueous Ferrofluids and Evaluation of Radio Frequency Power Loss for Magnetic Hyperthermia. *Nanoscale Research Letters* 2010, 5, 1706–1711.

117. Lee, S. J., Jeong, J. R., Shin, S. C., Huh, Y. M., Song, H. T., Su, J. S., Chang, Y. H., Jeon, B. S., Kim J-D. Intracellular translocation of superparamagnetic iron oxide nanoparticles encapsulated with peptide-conjugated poly. D, L lactide-co-glycolide. *Journal of Applied Physics* 2005, 97, 10Q913.

118. Yan, W., Guangfu, Y., Chuying, M., Zhongbing, H., Xianchun, C., Xiaoming, L., Yadong, Y., Hao, Y. Synthesis and Cellular Compatibility of Biomineralized Fe_3O_4 Nanoparticles in Tumor Cells Targeting Peptides. *Colloids Surf B Biointerfaces* 2013, 107, 180–188.

119. Gallarate, M., Trotta, M., Battaglia, L., Chirio, D. J., Preparation of solid lipid nanoparticles from W/O/W emulsions: preliminary studies on insulin encapsulation. *Microencapsul.* 2009, 26(5), 394–402.

120. Gallarate, M., Battaglia, L., Peira, E., Trotta, M., Peptide-Loaded Solid Lipid Nanoparticles Prepared through Coacervation Technique. *International Journal of Chemical Engineering.* 2011, 6.

121. Mahapatro, A., Singh, K. D., Biodegradable nanoparticles are excellent vehicle for site directed in-vivo delivery of drugs and vaccines. Journal of Nanobiotechnology 2011, 9, 55.

122. Reis, P. C., Neufeld, J. R., Ribeiro, A. J., Veiga, F., Nanoencapsulation, I., Methods for preparation of drug-loaded polymeric nanoparticles. *Nanomedicine: Nanotechnology, Biology, and Medicine.* 2006, 2, 8–21.

123. Chau, Y., Padera, F. R., Dang, M. N., Langer, R., Antitumor efficacy of novel polymer-peptide-drug conjugate in human tumor xenograft models. *Int, J., Cancer.* 2006, 118, 1519–1526.

124. Mahato, I. R., Narang, S. A., Thoma, L., Miller, D. D., Emerging Trends in Oral Delivery of Peptide and Protein Drugs Critical Reviews. *Therapeutic Drug Carrier System* 2003, 20(2–3), 153–214.

125. Aubin-Tam, M. E., Hamad-Schifferli, K., Structure and function of nanoparticle-protein conjugates. *Biomed Mater.* 2008, 3(3), 034001.

126. Wagner, S. C., Roskamp, M., Cölfen, H., Böttcher, C., Schlecht, S., Koksch, B., Switchable electrostatic interactions between gold nanoparticles and coiled coil peptides direct colloid assembly. *Org Biomol Chem.* 2009, 7(1), 46–51.

127. Majzik, A., Fülöp, L., Csapó, E., Bogár, F., Martinek, T., Penkea, B., Bíró, G., Dékány, I., Functionalization of gold nanoparticles with amino acid, -amyloid peptides and fragment. *Colloids and Surfaces B: Biointerfaces* 2010, 81, 235–241.

128. Pensa, E., Cortés, E., Corthey, G., Carro, P., Vericat, C., Fonticelli, H. M., Benítez, G., Rubert AA., Salvarezza, C. R., The Chemistry of the Sulfur–Gold Interface: In Search of a Unified Model. *Acc. Chem. Res.* 2012, 45(8), 1183–1192.

129. Park, J. H., Self-assembled nanoparticles based on glycol chitosan bearing 5 beta-cholanic acid for RGD peptide delivery. *Journal of Controlled Release.* 2004, 95(3), 579–88.

130. Wagner, C. S., Roskamp, M., Olfen, C. H., Ottcher, B. C., Schlecht, S., Koksch, B., Switchable electrostatic interactions between gold nanoparticles and coiled coil peptides direct colloid assembly. *Org. Biomol. Chem.* 2009, 7, 46–51.

131. Hildebrandt, N., Hermsdorf, D., Signorell, R., Schmitz, S. A., Diederichsen Ulf. Superparamagnetic iron oxide nanoparticles functionalized with peptides by electrostatic interactions. *Commemorative Issue in Honor of Prof. Lutz, F., Tietze on the occasion of his 65th anniversary* 2007, 5, 79–90.

132. Majzik, A., Fülöp, L., Csapó, E., Bogár, F., Martinek, T., Penke, B., Bíró, G., Dékány, I., Colloids and Surfaces B: Biointerfaces Functionalization of gold nanoparticles with amino acid, -amyloid peptides and fragment. *Colloids and Surfaces B: Biointerfaces* 2010, 81, 235–241.

133. Liu, B. R., Li, J. F., Lu, S. W., Leel, H. J., Huang, Y. W., Shannon, K. B., Aronstam, R. S., Cellular internalization of quantum dots noncovalently conjugated with arginine-rich cell-penetrating peptides. *J Nanosci Nanotechnol.* 2010, 10(10), 6534–43.

134. Hennrich, L. M., Boersema, J. P., Toorn, H. V. D., Mischerikow, N., Heck, J. R. A., Mohammed, S., Effect of Chemical Modifications on Peptide Fragmentation Behavior upon Electron Transfer Induced Dissociation. *Anal. Chem.* 2009, 81(18), 7814–7822.

135. Haliza, K., Nik NSND, Shariza, S., Synthesis of a New Potential Conjugated TAT-Peptide-Chitosan Nanoparticles Carrier via Disulfide Linkage. *Journal of Nanomaterials.* 2012, 7.

136. Dennis, M. A., Sotto, C. D., Mei, C. B., Medintz, L. I., Mattoussi, H., Bao, G., Surface Ligand Effects on Metal-Affinity Coordination to Quantum Dots: Implications for Nanoprobe *Self-Assembly*. *Bioconjug Chem.* 2010, 21(7), 1160–1170.

137. Delehanty, B. J., Medintz, L. I., Pons, T., Dawson, E. P., Brunel, M. F., Mattoussi, H., Self-Assembled Quantum Dot-Peptide Bioconjugates for Selective Intracellular Delivery. *Bioconjug Chem.* 2006, 17(4), 920–927.

138. Lee, B. C., Zuckermann, N. R., Templated display of biomolecules and inorganic nanoparticles by metal ion-induced peptide nanofibers. *Chem. Commun.* 2010, 46, 1634–1636.

139. Gemmill, B. K., Delehanty, B. J., Stewart, H., Susumu, K. M., Further progress in cytosolic cellular delivery of quantum dots. In: Colloidal Nanocrystals for Biomedical Applications, V. I. I., *Proc. SPIE* 2012, 82320B, 8232.

140. Sun, W., Fletcher D, van Heeckeren, R. C., Davis, P. B., Non-covalent ligand conjugation to biotinylated DNA nanoparticles using TAT peptide genetically fused to monovalent streptavidin. *J Drug Target.* 2012, 20(8), 678–90.

141. Holmberg, A., Blomstergren, A., Nord, O., Lukacs, M., Lundeberg, J., Uhlén, M., The biotin-streptavidin interaction can be reversibly broken using water at elevated temperatures. *Electrophoresis* 2005, 26(3), 501–10.

142. Langer, K., Coester, C., Weber, C., Von Briesen, H., Kreuter Preparation of avidin-labeled protein nanoparticles as carriers for biotinylated peptide nucleic acid. *Eur J Pharm Biopharm.* 2000, 49(3), 303–7.

143. Chen, W., Zhan, C., Gu, B., Meng, Q., Wang, H., Lu, W., Huimin, H., Targeted brain delivery of itraconazole via RVG29 anchored nanoparticles. *Journal of Drug Targeting* 2011, 19(3), 228–234.

144. Perrault, D. S., Chan, C. W. W., *In vivo* assembly of nanoparticle components to improve targeted cancer imaging. PNAS 2010, 107(25), 11194–11199.

145. Coester, C., Kreuter J, von Briesen, H., Langer, K., Preparation of avidin-labeled gelatin nanoparticles as carriers for biotinylated peptide nucleic acid (PNA). *Int J Pharm.* 2000, 196(2), 147–9.

146. Heitzmann, H., Richards, M. F., Use of the Avidin-Biotin Complex for Specific Staining of Biological Membranes in Electron Microscopy. *Proc. Nat. Acad. Sci. USA.* 1974, 71(9), 3537–3541.

147. Anders, H., Anna, B., Olof, N., Morten, L., Joakim, L., Mathias, U., The biotin-streptavidin interaction can be reversibly broken using water at elevated temperatures. *Electrophoresis* 2005, 26, 501–510.

148. Toti, S. U., Raja, G. B., Grill, E. A., Panyam, J., Interfacial Activity Assisted Surface Functionalization: A Novel Approach to Incorporate Maleimide Functional Groups and cRGD Peptide on Polymeric Nanoparticles for Targeted Drug Delivery. *Mol Pharm.* 2010, 7(4), 1108–1117.

149. Zou, J., Zhang, Y., Zhang, W., Ranjan, S., Sood, R., Mikhailov, A., Kinnunen, P., Pyykkö, I., Internalization of liposome nanoparticles functionalized with TrkB ligand in ratcochlear cell populations. *European Journal of Nanomedicine.* 2008, 3, 8–16.

150. Yaming, S., Liping, W., Yuhua, S., Hao, Z., Hongmei, L., Hanzhi, L., Bai, Y., Tianyu, L., Xuexun, F., Wei, L., NHS-mediated QDs-peptide/protein conjugation and its application for cell labeling. *Talanta.* 2008, 75, 1008–1014.

151. Park, S. N., Park, J. C., Kim, H. O., Song, J. M., Suh, H., Characterization of porous collagen/hyaluronic acid scaffold modified by 1-ethyl-3-(3-dimethylaminopropyl) carbodiimide cross-linking. *Biomat.* 2002, 23, 1205–1212.

152. Van, B. S. S., Van, E. M. B., Van, H. J. C., Staudinger ligation as a method for bioconjugation. *Angew Chem Int Ed Engl.* 2011, 50(38), 8806–27.

153. Sanne, W. A. R., Ingrid, V. B., Jos, M. H. R., Maarten MBMC. Efficient, chemoselective synthesis of immunomicelles using single-domain antibodies with a C-terminal thioester. *Biotechnol.* 2009, 9, 66.

154. Lin, R. Y., Dayananda, K., Chen, T. J., Chen, C. Y., Liu, G. C., Lin, K. L., Wang, Y. M., Targeted RGD nanoparticles for highly sensitive in vivo integrin receptor imaging. *Contrast Media Mol Imaging.* 2012, 7(1), 7–18.

155. Dominique, A. R., Harry, B., Conlin, P. O., Jeffrey, A. H., Biofunctional polymer nanoparticles for intra-articular targeting and retention in cartilage. *Nature Materials.* 2008, 7, 248–254.

156. Jiao, L., Meng, S., Molly, S., Shoichet. Click Chemistry Functionalized Polymeric Nanoparticles Target Corneal Epithelial Cells through RGD-Cell Surface Receptors. *Bioconjugate Chem.* 2009, 20, 87–94.

157. Maltzahn, V. G., Ren, Y., Park, J. H., Min, D. H., Ramana, K. V., Jayakumar, J., Fogal, V., Sailor J M, Ruoslahti, E., Bhatia, N. S., *In Vivo* Tumor Cell Targeting with "Click" Nanoparticles. *Bioconjugate Chem.* 2008, 19, 1570–157.

158. Nguyen, T. K. T., Luke, A. W., Functionalization of nanoparticles for biomedical applications. *Green Nano Today.* 2010, 5, 213–230.

159. Xiaoling Li, Bhaskara, R. J., Ligand based targeting Approaches to Drug Delivery. In: Design of Controlled Release Drug Delivery Systems. *McGraw-Hill publishers.* 2012, 12, 376–398.

160. Best, C. H., Scott, D. A., The Preparation of Insulin. *J. Biol. Chem.* 1923, 57, 709–723.

161. http://www.genengnews.com/gen-articles/focusing-on-task-of-reinventing-peptide-drugs/2538/.

162. Ingemann, M. F. S., Hovgaard, L., Bronsted, H., Peptide and Protein drug delivery Systems for Non-parenteral Routes of Administration. In Frokjaer, S., Hovgaard, L. (Ed.) 1986 Drug Metabolism Review.

163. Ma, W., Chen, M., Kaushal, S., McElroy, M., Zhang, Y., Ozkan, C., Bouvet, M., Kruse, C., Grotjahn, D., Ichim, T., Minev, B., PLGA nanoparticle-mediated delivery of tumor antigenic peptides elicits effective immune responses. *Int J Nanomedicine.* 2012, 7, 1475–1487.

164. Laroui, H., Dalmasso, G., Nguyen, H. T., Yan, Y., Sitaraman, S. V., Merlin, D., Drug-loaded nanoparticles targeted to the colon with polysaccharide hydrogel reduce colitis in a mouse model. *Gastroenterology.* 2010, 138(3), 843–53.

165. Siling, W., Tongying, J., Mingxin, M., Yanchen, H., Jinghai, Z., Preparation and evaluation of anti-neuroexcitation peptide (ANEP) loaded N-trimethyl chitosan chloride nanoparticles for brain-targeting. *International Journal of Pharmaceutics.* 2010, 386, 249–255.

166. Bartlett, L. R., Sharma, S., Panitch, B. S. A. Cell-penetrating peptides released from thermosensitive nanoparticles suppress pro-inflammatory cytokine response by specifically targeting inflamed cartilage explants. *Nanomedicine: Nanotechnology, Biology, and Medicine Nanomedicine* 2013, 9(3), 419–427.

167. Abdallah, M., Shiho, F., Yuichi, T., Hirofumi, T., *In vitro* and in vivo evaluation of WGA–carbopol modified liposomes as carriers for oral peptide delivery. *European Journal of Pharmaceutics and Biopharmaceutics* 2011, 77, 216–224.

168. Kiziltepe, T., Ashley, J. D., Stefanick, J. F., Qi, Y. M., Alves, N. J., Handlogten, M. W., Suckow, M. A., Navari, R. M., Bilgicer, B., Rationally engineered nanoparticles target multiple myeloma cells, overcome cell-adhesion-mediated drug resistance, and show enhanced efficacy in vivo. *Blood Cancer Journal.* 2012, 2:e64.

169. António, J. A., Eliana, S., Solid lipid nanoparticles as a drug delivery system for peptides and proteins. *Advanced Drug Delivery Reviews.* 2007, 59, 478–490.

170. Forssena, E., Willis, M., Ligand-targeted liposomes. *Advanced Drug Delivery Reviews* 1998, 29, 249–271.

171. Baillie, C. T., Winslet, M. C., Bradley NJ. Tumor vasculature: a potential therapeutic target. *British Journal of Cancer.* 1995, 72, 257–267.

172. Green, M., Loewenstein, P. M., Autonomous functional domains of chemically synthesized human immunodeficiency virus Tat trans-activator protein. *Cell.* 1988, 55(6), 1179–1188.

173. Frankel, A. D., Pabo, C. O., Cellular uptake of the Tat protein from human immuno-deficiency virus. *Cell.* 1988, 55(6), 1189–1193.

174. Dingwall, C., Ernberg, I., Gait, J. M., Green MS., Heaphy, S., Karn, J., Lowe, D. A., Singh, M., Skinner, A. M., Valerio, R., Human Immunodeficiency Virus 1 Tat Protein Binds Trans-activation-responsive Region(TAR) RNA *In Vitro.* *Proc. Nati. Acad. Sci.* 1989, 86, 6925–6929.

175. Subrizi, A., Tuominen, E., Bunker, A., Róg, T., Antopolsky, M., Urtti, A., Tat (48–60) peptide amino acid sequence is not unique in its cell penetrating properties and cell-surface glycosaminoglycans inhibit its cellular uptake. J Control Release. 2012, 158(2), 277–85.

176. Brooks, H., Lebleu, B., Vivès, E. L. P., Qianjun He, Jianan Liu, Yu Chen, Ming Ma, Linlin Zhang, Jianlin Shi. Tat peptide-mediated cellular delivery: back to basics. *Adv Drug Deliv Rev.* 2005, 57(4), 559–77.

177. Pan, L., He, Q., Liu, J., Chen, Y., Ma, M., Zhang, L., Shi, J., Nuclear-Targeted Drug Delivery of TAT Peptide-Conjugated Monodisperse Mesoporous Silica Nanoparticles. *J. Am. Chem. Soc.* 2012, 134(13), 5722–5725.

178. Nitin, N., Leslie, L., Won, J. R., Gang, B., Tat peptide is capable of importing large nanoparticles across nuclear membrane in digitonin permeabilized cells. *Ann Biomed Eng.* 2009, 37(10), 2018–2027.

179. Vives, E., Richard, J. P., Rispal, C., Lebleu, B., TAT peptide internalization: seeking the mechanism of entry. *Curr. Protein Pept. Sci.* 2003, 4, 125–132.

180. Wadia, J. S., Stan, R. V., Dowdy, S. F., Transducible TAT-HA fusogenic peptide enhances escape of TAT-fusion proteins after lipid raft macropinocytosis. *Nat. Med.* 2004, 10, 310–315.

181. Ray, A., Larson, N., Pike, B. D., Grüner, M., Naik, S., Bauer, H., Malugin, A., Greish, K., Ghandehari, H., Comparison of Active and Passive Targeting of Docetaxel for Prostate Cancer Therapy by HPMA Copolymer–RGDfK Conjugates. Mol. *Pharmaceutics* 2011, 8 (4), 1090–1099.

182. Omid, V., Forrest, M. K., Chen, F., Ni, M., Soumen, J., Matthew, C. L., Hyejung, M., Richard, G, E, James, O. P., Miqin, Z., Chlorotoxin bound magnetic nanovector

tailored for cancer cell targeting, imaging, and siRNA delivery. Biomaterials 2010, 31, 8032–8042.

183. Li-Yong, J., Li-Li, C., Le-Jian, W., Xiao-Ying, Y., Ri-Sheng, Y., Min-Ming, Z., Yong-Zhong, D., Actively-targeted LTVSPWY peptide-modified magnetic nanoparticles for tumor imaging. *International Journal of Nanomedicine.* 2012, 3981–3989.

184. Shuxian, S., Dan, L., Jinliang, P., Ye, S., Zonghai, L., Jian-Ren, G., Yuhong, X., Peptide ligand-mediated liposome distribution and targeting to EGFR expressing tumor in vivo. *International Journal of Pharmaceutics.* 2008, 363, 155–161.

185. Bing-Xiang, Z., Yang, Z., Yue, H., Lin-Min, L., Ping, S., Xin, W., Su, C., Ke-Fu, Y., Xuan, Z., Qiang, Z., The efficiency of tumor-specific pH-responsive peptide-modified polymeric micelles containing paclitaxel. *Biomaterials.* 2012, 33, 2508–2520.

186. Yao, L., Danniels, J., Moshnikova, A., Kuznetsov, S., Ahmed, A., Engelman, D. M., Reshetnyak, Y. K., Andreev OA pHLIP peptide targets nanogold particles to tumors. *Proc Natl Acad Sci USA.* 2013, 110(2), 465–70.

187. Zhaohui, W., Yang, Y., Wenbing, D., Jingkai, L., Jingrong, C., Hounan, W., Lan, Y., Hua, Z., Xueqing, W., Jiancheng, W., Xuan, Z., Qiang, Z., The use of a tumor metastasis targeting peptide to deliver doxorubicin-containing liposomes to highly metastatic cancer. *Biomaterials.* 2012, 33, 8451–8460.

188. Yunching, C., Jinzi, J. W., Leaf, H., Nanoparticles targeted with NGR motif deliver c-myc siRNA and doxorubicin for anticancer therapy. *Mol Ther.* 2010, 18(4), 828–834.

189. National Institutes of Health Autoimmune Disease Coordinating Committee Report. 2002. Bethesda (MD): The Institutes.

190. Larche, M., Peptide therapy for allergic diseases: basic mechanisms and new clinical approaches. *Pharmacology and Therapeutics.* 2005, 108, 353–361.

191. Larché, M., Allergen-derived T cell peptides in immunotherapy. Revue française d'allergologie et d'immunologie clinique 2003, 43, 59–63.

192. Chen, K., Chen, X., Integrin Targeted Delivery of Chemotherapeutics. Theranostics 2011, 1, 189–200.

193. Virginie, F., Laurence, P., Anne des, R., Vincent, P., Hélène, F., Valentine, W., Marie-Lyse, V., Christine, J., Alain, V., Jacqueline, M. B., Yves, J. S., Véronique, P., Targeting nanoparticles to M cells with non-peptidic ligands for oral vaccination. *European Journal of Pharmaceutics and Biopharmaceutics.* 2009, 73(1), 16–24.

194. Scheinman, I. R., Trivedi, R., Vermillion, S., Kompella, B. U., Functionalized STAT1 siRNA nanoparticles regress rheumatoid arthritis in a mouse model. Nanomedicine 2011, 6(10), 1669–82.

195. Zhang, N., Chittapuso, C., Ampassavate, C., Siahaan, T. J., Berkland, C., PLGA Nanoparticle-Peptide Conjugate Effectively Targets Intercellular Cell-Adhesion Molecule-1. *Bioconjug Chem.* 2008, 19(1), 145–152.

196. Garinot, M., Fiévez, V., Pourcelle, V., Stoffelbach, F., Rieux, A. D., Plapied, L., Theate, I., Freichels, H., Jérôme, C., Marchand-Brynaert, J., Schneider, Y. J., Préat, V., PEGylated PLGA-based nanoparticles targeting M cells for oral vaccination. *Journal of Controlled Release* 2007, 120, 195–204.

197. Sneh-Edri Hadas, Likhtenshtein Diana, Stepensky David. Intracellular targeting of PLGA nanoparticles encapsulating antigenic peptide to the endoplasmic reticulum of dendritic cells and its effect on antigen cross-presentation in vitro. *Mol Pharm* 2011, 8(4), 1266–1275.

198. Rubinstein, I., Sadikot, R., A novel peptide nanomedicine against acute lung injury: GLP-1 in phospholipid micelles. *Pharmaceutical research.* 2011, 28(3), 662–72.

199. Allon, N., Saxena, A., Chambers, C., Bhupendra, P., A new liposome-based gene delivery system targeting lung epithelial cells using endothelin antagonist. *Journal of Controlled Release.* 2012, 160, 217–224.

200. Gadi, S., Tal, D., Moshe, E., Ron, N. A., Smadar, C., Targeting of polymeric nanoparticles to lung metastases by surface-attachment of YIGSR peptide from laminin. *Biomaterials.* 2011, 32, 152–161.

201. Anna, H., Christine, R., Elisabeth, V. G., Matthias, S. L., Radolf, T., Thurner, G., Andreae, F., Prassl, R., Decristoforo, C., Targeting properties of peptide-modified radiolabeled liposomal nanoparticles. *Nanomedicine: Nanotechnology, Biology, and Medicine.* 2012, 8, 112–118.

202. Christopher, J. M., Smith, W. M., Griffiths, C. P., McKeown, B. N., Gumbleton, M., Enhanced pulmonary absorption of a macromolecule through coupling to a sequence-specific phage display-derived peptide. *Journal of Controlled Release.* 2011, 151, 83–94.

203. Melis, M., Spatafora, M., Melodia, A., Pace, E., Gjomarkaj, M., Merendino, A. M., Bonsignore, G., ICAM-1 expression by lung cancer cell lines: effects of upregulation by cytokines on interaction with LAK cell. *Eur Respir J* 1996, 9, 1831–1838.

204. Chuda, C., Sheng-XX, Abdulgader, B., Tatyana, Y., Teruna, J. S., Cory, J. B., ICAM-1 targeting of doxorubicin-loaded PLGA nanoparticles to lung epithelial cells. *European Journal of Pharmaceutical Sciences.* 2009, 37, 141–150.

205. Jean-Sébastien, T., Béatrice, H., Steffen, W., Mélanie, B., Julien, B., Sylvie, F., Francis, S., Benoît, F., Antitumor activity of liposomal ErbB2/HER2 epitope peptide-based vaccine constructs incorporating TLR agonists and mannose receptor targeting. *Biomaterials* 2011, 32, 4574–4583.

206. Bartneck, M., Ritz, T., Keul, H. A., Wambach, M., Bornemann, J., Gbureck, U., Gassler, N., Groll, J., Trautwein, C., Tacke, F., Effects of Nanorod-Conjugated Peptides For Cell Targeting Of Hepatic Macrophages In Acute And Chronic Murine Liver Injury. *Journal of Hepatology.* 2012, 56(2):S146.

207. Hayashi, Y., Yamauchi, J., Khalil, A. I., Kajimoto, K., Akita, H., Harashima, H., Cell penetrating peptide-mediated systemic siRNA delivery to the liver. *International Journal of Pharmaceutics.* 2011, 419, 308–313.

208. Friedman, L. S., Hepatic Stellate Cells: Protean, Multifunctional, and Enigmatic Cells of the Liver. *Physiol Rev.* 2008, 88(1), 125–172.

209. Feng, L., Jian-yong, S., Ji-yao, W., Shi-lin, D., Wei-yue, L., Min, L., Chao, X., Jian-ying, S., Effect of hepatocyte growth factor encapsulated in targeted liposomes on liver cirrhosis. *Journal of Controlled Release* 2008, 131, 77–82.

210. Du, S. L., Wang, J. Y., Pan, H., Lu, W. Y., Wang, J., Antifibrotic effects of cyclic RGD-peptide mediated liposomal interferon: an experimental on rats. *Zhonghua Yi Xue Za Zhi.* 2005, 85(15), 1015–20.

211. Takeshi, T., Mieko, I., Shigeru, K., Fumiyoshi, Y., Mitsuru, H., Novel PEG-matrix metalloproteinase-2 cleavable peptide-lipid containing galactosylated liposomes for hepatocellular carcinoma-selective targeting. *Journal of Controlled Release.* 2006, 111, 333–342.

212. Liang, H., Rongqin, H., Jianfeng, L., Shuhuan, L., Shixian, H., Chen, J., Plasmid pORF-hTRAIL and doxorubicin co-delivery targeting to tumor using peptide-conjugated polyamidoamine dendrimer. *Biomaterials.* 2011, 32, 1242–1252.

213. Shuhuan, L., Yubo, G., Rongqin, H., Jianfeng, L., Shixian, H., Yuyang, K., Liang, H., Chen, J., Gene and doxorubicin co-delivery system for targeting therapy of glioma. Biomaterials 2012, 33, 4907–4916.

214. Liang, H., Jianfeng, L., Shixian, H., Rongqin, H., Shuhuan, L., Xing, H., Peiwei, Y., Dai, S., Xuxia, W., Hao, L., Chen, J., Peptide-conjugated polyamidoamine dendrimer as a nanoscale tumor-targeted T1 magnetic resonance imaging contrast agent. *Biomaterials* 2011, 32, 2989–2998.

215. Yong-Zhong, D., Li-Li, C., Ping, L., Jian, Y., Hong, Y., Fu-Qiang, H., Tumor cells-specific targeting delivery achieved by A54 peptide functionalized polymeric micelles. *Biomaterials.* 2012, 33, 8858–8867.

216. Kelly, K. A., Bardeesy, N., Anbazhagan, R., Gurumurthy, S., Berger, J., Targeted Nanoparticles for Imaging Incipient Pancreatic Ductal Adenocarcinoma. *PLoS Med.* 2008, 5(4):e85.

217. Hongliang, X., Xinyi, J., Jijin, G., Xianyi, S., Liangcen, C., Kitki, L., Yanzuo, C., Xiao, W., Ye, J., Xiaoling, F., Angiopep-conjugated poly(ethylene glycol)-co-poly(e-caprolactone) nanoparticles as dual-targeting drug delivery system for brain glioma. *Biomaterials.* 2011, 32, 4293–4305.

218. Gao, H., Pan, S., Yang, Z., Cao, S., Chen, C., Jiang, X., Shen, S., Pang, Z., Hu, Y., A cascade targeting strategy for brain neuroglial cells employing nanoparticles modified with angiopep-2 peptide and EGFP-EGF1 protein. *Biomaterials.* 2011, 32, 8669–8675.

219. Weilun, K., Kun, S., Rongqin, H., Liang, H., Yang, L., Jianfeng, L., Yuyang, K., Liya, Y., Jinning, L., Chen, J., Gene delivery targeted to the brain using an Angiopep-conjugated polyethyleneglycol-modified polyamidoamine dendrimer. *Biomaterials* 2009, 30, 6976–6985.

220. Jinfeng, R., Shun, S., Dangge, W., Zhangjie, X., Liangran, G., Zhiqing, P., Yong, Q., Xiyang, S., Xinguo, J., The targeted delivery of anticancer drugs to brain glioma by PEGylated oxidized multi-walled carbon nanotubes modified with angiopep-2. *Biomaterials.* 2012, 33, 3324–3333.

221. Kun, S., Rongqin, H., Jianfeng, L., Liang, H., Liya, Y., Jinning, L., Chen, J., Angiopep-2 modified PE-PEG based polymeric micelles for amphotericin B delivery targeted to the brain. *Journal of Controlled Release.* 2010, 147, 118–126.

222. Hongliang, X., Xianyi, S., Xinyi, J., Wei, Z., Liangcen, C., Xiaoling, F., Anti-glioblastoma efficacy and safety of paclitaxel-loading Angiopep-conjugated dual targeting PEG-PCL nanoparticles. *Biomaterials* 2012, 33, 8167–8176.

223. Jinfeng, R., Shun, S., Dangge, W., Zhangjie, X., Liangran, G., Zhiqing, P., Yong, Q., Xiyang, S., Xinguo, J., The targeted delivery of anticancer drugs to brain glioma by PEGylated oxidized multi-walled carbon nanotubes modified with angiopep-2. *Biomaterials.* 2012, 33, 3324–3333.

224. Hongliang, X., Xianyi, S., Xinyi, J., Wei, Z., Liangcen, C., Xiaoling, F., Anti-glioblastoma efficacy and safety of paclitaxel-loading Angiopep-conjugated dual targeting PEG-PCL nanoparticles. *Biomaterials.* 2012, 33, 8167–8176.

225. Francesca, R., Ilaria, C., Silvia, S., Elisa, S., Alfredo, C., Mario, S., Pierre-Olivier, C., Moghimi, S. M., Massimo, M., Giulio Si. Functionalization with ApoE-derived peptides enhances the interaction with brain capillary endothelial cells of nanoliposomes binding amyloid-beta peptide. *Journal of Biotechnology.* 2011, 156, 341–346.

226. Francesca, R., Ilaria, C., Cristiano, Z., Silvia, S., Maria, G., Roberta, R., Barbara, L. F., Francesco, N., Gianluigi, F., Alfredo, C., Mario, S. M. M., Giulio, S., Functionalization of liposomes with ApoE-derived peptides at different density affects cellular uptake and drug transport across a blood-brain barrier model. *Nanomedicine: Nanotechnology, Biology, and Medicine.* 2011, 7, 551–559.

227. Yu, X., Liang, L., Xueqing, W., Jiancheng, W., Xuan, Z., Qiang, Z., Chloride channel-mediated brain glioma targeting of chlorotoxin-modified doxorubicin-loaded liposomes. *Journal of controlled release.* 2011, 152, 402–410.

228. Kievit, M. F., Veiseh, O., Fang, C., Bhattarai, N., Lee, D., Ellenbogen, G. R., Zhang, M., Chlorotoxin Labeled Magnetic Nanovectors for Targeted Gene Delivery to Glioma. *ACS Nano.* 2010, 4(8), 4587–4594.

229. Yu, X., Liang, L., Xueqing, W., Jiancheng, W., Xuan, Z., Qiang, Z., Chloride channel-mediated brain glioma targeting of chlorotoxin-modified doxorubicin-loaded liposomes. *Journal of Controlled Release.* 2011, 152, 402–410.

230. Xin-hua, T., Feng, W., Tian-xiao, W., Dong, W., Jun, W., Xiao-ning, L., Peng, W., Lei, R., Blood–brain barrier transport of Tat peptide and polyethylene glycol decorated gelatin–siloxane nanoparticle. *Materials Letters.* 2012, 68, 94–96.

231. Malhotra, M., Tomaro, D. C., Prakash, S., Synthesis of TAT peptide-tagged PEGylated chitosan nanoparticles for siRNA delivery targeting neurodegenerative diseases. *Biomaterials.* 2013, 34, 1270–1280.

232. Yao, Q., Huali, C., Qianyu, Z., Xiaoxiao, W., Wenmin, Y., Rui, K., Jie, T., Li, Z., Zhirong, Z., Qiang, Z., Ji, L., Qin, H., Liposome formulated with TAT-modified cholesterol for improving brain delivery and therapeutic efficacy on brain glioma in animals. *International Journal of Pharmaceutics.* 2011, 420, 304–312.

233. Ja-Young, K., Won, I. L. C., Young, H. K., Giyoong, T., Brain-targeted delivery of protein using chitosan- and RVG peptide-conjugated, pluronic-based nano-carrier. *Biomaterials.* 2013, 34, 1170–1178.

234. Son, S., Hwang, D. W., Singha, K., Jeong, J. H., Park, T. G., Lee, D. S., Kim, W. J., RVG peptide tethered bioreducible polyethylenimine for gene delivery to brain. *Journal of Controlled Release.* 2011, 155, 18–25.

235. Cheng, G., Xiangning, L., Lingling, X., Yu-Hui, Z., Target delivery of a gene into the brain using the RVG29-oligoarginine peptide. *Biomaterial.* 2012, 33, 3456–3463.

236. Yang, L., Rongqin, H., Liang, H., Weilun, K., Kun, S., Liya, Y., Jinning, L., Chen, J., Brain-targeting gene delivery and cellular internalization mechanisms for modified rabies virus glycoprotein RVG29 nanoparticles. *Biomaterial.* 2009, 30, 4195–4202.

237. Tosi, G., Vergoni, A. V., Ruozi, B., Bondioli, L., Badiali, L., Rivasi, F., Costantino, L., Forni, F., Vandelli, M. A., Sialic acid and glycopeptides conjugated PLGA nanoparticles for central nervous system targeting: *In vivo* pharmacological evidence and biodistribution. Journal of Controlled Release 2010, 145, 49–57.

238. Elina, T., Jenni, L., Andrey, M., VPH, Minna, K., Peptide-functionalized chitosan–DNA nanoparticles for cellular targeting. *Carbohydrate Polymers* 2012, 89, 948–954.

239. Quanyin, H., Guangzhi, G., Zhongyang, L., Mengyin, J., Ting, K., Deyu, M., Yifan, T., Zhiqing, P., Qingxiang, S., Lei, Y., Huimin, X., Hongzhan, C., Xinguo, J., Xiaoling, G., Jun, C., F3 peptide-functionalized PEG-PLA nanoparticles co-administrated with tLyp-1 peptide for anti-glioma drug delivery. *Biomaterials.* 2013, 34, 1135–1145.

240. Mathew, A., Fukuda, T., Nagaoka, Y., Hasumura, T., Morimoto, H., Curcumin Loaded-PLGA Nanoparticles Conjugated with Tet-1 Peptide for Potential Use in Alzheimer's Disease. *PLoS ONE.* 2012, 7(3): e32616.

241. Lippert, W. J., Vascular disrupting agents. *Bioorganic and Medicinal Chemistry* III 2007, 15, 605–615.

242. Xiang, C., Xianhuo, W., Yongsheng, W., Li, Y., Jia, H., Wenjing, X., Afu, F., Lulu, C., Xia, L., Xia, Y., Yalin, L., Wenshuang, W., Ximing, S., Yongqiu, M., Yuquan, W., Lijuan, C., Improved tumor-targeting drug delivery and therapeutic efficacy by cationic liposome modified with truncated bFGF peptide. *Journal of Controlled Release.* 2010, 145, 17–25.

243. Quanyin, H., Guangzhi, G., Zhongyang, L., Mengyin, J., Ting, K., Deyu, M., Yifan, T., Zhiqing, P., Qingxiang, S., Lei, Y., Huimin, X., Hongzhan, C., Xinguo, J., Xiaoling, G., Jun, C., F3 peptide-functionalized PEG-PLA nanoparticles co-administrated with tLyp-1 peptide for anti-glioma drug delivery. *Biomaterials.* 2013, 34, 1135–1145.

244. Zhang, H., Luo, J., Yuanpei, L., Henderson, P. T., Wang, Y., Wachsmann, H. S., Zhao, W., Lam KS., Pan, C. X., Characterization of high-affinity peptides and their feasibility for use in nanotherapeutics targeting leukemia stem cells. *Nanomedicine: Nanotechnology, Biology, and Medicine.* 2012, 8, 1116–1124.

245. Winer, I., Wang, S., Lee, Y. E. K., Fan, W., Gong, Y., Burgos-OD, Spahlinger, G., Kopelman, R., Buckanovich, R. J., F3-Targeted Cisplatin-Hydrogel Nanoparticles as an Effective Therapeutic That Targets Both Murine and Human Ovarian Tumor Endothelial Cells *In vivo. Cancer Res.* 2010, 70(21), 8674–83.

246. Shahin, M., Ahmed, S., Kaur, K., Lavasanifar, A., Decoration of polymeric micelles with cancer-specific peptide ligands for active targeting of paclitaxel. *Biomaterials* 2011, 32, 5123–5133.

247. Fan, Z., Xinglu, H., Lei, Z., Ning, G., Gang, N., Magdalena, S., Seulki, L., Hong, X., Andrew, Y. W., Khalid, A. M., Michael, G. R., Guangming, L., Xiaoyuan, C., Noninvasive monitoring of orthotopic glioblastoma therapy response using RGD-conjugated iron oxide nanoparticles. *Biomaterials.* 2012, 33, 5414–5422.

248. Zitzmann, S., Ehemann, V., Manfred, S., Arginine-Glycine-Aspartic Acid (RGD)-Peptide Binds to Both Tumor and Tumor-Endothelial Cells *in Vivo. Cancer Res.* 2002, 62, 5139.

249. Ruoslahti, E., RGD and other recognition sequences for integrins. *Annu Rev Cell Dev Biol.* 1996, 12, 697–715.

250. Zhang, L., Hoffman, J. A., Ruoslahti, E., Molecular Profiling of Heart Endothelial Cells. *Circulation.* 2005, 112(11), 1601–11.

251. Zhang, H., Kusunose, J., Kheirolomoom, A., Seo, J. W., Jinyi, Q., Watson, K., Lindfors, H., Ruoslahti, E., Sutcliffe, J. L., Ferrara, K. W., Dynamic imaging of arginine-rich heart-targeted vehicles in a mouse model Biomaterials. 2008, 29(12), 1976–1988.

252. Zhang, H., Li, N., Sirish, P., Mahakian, L., Ingham, E., Curry, F. R., Yamada, S., Chiamvimonvat, N., Ferrara, K. W., The cargo of CRPPR-conjugated liposomes crosses the intact murine cardiac endothelium. *Journal of Controlled Release.* 2012, 163, 10–17.

253. Lestini, B. J., Sagnella SM., Zhong, X., Shive MS., Richter NJ., Jayaseharan, J., Case, A. J., Kottke, M. K., Anderson, J. M., Marchant, R. E., Surface modification of liposomes for selective cell targeting in cardiovascular drug delivery *Journal of Controlled Release.* 2002, 78, 235–247.

254. Singh, S. R., Grossniklaus, H. E., Kang, S. J., Edelhauser, H. F., Ambati, B. K., Kompella, U. B., Intravenous transferrin, RGD peptide and dual-targeted nanoparticles enhance anti-VEGF intraceptor gene delivery to laser-induced, C. N. V., *Gene Ther.* 2009, 16(5), 645–659.

255. Lu, J., Shi, M., Shoichet, M. S., Click Chemistry Functionalized Polymeric Nanoparticles Target Corneal Epithelial Cells through RGD-Cell Surface Receptors. *Bioconjugate Chem.* 2009, 20, 87–94.

256. Melamangalam, S. J., Dhruba, J. B., Thangirala, S., Shaker, A. M., Nano chitosan peptide as a potential therapeutic carrier for retinal delivery to treat age-related macular degeneration. *Molecular Vision.* 2012, 18, 2300–2308.

257. Hossen, M. N., Kajimoto, K., Hidetaka, A., Hyodo, M., Ishitsuk, T., Harashim, H., Ligand-based targeted delivery of a peptide modified nanocarrier to endothelial cells in adipose tissue. *Journal of Controlled Release.* 2010, 147, 261–268.

258. Lu, S., Xu, X., Zhao, W., Wu, W., Yuan H, et al. Targeting of Embryonic Stem Cells by Peptide-Conjugated Quantum Dots. *PLoS ONE.* 2010, 5(8): e12075.

259. Zhang, X. Y., Chen, J., Zheng, Y. F., Gao, X. L., Kang, Y., Liu, J., Cheng, M., Sun, H., Xu, C., Follicle-Stimulating Hormone Peptide Can Facilitate Paclitaxel Nanoparticles to Target Ovarian Carcinoma *In Vivo. Cancer Res.* 2009, 69(16):OF1–9.

260. Jin, Y., Song, Y., Zhu, X., Zhou, D., Chen, C., Zhang, Z., Huang, Y., Goblet cell-targeting nanoparticles for oral insulin delivery and the influence of mucus on insulin transport. *Biomaterials.* 2012, 33(5), 1573–82.

261. Jianian, C., Shaoshun, L., Qi, S., Folic acid and cell-penetrating peptide conjugated PLGA–PEG bifunctional nanoparticles for vincristine sulfate delivery. *European Journal of Pharmaceutical Sciences.* 2012, 47, 430–443.

262. David, A., Kopeckova, P., Minko, T., Rubinstein, A., Kopecek, J., Design of a multivalent galactoside ligand for selective targeting of HPMA copolymer–doxorubicin conjugates to human colon cancer cells. *European Journal of Cancer.* 2004, 40, 148–157.

263. Wendy, U. W. J., Astrid, J. V. B., Bruijns SC, et al. Glycan-modified liposomes boost CD4+ and CD8+ T-cell responses by targeting DC-SIGN on dendritic cells. *Journal of Controlled Release* 2012, 160, 88–95.

264. Medha, D. J., Wendy, W. J. U., Astrid JVB, et al. DC-SIGN mediated antigen-targeting using glycan-modified liposomes: Formulation considerations. *International Journal of Pharmaceutics.* 2011, 416, 426–432.

265. Jun, Y., Kamihira, M., Iijimajournalo, S., Fermentationan, F. D., Enhancement of Transfection Efficiency Using Ligand-Modified Lipid Vesicles. *Bioengineering.* 1998, 85(5), 525–528.

266. Jain, A., Jain, S. K., *In vitro* and cell uptake studies for targeting of ligand anchored nanoparticles for colon tumors. *European Journal of Pharmaceutical Sciences.* 2008, 35, 404–416.

267. Christiane, B., Claus, M. L., Woodley, J. F., Lectin-mediated drug targeting: history and applications. *Advanced Drug Delivery Reviews* 2004, 56, 425–435.

268. Abdallah, M., Shiho, F., Yuichi, T., Hirofumi, T., *In vitro* and in vivo evaluation of WGA–carbopol modified liposomes as carriers for oral peptide delivery. *European Journal of Pharmaceutics and Biopharmaceutics.* 2011, 77, 216–224.

269. Abu-Daha, R., Schafer Ulrich, F., Lehr, C. M., Lectin-functionalized liposomes for pulmonary drug delivery: effect of nebulization on stability and bioadhesion. *European Journal of Pharmaceutical Sciences.* 2001, 14, 37–46.

270. Na, Z., Qineng, P., Guihua, Wenfang, X., Yanna, C., Xiuzhen, H., Lectin-modified solid lipid nanoparticles as carriers for oral administration of insulin. *International Journal of Pharmaceutics.* 2006, 327(1–2), 153–159.

271. Li, J., Wu, H., Hong, J., Xu, X., Yang, H., Odorranalectin is a Small Peptide Lectin with Potential for Drug Delivery and Targeting. *PLoS ONE.* 2008, 3(6): e2381.

272. Heidel, J. D., Yu, Z., Liu, J. Y. C., Rele, S. M., Liang, Y., Zeidan, R. K., Kornbrust, D. J., Davis, M. E., Administration in non-human primates of escalating intravenous doses of targeted nanoparticles containing ribonucleotide reductase sub-unit M2 siRNA. *Proc. Nat. Acad. Sci. USA.* 2007, 104, 5715.

273. Davis, M. E., Zuckerman, E. J., Choi JCH, et al. Evidence of RNAi in humans from systemically administered siRNA via targeted nanoparticles. *Nature.* 2010, 464(7291), 1067–1070.

274. Jiang, C., Youssef, J., Maya, K., Xu-bo, Y., Wei, F., Chun-sheng, K., Pei-yu, P., Didier, B., Characterization of endocytosis of transferrin-coated PLGA nanoparticles by the blood–brain barrier. *International Journal of Pharmaceutics.* 2009, 379, 285–292.

275. Huang, R. Q., Qu, Y. H., Ke, W. L., Zhu, J. H., Pei, Y. Y., Jiang, C., Efficient gene delivery targeted to the brain using a transferrin-conjugated polyethyleneglycol-modified polyamidoamine dendrimer. *FASEB J.* 2007, 21.

276. Gua, M., Yuan, X., Kang, C., Zhao, Y., Tian, N., Pu, P., Sheng, J., Surface biofunctionalization of PLA nanoparticles through amphiphilic polysaccharide coating and ligand coupling: Evaluation of biofunctionalization and drug releasing behavior. *Carbohydrate Polymers.* 2007, 67, 417–426.

277. Sungho, B., Kyungwan, M., Tae, H. K., Eun, S. L., Kyung, T. O., Eun-Seok, P., Kang, C. L., Yu, S. Y., Doxorubicin-loaded human serum albumin nanoparticles surface-modified with TNF-related apoptosis-inducing ligand and transferrin for targeting multiple tumor types. *Biomaterials.* 2012, 33, 1536–1546.

278. Salvati, A., Pitek, S. A., Monopoli PM, et al. Transferrin-functionalized nanoparticles lose their targeting capabilities when a biomolecule corona adsorbs on the surface. *Nature Nanotechnology.* 2013, 8, 137–143.

279. Huang, R., Ke, W., Liu, Y., Jiang, C., Pei, Y., The use of lactoferrin as a ligand for targeting the polyamidoamine-based gene delivery system to the brain. *Biomaterials.* 2008, 29(2), 238–46.

280. Aggarwal, A., Saraf, S., Asthana, A., Gupta, U., Gajbhiye, V., Jain, N. K., Ligand based dandritic system for tumor targeting. *Int J Pharm.* 2008, 350, 3–13.

281. Lee, H., Jang I.H., Ryu S.H., Park T.G. N-terminal site specific mono-PEGylation of epidermal growth factor. *Pharm Res.* 2003, 20, 818–825.
282. Lee, H., Park, T. G., Preparation and characterization of monoPEGylated epidermal growth factor: evaluation of in vitro biological activity. *Pharm Res.* 2002, 19, 845–851.
283. Bhirde, A. A., Patel, V., Gavard, J., Zhang, G., Sousa, A. A., Masedunskas, A., Leapman, R. D., Weigert, R., Gutkind, J. S., Rusling, J. F., Targeted killing of cancer cells in vivo and in vitro with EGF-directed carbon nanotube-based drug delivery. *ACS Nano.* 2009, 3, 307–316.
284. Klaus, U., Ulrich, S., Katrin, B., Alexander, W., Josef, A., Carbon nanoparticle-induced lung epithelial cell proliferation is mediated by receptor-dependent Akt activation. *AJP – Lung Physiol.* 2008, 294, L358–L367.
285. Kreuter, J., Shamenkov, D., Petrov, V., Ramge, P., Cychutek, K., Koch-Brandt, C., Alyautdin, R., Apolipoprotein-mediated transport of nanoparticle-bound drugs across the blood-brain barrier. *J Drug Target* 2002, 10(4), 317–25.
286. Michaelis, K., Hoffmann, M. M., Dreis, S., Herbert, E., Alyautdin, R. N., Michaelis, M., Kreuter, J., Langer, K., Covalent Linkage of Apolipoprotein E to Albumin Nanoparticles Strongly Enhances Drug Transport into the Brain. *JPET* 2006, 317(3), 1246–1253.
287. Cruz, L. J., Paul, J. T., Remco, F., Figdor, C. G., The influence of PEG chain length and targeting moiety on antibody-mediated delivery of nanoparticle vaccines to human dendritic cells. *Biomaterials.* 2011, 32, 6791–6803.
288. Hatakeyama, H., Akita, H., Ishida E, et al. Tumor targeting of doxorubicin by anti-MT1-MMP antibody-modified PEG liposomes. *International Journal of Pharmaceutics.* 2007, 342, 194–200.
289. Saul, J. M., Annapragada, V. A., Bellamkonda, R. V., A dual-ligand approach for enhancing targeting selectivity of therapeutic nanocarriers. *Journal of Controlled Release.* 2006, 114, 277–287.
290. Gupta, U., Dwivedi, S. K. D., Bid, H. K., Konwar, R., Jain, N. K., Ligand anchored dendrimers based nanoconstructs for effective targeting to cancer cells. *International Journal of Pharmaceutics.* 2010, 393, 185–196.
291. Liao, D., Liu, Z., Wrasidlo, W., Chen, T., Luo, Y., Xiang, R., Reisfeld, R. A., Synthetic enzyme inhibitor: a novel targeting ligand for nanotherapeutic drug delivery inhibiting tumor growth without systemic toxicity. *Nanomedicine.* 2011, 7(6), 665–73. http://www.sciencedirect.com/science/article/pii/S1549963411000980 – item1http://www.sciencedirect.com/science/article/pii/S1549963411000980 – item2http://www.sciencedirect.com/science/article/pii/S1549963411000980 – item3.
292. Liu, C., Chengzao, S., Huang, H., Janda, K., Edgington, T., Overexpression of Legumain in Tumors Is Significant for Invasion/Metastasis and a Candidate Enzymatic Target for Prodrug Therapy. *Cancer Res.* 2003, 63, 2957.
293. Lulu, C., Neng, Q., Xia, L., Kaili, L., Xiang, C., Li, Y., GuHe, Y. W., Lijuan, C., A novel truncated basic fibroblast growth factor fragment-conjugated poly (ethylene glycol)-cholesterol amphiphilic polymeric drug delivery system for targeting to the FGFR-overexpressing tumor cells. *International Journal of Pharmaceutics* 2011, 408, 173–182.

294. Xi, L., Hongyu, Z., Lei, Y., Guoqing, D., Atmaram, S., Pai-Panandiker, Xuefei Huang, Bing Yan. Enhancement of cell recognition in vitro by dual-ligand cancer targeting gold nanoparticles. *Biomaterials.* 2011, 32, 2540–2545.

295. Petra, K., Nataša, O., Mateja, C., Janko, K., Julijana, K., Targeting cancer cells using PLGA nanoparticles surface modified with monoclonal antibody. *Journal of Controlled Release.* 2007, 120, 18–26.

296. Murad, J. P., Espinosa, E. V. P., Ting, H. J., McClure, D., Khasawneh, F. T., A novel antibody targeting the ligand binding domain of the thromboxane A2 receptor exhibits antithrombotic properties in vivo. *Biochemical and Biophysical Research Communications.* 2012, 421, 456–461.

297. Shy, C. W., Koon, G. N., En-Tang, K., Daniel, W. P., Deborah, E. L., HER-2-mediated endocytosis of magnetic nanospheres and the implications in cell targeting and particle magnetization. *Biomaterials.* 2008, 29, 2270–2279.

298. Maya, S., Kumar, L. G., Sarmento, B., Rejinold, N. S., Menon, D., Nair, V. S., Jayakumar, R., Cetuximab conjugated O-carboxymethyl chitosan nanoparticles for targeting EGFR overexpressing cancer cells. *Carbohydrate Polymers.* 2010, 1–32.

299. Zhang, C., Zhao, L. Q., Dong, Y. F., Zhang, X. Y., Lin, J., Chen, Z., Folate-mediated poly (3-hydroxybutyrate-co-3-hydroxyoctanoate) nanoparticles for targeting drug delivery. *European Journal of Pharmaceutics and Biopharmaceutics.* 2010, 76, 10–16.

300. Gabizon, A., Shmeeda, H., Horowitz, T. A., Zalipsky, S., Tumor cell targeting of liposome-entrapped drugs with phospholipid-anchored folic acid–PEG conjugates. *Advanced Drug Delivery Reviews.* 2004, 56, 1177–1192.

301. Kesharwani, P., Tekade, K. R., Gajbhiye, V., Jain, K., Jain, N. K., Cancer targeting potential of some ligand-anchored poly(propylene imine) dendrimers: a comparison. *Nanomedicine: Nanotechnology, Biology, and Medicine* 2011, 7, 295–304.

302. Gupta, Y., Kohli, D. V., Jain, S. K., Vitamin B12-mediated transport: a potential tool for tumor targeting of antineoplastic drugs and imaging agents. *Crit Rev Ther Drug Carrier Syst.* 2008, 25(4), 347–79.

303. Nie, Z., Liu, K. J., Zhong, C. J., Wang, L. F., Yang, Y., Tian, Q., Liu, Y., Enhanced radical scavenging activity by antioxidant-functionalized gold nanoparticles: A novel inspiration for development of new artificial antioxidants. *Free Radical Biology and Medicine* 2007, 43, 1243–1254.

304. Lin, J. J., Chen, J. S., Huang, S. J., Ko, J. H., Wang, Y. M., Chen, T. L., Wang, L. F., Folic acid–Pluronic F127 magnetic nanoparticle clusters for combined targeting, diagnosis, and therapy applications. *Biomaterials.* 2009, 30, 5114–5124.

305. Farokhzad, C. O., Jon, S., Khademhosseini, A., Tran, T. T., LaVan, A. D., Langer Rt. Nanoparticle-Aptamer Bioconjugates: A New Approach for Targeting Prostate Cancer Cells. *Cancer Research* 2004, 64, 7668–7672.

306. Chen, T., Shukoor, I. M., Chen, Y., Yuan, Q., Zhu, Z., Zhao, Z., Gulbakan, B., Tan, W., Aptamer-conjugated nanomaterials for bioanalysis and biotechnology applications. *Nanoscale.* 2011, 3, 546–556.

307. Lamuraglia, M., Chami, L., Schwartz, B., Leclere, J., Lassau, N., Cancer nanotechnology: drug encapsulated nanoparticle-aptamer bioconjugates for targeted delivery to prostate cancer cells. *European Journal of Cancer Supplements* 2005, 3(2), 229–230.

308. Kashefi-Kheyrabadi, L., Mehrgardi, A. M., Aptamer-conjugated silver nanoparticles for electrochemical detection of adenosine triphosphate. *Biosensors and Bioelectronics.* 2012, 37, 94–98.

309. Yuana, Q., Lua, D., Zhanga, X., Chena, Z., Tan, W., Aptamer-conjugated optical nanomaterials for bioanalysis. *Trends in Analytical Chemistry.* 2012, 39, 72–86.

310. Lee, H. J., Yigit, V. M., Mazumdar, D., Lu, Y., Molecular diagnostic and drug delivery agents based on aptamer-nanomaterial conjugates. *Advanced Drug Delivery Reviews.* 2010, 62, 592–605.

CHAPTER 3

CPP AND CTP IN DRUG DELIVERY AND CELL TARGETING

CONTENTS

ABSTRACT

Current exploration of novel potential therapeutics that doesn't enter the clinic owing to poor delivery and low bioavailability has made their foundation in therapeutic development. Various technologies have been employed to improve cellular uptake of these therapeutic molecules, including cell-penetrating peptides (CPPs). These potential therapeutic molecules were discovered 20 years ago. Their discovery was based on the potency of several proteins to enter cells. Till now several CPPs have been investigated which can be categorized into two major classes (those require chemical linkage with the drug for cellular internalization and second includes formation of stable, non-covalent complexes with cargos). Current research explored the potential of CPPs in non-invasive cellular import of cargos. CPPs are very capable in delivering various molecules into cells. In addition they have been successfully employed for ex vivo and in vivo delivery of therapeutic molecules. Nevertheless, apart from some particular cases, their lack of cell specificity remains the major limitation for their clinical development. During this time several peptides with precise binding activity for a known cell line (cell-targeting peptides) have also been demonstrated in the literature. One of the primary objective of peptide mediated drug delivery based research is to optimize the tissue and cell delivery of therapeutic molecules by means of peptides which unite both targeting and internalization benefits.

3.1   INTRODUCTION

Peptides and proteins are now recognized as potential biological molecules to act as targeted drug therapies in the treatment of a variety of diseases. Nevertheless limitations such as low stability and cell permeability are often hinder their utilization in vivo. These are the major obstacles in pharmaceutical drug design and controlled drug delivery is the cellular uptake of biologically active molecules. Utilization of viral vectors and certain methods such as microinjection, electroporation, and liposome encapsulation with many therapeutic agents like proteins, peptides, and oligonucleotides have been successfully used to target cells. However these internalization strategies have a number of limitations such limitations includes high variability of drug expression among target cells, inefficient drug delivery, cellular damage and toxicity and restrictions due to drug and cell type. Owing to the high rate of cell mortality caused by the high-voltage pulses used during electroporation and the unavailability of specialized equipment for microinjection, physical methods are usually restricted. The first barrier which prevents the direct translocation of hydrophilic macromolecules into the intracellular environment is plasma membrane. Various processes such as endocytosis are exploited to transport a macromolecule into a cell in vivo, but the fate of an endocytosed macromolecule is unpredictable. There are possibilities of entrapment of such molecules within endosomes. This may cause degradation due to the low pH and digestive enzymes located within such vesicles. Therefore there is a urgent requirement of efficient and safe peptide/protein delivery systems that can overcome these problems and increase a therapy's bioavailability. This necessity leads to the exploration of potential vector compounds known as cell-penetrating peptides or protein transduction domains. Cell-penetrating peptides are the most effective transporter that are utilized to increase the uptake of various biologically active peptide/protein cargos upon fusion or attachment to its sequences.

Recent research has highlighted the potential of peptides especially cell-penetrating peptides in terms of their efficiency in delivering various molecules into cells. In addition to CPP, several peptides those are having specific binding activity for a given cell line have also been reported. Nevertheless in spite of few peptides, most of the peptides lack cell specificity and some show their binding efficiency with a given cell line. This problem can be eliminated by optimization of various factors involved during the binding or targeting of peptide to cell or tissue for the delivery of delivery of therapeutic molecules. In this chapter we describe the prominent strategies that are presently employed or likely to be employed in the near future to both target and deliver peptides and its conjugates at desired site. Cell-penetrating peptides are the unique class of diverse peptides that are typically made up of 5 to 30 amino acids and are efficient in crossing the cellular membrane. So far CPPs have been used for a variety of applications, for example, can act as vectors for siRNA nucleotides, small molecules, proteins, and for other peptides, both in vitro and in vivo. Role of CPPs is not only to carry a functional peptide inside the cell, but it can also incorporate a functional motif. Utilization of CPP is to mediate drug delivery into a specific cell type. This can be achieved by masking temporally before being exposed at a particular site of action. Exposed CPP can be activated by several factors such as local pH, the presence of specific enzymes or the mechanical release from a scavenging structure. Owing to non-specific interactions with ubiquitous cellular components, drug-CPP chimeras disperse all over the body. This limitation can be overcome by dosing high level of material with potential secondary effects. In spite of these limitations some excellent biological effects have been discovered such as *vivo* delivery of drugs loaded on CPPs. However peptide mediated drug delivery in to the desirable cell via peptide-recognition motif has not been broadly quantified in terms of efficacy. Reported doses of these materials are designed to obtain the expected biological responses but such high doses hardly represented in most of the studies. On the basis of several researches, it has been observed that future of peptide mediated drug delivery lie in the development of modular units in which CPPs and CTPs will be assembled and disassembled sequentially.

Generally mechanism of action of peptides can be described in two steps: first step involves in elevating the concentrate at the target place. This can be achieved by coating of the first part of a unit, which could be a polymeric structure with a cell-recognition peptide motif. During the second step drug loaded-CPP become free in close vicinity of the target by dismantled. This is achieved by using a localized device, such as hyperthermia or ultrasounds. After following these steps drug could be efficiently concentrated at the tumor site and then efficiently delivered into the cancer cells. Primary objective of peptide-based research is to prepare multi-unit entities peptide-based delivery system, which will be very precise in its function. In addition they can be triggered at the right moment in order to provide a more efficient and/or more tolerated therapeutic delivery strategy. Because of following advantages these peptide-based delivery vehicles remain very interesting to develop for drug delivery purpose:

Advantages of peptide-based delivery vehicles

- Peptides can be modified accordingly with carbohydrates, lipids or phosphate groups to of chemical groups from drugs or from structures aimed to transport drug.
- Peptides can be stored freeze-dried, thus limiting the problems of long-term storage.
- Their production can be carried out easily on a large scale (see example of the anti-HIV T20 peptide).
- Well-established analytical techniques such as analytical chromatographies and mass undesired effects due to side sequences, transport and distribution.

3.2   CELL-PENETRATING PEPTIDES

Recently ten new peptides have entered in the market. Out of these ten peptides four were marketed with global sales over US$3.18 billion for Copaxone, Lupron (US$2.12 billion), Zoladex (US$1.14 billion) and Sandostatin (US$1.12 billion). Additionally various peptides have entered clinical studies, but none of the peptides on the market today targets intracellular proteins, thus limiting the potential therapeutic space. It has been estimated that 10% of the druggable genome can be targeted by traditional rule-of-five small molecules, thereby leaving a large number of targets still untapped. However various peptides have been discovered to target protein–protein interactions. Out of these peptides few have made significant progress in clinical trials. Based on amino acid composition and 3D structure, variety of CPPs are present. Cationic, anionic, and neutral sequences based peptides is the most common classification of peptides. Variation in amino acid sequences has shown varying degrees of hydrophobicity and polarity. However specific groups of CPPs are related by high sequence identity and common structural features. Generally CPPs have no sequence homology, hence structural diversity results in different modes of uptake, and different levels of uptake depending on the cell line, and other conditions. In addition some other factor such as the type of cargo carried by a CPP, which can be covalently or non-covalently attached to the CPP, can also affect profoundly mode and levels of uptake. To gain entry into the cell, CPPs generally follow endocytosis and direct translocation through the cellular membrane. Endocytosis mediated delivery system occurs by various mechanisms such as clathrin-dependent endocytosis (CDE) and clathrin-independent (CDI). Clathrin-dependent endocytosis mediated delivery is mediated by the cytoplasmic domains of plasma membrane proteins which are recognized by adaptor proteins and packaged into clathrin-coated vesicles that are brought into the cell. Clathrin-dependent endocytosis can be further classified in to many forms such as such as macropinocytosis and caveolae and/or lipid raft-mediated endocytosis. It has been investigated that these pathways are involved in the uptake of CPPs.

In a usual process cellular internalization and recycling is equivalents to their cell surface one to five times per hour. Mechanism that supports cellular internalization facilitates the entry of peptides. Endocytic mechanism is usually followed by peptides to support the continuous internalization which promotes the strong affinity for the cell membrane. On the other hand, peptides having stable physical chemical properties might be able to cross the cell membrane directly, as similar to small molecules. It is extensively studied that the cell membrane is not a homogeneous double layer, for example, density of some regions are not uniform (some regions are denser; others are more fluid, owing to different lipid composition and lipid density). Varying composition, density and dynamics of lipids is dependent certain regions of the membrane and a variety of cellular pathways. Such a lipid profile variation or heterogeneity leads to different levels and modes of uptake. This may further depend on the conditions used for testing CPPs. Concentration dependent uptake of many cationic CPPs works in such a fashion so that CPPs concentration above concentration threshold results in rapid cytosolic uptake, signifies direct translocation, while at lower concentrations the uptake is primarily endocytic. Such type of behavior was observed in several cationic CPPs, including Penetratin™, Tat-derived peptides, R9, S413-PVrev and lactoferrin. Variation in CPP dependent concentration threshold was observed however it is generally present in the low micromolar range. In contrast it was documented that Penetratin have shown that translocation in CHO-K1 cells take place only below 2 mM. This research suggested that the cell-type (and thus the membrane composition) affects the balance between different internalization pathways. Cationic CPPs primarily entered in to cell by formation of electrostatic interactions with cell surface glycosaminoglycans (GAGs). Because of this electrostatic interactions clustering of GAGs at the cell surface occurs which may further trigger the activation of intracellular signals followed by the actin remodeling and cellular entry through a variety of internalization pathways ranging from direct translocation to endocytosis. During the investigation it was observed that the acid sphingomyelinase activation, followed by a change in the cell membrane lipid composition might be responsible for direct translocation of cationic CPPs. Extracellular concentrations above 10 mM often require for cell-based activity of peptides carried through cationic CPPs. This information supports the concept that certain endocytic pathways do not lead to therapeutically useful concentrations in the cytosol, unless the peptide can access some endosomal escape routes. Peptide entrapment in endosomes results in degradation, with their lack of ability to reach cytosolic targets. For achieving the therapeutic application the cytosolic concentration higher than 10 mM is not satisfactory which causes a major limitation for many CPPs. Just like other peptides CPPs also suffer from various limitations such as:

- *Short duration of action*: This is caused by proteolysis and rapid renal clearance. Proteolysis can be overcome by unnatural amino acids and conformational stabilization of the 3D structure, whereas reduction in the amount of free peptide in the plasma with various methods ('depot' formation in the site of injection and association with carrier proteins) prevents renal clearance.

- *Lack of oral bioavailability*: Novel routes of administration intranasal, inhalation and injectable depot formulations can be utilized to avoid the lack of oral bioavailability.

For achieving the efficient delivery across the plasma membrane administration of high quantities of drugs is required. This is done to obtain the expected intracellular biological effect. Hence modification of the translocation process across the plasma membrane will significantly reduce the quantity of drug to be administered, and the side effects on healthy tissues that are currently observed in most of the cases. It was discovered in 1988, Tat, the HIV transactivator of transcription protein, was the first sequence found to be capable of translocating cell membranes and gaining intracellular access. During the earlier times various proteins (when dissolved in extracellular medium) were translocated spontaneously through the plasma membrane. Among all, two proteins were extensively studied: Tat protein from the HIV-1 virus and the *Drosophila melanogaster Antennapedia homeodomain*, made up of 101 and 60 amino acids, respectively, and the minimal domain needed for translocation was defined. These sequences were further reduced to short sequences of 10 to 16 amino acids. Such short sequences has opened opportunities for the chemical synthesis of different mutants and analogs that are called cell-penetrating peptides (CPPs) or "protein transduction domains" (PTDs). Upon further investigation, it was shown that this sequence could be shortened to a few amino acids, referred to as the Tat peptide, without sacrificing translocation capacity. Such an achievement in peptide science has brought revolution which was further followed by the first use of CPPs as vectors when penetrating was employed for the delivery of a small exogenous peptide in 1994. Later on various CPPs have developed dramatically and the number continues to increase. Though selection of CPP often depends on the application at hand, some of the most commonly used peptides include Tat, penetrating, polyarginine and transportan. These CPP has the potential in delivering several proteins, nucleic acids, small molecule therapeutics, quantum dots, and MRI contrast agents and few types of cargo. Moreover, this potential and highly efficient translocation system has been observed in a variety of cell lines with minimal toxicity, overcoming challenges often faced with other delivery methods.

Several cellular biological investigations are evident for the fact that classical endocytosis pathway might be responsible for cellular internalization of hydrophilic macromolecules. In contrast several peptides have been studied for their translocation across the eukaryotic cells plasma membrane by a seemingly energy in dependent pathway. Such type of peptide can be effectively used for the intracellular delivery of macromolecules having molecular weights several times greater than their own. CPPs are efficient for a range of cell types and therefore exhibit potential therapeutic application. Owing to the low biomembrane permeability and their relatively rapid degradation, polypeptides and oligonucleotides are generally considered to be of limited therapeutic value. This pauses an obstacle in pharmaceutical industry. Certain chances in manipulation of the biological targets would increase, in case large hydrophilic compounds could be administered intracellularly. The ultimate goal to transport hydrophilic macromolecules into the cytoplasmic and nuclear compartments of living cells without disrupting the plasma membrane is not successfully achieved yet. Based on the physical chemical properties CPPs can be broadly classified according to net average charge and classified in to cationic, amphipathic and hydrophobic peptides. Anionic CPPs can be classified as hydrophobic or amphipathic CPPs. By contrast, many cationic CPPs are highly charged peptides, without any amphipathic arrangement or hydrophobic character.

3.3   DELIVERY OF CPPs INTO CELLS

Complications during the cellular uptake of biologically active molecules are the major obstacles that restrict the development of several drugs. Various techniques such as electroporation, microinjection, and liposome encapsulation have been used for delivering many therapeutic agents (like proteins, peptides, and oligonucleotides) that are used to target cells, but these internalization strategies have a number of limitations. Owing to several limitations such as high variability of drug expression among target cells, inefficient drug delivery, cellular damage and toxicity and restrictions due to drug and cell type these techniques are restricted. Additionally utilization of physical methods such as electroporation and microinjection, restricted due to the high rate of cell mortality (caused by the high-voltage pulses used during electroporation) and the unavailability of particular equipment required for microinjection. The first barrier that prevents the direct translocation of hydrophilic macromolecules into the intracellular environment is called as plasma membrane. Translocation of hydrophilic macromolecules into the intracellular environment is usually prevented by first barrier known as plasma membrane. These hydrophilic macromolecules are easily translocated into a cell in vivo through the most common pathway "Endocytosis," but the destiny of an endocytosed macromolecule is unpredictable. These molecules are initially trapped within endosomes and go through degradation. Level of degradation is dependent on the extent of 1 pH variation (especially degraded at acidic pH) and concentration of digestive enzymes located within such vesicles. Such difficulties can be overcome by the development of proteins, which can be easily travel across mammalian cell membranes. Development of protein is based on the amino acid sequences especially PTD; a region also known as the CPP. CPPs are a class of short (20–40 amino acids) cationic peptides that have the ability to permeate the cell membranes of many different mammalian cells types. Protein transduction domain has been successfully utilized to transport different biomolecules into cells. After internalization, CPPs maintained their activities to translocate biologically active molecules into cells, which make these peptides most promising candidates for drug delivery applications. The most prominent examples of CPPs are Penetratin (a 16 amino acid domain from the *Drosophila* Antennapedia protein), a flock house virus (FHV) coat peptide (sequence 35–49), and oligoarginines, and TAT peptide from HIV. Recent report suggested the utilization of amphipathic peptide carriers for intracellular protein delivery. Noncovalent binding of amphipathic peptides to their protein cargos, chiefly through hydrophobic interactions is the most simple, yet effective, aspect of these delivery systems. Ideally a synthetic carrier should fulfill the following criteria:

- Be able to protect proteins from protease degradation;
- Be able to release the protein efficiently within target cells;
- Be biodegradable and biocompatible;
- Be compatible with easy, large scale synthesis;
- Be functional for tissue or cell-type specific delivery;
- Be relatively stable for a long shelf life;
- Capable of interacting with protein cargos without the need for sophisticated procedures;
- Have a high drug loading capacity;
- Have a low toxicity and low immunogenicity.

Considering these points CPPs, we come to conclusion that CPPs, are emerging as superior carrier. Generally CPPs are short amino acid sequences, which are synthesized without the need for complicated chemistry or instrumental procedures. Arrangement of amino acids or their surface or their backbone can be chemically modified during synthesis to prevent degradation and improve bioavailability. Additionally the excellence of synthesized CPPs can be accurately restricted. Owing to the biocompatible and biodegradable properties, CPPs degrade in the body into naturally occurring molecules, thus producing relatively little toxicity. It has been reported these low molecular weight peptides are less immunogenic than other non-viral and viral carriers, therefore they are more suitable for repeated administration and can thus be used in the case of long-term therapies. Furthermore it has been investigated that, the specific tissue or cell-type delivery of therapeutic compounds can also be achieved by designing CPPs with different functional domains. These functional active peptide-based carriers are rationally designed with an objective to functionally deliver protein-based therapeutic agents to intracellular targets.

3.3.1 CPPs: CLINICAL DEVELOPMENT

Since 1988 several reports based on initial characterization of the protein transduction domain based on have been explored. Research conducted during this period bring these compounds at preclinical and clinical level however no CPP or CPP conjugate has passed the FDA hurdle and reached the clinics. First clinical trial on CPP was conducted for the topical treatment of psoriasis with cyclosporine–polyarginine conjugate. During this study it was observed that oligoarginine chimeric transporter facilitated the full penetration of cyclosporine into cells throughout the epidermis and dermis of human skin. In 2003 it was entered in Phase II clinical trials, but finally discontinued from the clinical trials. The first corporation that was succeeded in completing Phase II clinical trial (RT-001) was Revance Therapeutics. They have used a TATp cell penetrating-based platform technology, which allows topical delivery of botulinum toxin across the skin. Similarly many other products that are able to reach Phase II trials have been investigated during this period. Cell-penetrating peptides based AZX-100, which was explored by Capstone Therapeutics, mimics heat shock protein (HSP20) and bypasses the signaling pathways, leading to smooth muscle relaxation after topical application, evaluated in Phase II trials for prevention of dermal/keloid scarring. Afterwards various pharmaceuticals (KAI Pharmaceuticals, Avi Biopharma, Drais Pharmaceuticals, Novartis) were started involving in the production of CPP based therapeutics. These pharmaceuticals have successfully produced Cd inhibitor-TAT (47–57) conjugates for myocardial infarction, 6-aminohexanoic acid spaced oligoarginine [(R-Ahx-R)4], CPP–PMO conjugate for Duchenne muscular dystrophy and prodrug DTS-108 for cancer treatment. These products are still under clinical investigation

3.3.2 CPPs DRUG DELIVERY IN VIVO: PROBLEMS AND LIMITATIONS

CPPs has potential in mediating the in vitro cell uptake of diverse molecules into most cell lines however in vivo utilization appears much more complex mainly because of lack of cell specificity. Owing to the lack of specificity CPPs and their attached therapeutic compounds are dispersed almost all over the body independently of the way of administration. This type of distribution can be limited by the utilization of CPP. Earlier report confirmed that the parenteral administration of short Tat cell-penetrating sequence fused to *β-galactosidase* was found in the liver, kidney, lung and other tissues. Moreover, unexpectedly, these fused particles were also found in the brain, demonstrating that this fusion protein could cross also the blood brain barrier. Translocation of Tat-fusion protein across barrier in brain was again confirmed by Bcl-xL, a well-characterized death-suppressing molecule of the Bcl-2 family, also following IP injection. In another CPP based work it was discovered that CPP that are related to the protegrins family has been also revealed to pass across the blood brain barrier. Nevertheless no report has been revealed so far to elucidate accurately how CPPs and their cargoes could cross this highly selective membrane. Interestingly for such CPP-mediated in vivo transports no complete bio-distribution has been provided. Intensive inputs have been applied in improving the cell uptake of antibodies coupled to various CPPs. This is because of the tumor-targeting efficiency of most anticancer antibodies is strictly limited by their poor penetration into the tumor mass. Since it has been previously discussed that the conjugation between cell-specific antibodies and CPPs more likely to enhance their cellular internalization in vitro. In contrast such type of conjugation significantly reduced their selectivity in vivo since non-targeted tissues took up the chimeric construct through non-specific internalization mediated by the CPP. Therefore it was concluded that increasing the concentration of CPP copies grafted on a cell-specific monoclonal antibody (Mab), decreases the specificity of the Mab for the native cell. According to earlier report the considerable localization and retention of CPP-Mab fragments in non-targeted tissues in vivo have been confirmed, especially with a high peptide to antibody ratio. This concentration was achieved by the cell-penetrating "driving force" predominated over the specific antigen binding of the Mab fragments. In the similar way, Niesner et al. observed that the tumor to normal tissue ratio did not improve upon conjugation of the anti-p21 Mab with the Tat peptide. In the subsequent studies, Stein et al. conjugated a longer version of the Tat peptide spanning along residues [37–72] of the HIV-1 Tat protein to a Fab fragment of anti-tetanus toxin antibodies either by thioether or disulfide linkage. According to this work, it was suggested that only the disulfide conjugates effectively neutralized the toxin. More interestingly it was discovered that conjugation with disulfide showed a higher nuclear accumulation when compared to the thioether conjugate. On another hand it was investigated that intracellular reducing environment should be strong enough to cause the disulfide bridge reduction as shown in a previous study with a CPP peptide conjugate. Additionally one report suggested the some key differences in the body distribution, retention and metabolic fate of the conjugates depending on the type of CPP used. In one more investigation it was observed that Antennapedia or Tat peptides were co-administrated with sc(Fv)2 fragments, where there was no stable link between the CPPs and the sc(Fv)2 fragment, but a higher tumor retention was observed with the Antennapedia peptide and, to a lower extent, with the Tat CPP than with the sc(Fv)2 fragment alone. In further investigation it was observed that Tat peptide used in this study lacked one arginine residue, therefore, the lower tumor retention observed with Tat in this work should be reconsidered as the replacement of just a single cationic residue in Tat can greatly decrease its cell uptake. In one more study it was CPPs and cargo molecules were used to improve internalization of replication-deficient viruses for the therapeutic gene delivery.

This study was conducted in both in vitro and in vivo conditions; however no bio-distribution study was performed. So in conclusion there are some evidences that CPPs provoke a strong non-specific binding by sticking any molecule attached to them to non-targeted cells. This may lead to a dramatic loss of very "precious" material. This one of the major drawbacks for the use of these peptides as in vivo delivery vehicles if alternative strategies are not developed to promote cell-specific delivery.

3.3.3 CATIONIC CPPs: TOXICITY

Owing to high cationic nature some CPPs could be as toxic as other cationic polymers such as poly-L-lysine or poly-ethylene imine (PEI). It has been discovered that cationic Tat peptide has been shown to mediate some neurotoxicity. Thus several groups were evaluated for determining in vitro and in vivo cellular toxicity. In contrast, reports are also available on no toxicity or low toxicity (e.g., minimal membrane-translocating Tat domain causes no toxicity in very extreme conditions on HeLa. Similarly in various other reports it was proven that cationic CPP causes no toxicity (e.g., Toro et al. confirmed that the short Tat peptide was not toxic for lymphocytes at dose up to 300 μM). While differentiating in vivo and in vitro toxicities it's apparent that a number of studies have assessed the toxicity of cell penetrating peptides in vitro conditions. In vivo data are difficult to compare due to the variety of CPPs used and the differences between the cargoes attached. In addition, the modes of administration of the CPP-cargoes to the animals and the type of animal used, further complicates a delineation of toxic effects During in vitro studies comparative toxicity studies support the division of CPP into subgroups, namely RRPs and amphipathic peptides. Thus, the induction of membrane leakage by amphipathic peptides could be correlated with the hydrophobic moment. The evaluation of the toxicity of unmodified CPPs using a LDH-leakage, DiBAC4(3)–(membrane depolarization) and hemolytic assay showed rather severe toxic effects of MAP and transportan 10 as representatives of the amphipathic CPPs, but only mild effects of the RRPs TAT and penetrating. Using mouse myoblasts, oligoarginines consisting of minimal five and maximal 12 amino acids at different concentrations were analyzed for transduction and concomitant toxicological effects. Nona-arginine was identified as the oligoarginine of choice, combining high transduction frequencies with low short and long termed toxicological effects. Among the RRPs the toxicity decreases in the series oligoarginine>penetrating>TAT. The toxicological properties can be dramatically changed also upon the attachment of low molecular weight cargoes, for example, labels or other peptides. The toxicity of TAT fused to the anti-apoptotic Nemo-binding domain peptide as well as to the scrambled variant increased the detected toxicity in several cell types 100-fold. As the attachment of high molecular weight cargoes shifts the uptake mechanism to an endocytic pathway only, a reduction in toxicity in these cases most likely reflects the lower amount of bioavailable intracellular CPP-cargo.

It has been also suggested that utilization of different cationic peptides causes no toxicity in Hela or Jurkat cells with up to 20–30 μM concentrations. At a specific ratio of negative/positive charges, Si RNA transfected in vivo with a poly-arginine peptide did not induce any cellular. Moreover, frequent injections doses of fusion protein in mice corrected the metabolic disorder and immune defects with no apparent toxicity. In another study it was observed that primary or immortalized human keratocytes incubation with the Antennapedia or Tat peptide (with dosage up to 200 μM and 400 μM, respectively) resulted in no evidence of toxicity. Similarly primary human foreskin fibroblasts, human trabecular meshwork cell line, Vero, and HeLa cells have shown the same results. Further investigation in the similar work was performed by applying peptides to the cornea 4 times daily for 7 days under in vivo conditions, at concentrations 1000 times the IC50 values. It was observed that Antennapedia peptide showed no toxicity, whereas Tat caused some mild eyelid swelling. One report demonstrated the equivalent cell translocation potency of a short protamine derivative rich in arginine residues. This was equivalent to that of the Tat peptide combined with the absence of toxicity. Later on one more concentration dependent toxicity study was investigated by Cardozo et al., which showed that concentrations of up to 100 μM of Tat [48–57] were essentially harmless in all cells tested, whereas Antp [43–58] was significantly more toxic. Moreover, it's more significant that the concentration of peptide used against toxicity tests are above the concentrations used for delivering various drugs into cells (which range from 1 to 10 μM).

3.3.4 LOCAL CPP-MEDIATED DELIVERY

Non specific distribution of CPP-coupled molecules could be circumvented if the chimerical constructs or fusion proteins could be applied directly on the targeted cells, or at least in their close vicinity. Fusion proteins or chimeric proteins are proteins created through the joining of two or more genes that originally coded for separate proteins. These proteins combine whole peptides and therefore contain all functional domains of the original proteins. There are several examples that clearly reflect local application of CPPs could be compared to an in vitro experiment. First chimera constructs that were entered a phase II clinical trial in 2003 for the treatment of psoriasis under the commercial name of PsorBan®. However it has been studied later that in spite of an efficient uptake of the chimera, the release of the free drug was not rapid enough to compete with clearance. This strategy has been applied for the local delivery of drugs through a CPP in the vitreous body and in the sub-retinal space of the eye following intraocular injection of CPPs. This strategy of localization of chimera constructs over target cell exhibit advantage then intact CPP that diffusion of the CPPs is apparently strongly reduced. Similarly intracoronary injections of a Tat-delta protein kinase C inhibitor chimera have also been studied for the treatment of acute myocardial infarction. Later on various studies were performed.

- *Local intracoronary administration of CPP;*
- *Delivery to the lungs of siRNAs bound to CPPs;*
- *siRNA designed to turn off the expression on an intracellular protein*: 30 to 45% knockdown of the corresponding RNA was recorded. Surprisingly, higher doses of the chimera did not induce a stronger response.

Therefore it was concluded that Tat and Antennapedia-coupled siRNAs induce different cellular responses, reflecting apparently a different intracellular fate of these two conjugates

3.3.5 STRATEGIES TO HOME CPPs

Based on the physiological or biological features of the targeted organ or cell type various strategies have been adopted to bypass the absence of cell specificity of CPPs and to design molecular systems to improve targeting. During targeting and while reaching towards the target area CPP domain is first somehow hidden and then later fully exposed to promote an efficient internalization. This is done to avoid unspecific uptake of CPP. These set of objectives are usually applied against cancer cells which are very often targeted for a their metabolism and therapeutic purpose.

3.3.5.1   Utilizing Matrix Metalloproteases

Utilization of activatable cell penetrating imaging probes is a promising strategy for the in vivo detection of proteolytic activity in pathological conditions. A radiolabeled activatable cell penetrating peptide (ACPP) sensitive toward matrix metalloproteinases (MMP)-2 and -9 was successfully employed for MMP detection in the course of remodeling post-myocardial infarction in vivo. Besides activation in infarcted heart tissue, the MMP-2/9 sensitive ACPP probe also showed a considerable degree of activation in the vascular compartment, leading to background uptake of the activated probe in basically all tissues. This study was carried out in 2004, and become the first targeting system, which uses an "activatable" CPP. Potential of this strategy is dependent on the ability of a specific CPP to interact intramolecularly with a polyanionic counterpart through ionic interactions. Both ionic parts of this construct were linked together via a cleavable matrix metalloprotease (MMP) sequence, forming a hairpin. No cleavage occurred (due to the insufficient level of circulating MMPs) when this prepared construct was administered in the blood stream. In tumor surroundings the concentration of MMP is significantly. This was happens because they are secreted by tumor cells. Therefore the linker was cleaved and the ionic parts of the chimera could dissociate, and ultimately CPP ability to bind to the surrounding cells was restored. This CPP based molecular construct promotes an indirect, but more specific delivery system in vivo.

3.3.5.2   Utilizing Peritumoral Acidic pH

The pH of solid tumors is acidic due to increased fermentative metabolism and poor perfusion. It has been hypothesized that acid pH promotes local invasive growth and metastasis. Tumor microenvironment is characterized by low pH_e. Almost all solid tumors have a neutral to alkaline intracellular pH (pH_i), but they develop an acidic pH_e. The average pH_e could be as low as 6.0. A pH gradient ($pH_i > pH_e$) exists across the cell membrane in tumors. This gradient is contrary to that found in normal tissues, in which pH_i is lower than pH_e (7.2–7.4). Diffusion of the H^+ ions along concentration gradients from tumors into adjacent normal tissues creates a peritumoral acid gradient. The mechanisms responsible for the low pH_e include anaerobic glycolysis because of hypoxia, aerobic glycolysis, increased metabolic CO_2 production associated with uncontrolled cell growth, and increased activity of ion pumps on the cell membrane.

With the advent in peptide recent research it's now possible to hide the cellular "sticky opportunism" of cationic CPPs during their transport towards the tumor site. For this purpose ionic interactions between CPPs and their anionic counterparts have been exploited. Such type of delivery is mediated by: a conventional hydrophobic core made of a polymer into which any chemotherapeutic molecule can be incorporated, and a peripheral hydrophilic layer composed of poly-ethylene glycol and the Tat peptide. After the preparation of such system, anionic and ultra-sensitive di-block copolymer is then complexed to the cationic Tat. Such ionic interactions based complexation is expected to protect the cationic charges until slightly acidic microenvironment of the tumor triggers the protonation of the anionic moiety. Acidic microenvironment induced ionic dissociation and the successive exposure of the Tat peptide sequence, allows the preferential internalization of the drug-loaded polymer into the surrounding tumor cells. This process is further continued by using very pH sensitive sulfonamide (PSD) group. It has been observed that pH sensitive sulfonamide is fully protonated at pH 7.4 and becomes neutral at pH 6.8. According to earlier reports peritumoral pH around tumor is found to be about 6.9±0.14 at the tumor–host interface. Additionally this pH level has been dropped from 7.10 to 7.15 at only 200 μm away from the tumor. Tat peptide is advantageous in exhibiting its potential activity in close proximity of the tumor area. On the another side it has limitation over its volume of distribution. It has small volume of distribution, where the pro-drug becomes effective, as compared to the whole circulation. Thus pro-drug has its obligation to reach rapidly in the peritumoral area. Quick translocation across the compartment is required to fulfill its therapeutic effect prior to its renal and/or hepatic elimination. More information is required to furnish all the kinetic parameters of this interesting strategy. One report suggested the enhanced in vivo transfection of DNA using pH-sensitive Tat-modified pegylated (PEG) liposomes. Fundamentally it has been observed that attachment of PEG chains on the surface of liposome generates steric hindrances to shield the surface-attached Tat peptides and hence prevent the non-specific liposome/cell interactions. Coupling of PEG long chains to the liposome surface through a pH-sensitive hydrazone linker increases the chances of cleavage of the linker only in the acidic environment of the tumor. Intra-tumoral administration of these liposomes in tumor-bearing mice showed three times more efficient transfection than with the corresponding pH-insensitive system.

Role of the acidic pH has also been described to improve the cytoplasmic delivery of cargo molecules performed with CPPs. For this investigation entitled role of the acidic pH for a peptide derived from the HA2 glycopolypeptide of the influenza virus hemagglutinin was described. In this study lysosomal pH falls to 5 pH units, induced a conformational change of the HA2 derived-peptide, leading to a structural change and exposure of a fusion property. There are various alike studies have been reported that clearly reflect the significant role of acidic pH in CPPs delivery.

3.3.5.3   Utilizing Biological State of Targeted Cells

Pathological status of targeted cell plays an important role in designing the CPP based drug delivery system. Pathological status includes infection, or of a metabolic change induced by a pathological disorder. In this strategy toxic molecule can be coupled with CPP as a harmless prodrug. This could be activated only once inside a specific type of cells. This strategy was first adopted by Dowdy's group in 1999. In his work he has fused the Tat peptide to a caspase-3 protein precursor that could be activated only upon cleavage by the HIV protease. Synthesized chemical construct is converted in to its activated form only by the HIV protease that is exclusively present in HIV-infected cells, ultimately resulted in to the caspase-3 induced apoptosis of these cells. Prepared construct was reported to transduce efficiently about 100% of cells. In addition to this work one promising research, based on well-controlled in vivo experiment in humanized T cell SCID mice is required to definitively validate this approach. In this work same group also used the Tat peptide fused to the p27 tumor suppressor to evaluate the effect of its transduction on tumor proliferation in vitro and in vivo. Results suggested that indicated that cell cycle arrest of tumor cells was obtained in culture and an inhibition of the tumor growth was observed in mouse models.

3.3.6 LOCAL RELEASE OF CPP-LOADED DEVICES

Since it's already discussed that cationic CPPs have the undesired features of entering most of the cells, hence once administered, most of the peptides-drug conjugates will be internalized in the "wrong" cells. To overcome these limitations new chemicals are produced, which are often complex, certainly rather difficult to synthesize and/or fully characterize, and to market. For overcoming these problems, it is also possible to synthesize cell-penetrating systems representing a certain form of specificity, by simply allowing a CPP-loaded device to diffuse very close to the environment of the tissues to be targeted. This can be only possible in the case of solid tumors, or directly accessible organs and other localized pathology. Moreover, it can also possible for the treatment of several cutaneous pathologies, and possibly lung diseases. Additionally various other delivery systems like sustained release systems, including gel matrix, nanocapsules, liposomal structures, to name but a few, have been mentioned in the reports. In this strategy some could provoke the discharge of the encapsulated drug specifically in the region requiring the therapy, though the trouble of the swift and efficient internalization of these drugs into the targeted cells would remain. This problem can be circumvented by attaching a CPP to the therapeutic molecule. Several reports validated the potential of this strategy in combining a local delivery device and a CPP internalization system still remains very marginal, although some examples of its application have been described.

3.3.6.1 Role of Thermal Sensitive Polymers

Several alternative strategies have been adapted to proficiently direct drugs into a specific cell type with CPPs, for example, a chemotherapeutic

CPP coupled doxorubicin conjugate has been included in a macromolecular carrier. This elastin-like polypeptide carrier is thermal sensitive with a phase transition occurring between 39 and 42°C. This carrier based conjugate is adequately above the normal body temperature which avoids unwanted systemic aggregation. Hence, with the aid of different devices such as pulsed-high intensity focused ultrasounds (to induce external and focused hyperthermia), the Tat-drug conjugate can be released in the tumor environment and preferentially taken up by the tumor cells. Moreover, a spacer known as a Gly-Phe-Leu-Gly tetrapeptide was joined to the Tat peptide and the doxorubicin molecule. Gly-Phe-Leu-Gly tetrapeptide communicate to the target sequence of a protease belonging to the cathepsin family. As CPP get internalized via endocytosis, the lysosomal proteases are anticipated to cleave and release active drug moieties within the cell. It has been reported that drug potential was increased up to 20-fold when aggregation of ELP was induced by localized hyperthermia. Recent report also suggested feasibility to link carbon nanotubes to CPPs, to improve their cellular uptake, or to targeting-peptides to concentrate them at the tumor site.

3.3.6.2   Role of Ultrasound Sensitive Particles

Potential of ultrasounds can also be utilized to deliver drugs at specific location in body. More recently ultrasound sensitive particles exhibit two major applications. First strategy is dependent on the release of drug or genes at specified location where the sonication is applied. Ultrasound waves can reversibly enhance the micro-permeation of the membrane, hence inducing cellular transfer of the extracellular medium. More potential drug delivery of the molecule can be expected locally in the area of sonication, if the blood stream is loaded with the drug to be transfected into the cells. Noticeably sonication effects do not last long and this strategy mimics an in vitro transfection performed in vivo, thus leading to a higher degree of drug delivery in the cells within the treated area. Earlier reports have proven the potential of sonication in delivering cytotoxic agent, leading to a significant reduction of the injected dose. Nevertheless if this strategy applied to small drugs, fast clearance from the blood stream can be predicted, this will ultimately decrease the time window of the sonication effects. Thus to overcome this problem several efforts have been made to transport in the blood stream stable molecular structures that are entirely sensitive to ultrasounds. Such an attempt facilitates the entrapment of various drugs prior sonication, inducing their release only in the targeted environment. In addition simultaneous utilization of drugs conjugated to CPPs and sonication should promote the local cellular delivery. But this strategy only facilitates hydrophobic molecules (e.g., doxorubicin), can be effectively inserted in ultrasound sensitive structures. In contrast to hydrophobic molecules, because of their high cationic content, CPPs are relatively hydrophilic which restricts their inclusion unless their coupling to doxorubicin reduces significantly their hydrophilicity, or more extended hydrophobization of the CPPs is executed.

3.4 TYPES OF CPPs

CPPs have a great sequence variety and are diversely present in various forms therefore its now possible to identify three major classes of CPPs that are: cationic, amphipathic and hydrophobic. This classification is based on the physical–chemical properties of CPPs. According to the earlier reports 83% of the CPPs in this set have a net positive charge. Anionic CPPs do not form a class of their own and they are assigned to different classes on a case-by-case basis. Amphipathic CPPs, comprise of both cationic and anionic peptides, form the largest class (44%). Rests 15% of peptides are classified as hydrophobic. CPP is considered cationic if it contains a stretch of positive charges that is essential for uptake, and if the 3D arrangement does not lead to formation of an amphipathic helix. The net average charge of cationic and hydrophobic CPPs is close to +0.2. Based on the length of CPPs some includes long CPPs and some contains shorter CPPs. Long CPPs includes cationic Fushi-tarazu, contain a relatively shorter stretch of positive charges essential for uptake. Shorter CPPs, for example, hydrophobic BIP (Bax-inhibiting peptides) pentapeptides, contain a single positively charged residue. Hence based on these considerations CPPs with the similar average net charge are classified accordingly.

3.4.1   PROTEIN-DERIVED CELL-PENETRATING PEPTIDES

3.4.1.1   Penetratins: Homeodomain-Derived Peptides

Homeodomain-derived peptides are the class of transcription factors that bind with DNA through a structure of 60 amino acids in length. This structure of attachment is called as homeodomain. First evidence of cell-to-cell translocation of these transcription factors and their secretion as well as internalization by live cells was first observed during internalization of the homeodomain of Antennapedia (a Drosophila homeoprotein). Through site-directed mutagenesis it's possible to modify homeodomain, which helps in further understanding the mechanism of internalization. Site-directed mutagenesis can be used to modified its third helix (amino acids 43–58) was necessary and sufficient for translocation. This modification allows the development of penetrating. Structural and functional features of penetrating allow the access to active analogs that may further facilitate formulation of the inverted micelle internalization model. Whole procedure of internalization of homeoproteins, homeodomains and third helices of homeodomains occurs at 37°C and 4°C. These biomolecules cannot be saturated at this temperature which advocates that internalization does not require a chiral receptor protein. Nevertheless it has been reported that a-2,8-polysialic acid enhances the internalization of homeodomain peptides by increasing the local concentration of the homeodomain or by stabilizing the interaction of the homeodomain with membranes. In contrast it has been found that homeodomain peptides are also internalized by cells that do not express a-2,8-PSA; therefore, a similar function could be proposed for other complex sugars or negative charges that are exposed at the cell surface. Several peptides have been synthesized for understanding the exact mechanism of internalization of the third helix of the Antennapedia homeodomain.

3.4.1.2   Tat-Derived Peptides

Tat-derived peptide is a class of transcription-activating factor of 86–102 amino acids in length. This length is dependent on the viral strain. Tat-derived peptide is structured in three different functional domains:

- basic region (49–58 amino acids): responsible for nuclear import.
- cysteine-rich DNA-binding region (22–37 amino acids) with a zinc-finger motif.
- an acidic N-terminal region: important for transactivation.

Generally an uptake mechanism is time and concentration-dependent and can be partially inhibited by decreasing the temperature. However it has been discovered that chloroquine potentially protects Tat from degradation. Moreover, in some cells it potentially provokes the uptake suggesting that internalization of Tat can occur by endocytosis. In contrast some cells are completely dependent on the temperature for Tat uptake. This advocate the survival of different and, perhaps, competing uptake mechanisms. In addition to temperature, addition of polyanions such as heparin or dextrane sulfates reduces internalization of Tat. This result of polyanions might be a result of competition for charged molecules on the cellular membrane. Unexpectedly, uptake of Tat can also be stimulated by the addition of basic peptides such as protamine or Tat fragments. Present reports demonstrate the utilization of proteins both in vivo and in vitro. Till date various Tat-derived short peptides have been proven to translocate into the interior of different cell types. It has been observed that internalization occurs within minutes and is not distorted by decreasing the temperature to 4°C.

3.4.1.3 Signal-Sequence-Based Peptides

Signal sequence is a short (5–30 amino acids long) peptide present at the N-terminus of the majority of newly synthesized proteins that are destined towards the secretary pathway. These sequences are recognized by acceptor proteins that assist in dealing with the pre-protein from the translation machinery into the membrane of appropriate intracellular organelles. These proteins include those that reside either inside certain organelles such as Endoplasmic reticulum, Golgi or Endosome, secreted from the cell, or secreted into most cell membranes. Signal sequences that express proteins to the same intracellular compartment or mitochondria share some structural traits. The ER MTS contains 17–52 amino acids organized in an N-terminal positively charged section, a hydrophobic intersegment and a C-terminal polar region with peptidase recognition sites.

3.4.2 CHIMERIC OR SYNTHETIC CELL-PENETRATING PEPTIDES

3.4.2.1 Role of Transportan

Transportan is a chimeric CPP constructed from the peptides galanin and mastoparan. It has ability to internalize living cells carrying a hydrophilic load structural investigation revealed that well-defined alpha helix in the n-terminal domain. These arigine rich CPPs can form stable complexes with plasmid DNA in non covalent manner. A dominant example of transportan is galparan, which is a fusion between the neuropeptide galanin-1–13 and the wasp venom peptide mastoparan. This was developed under a program that was aimed at creating galanin receptor antagonists by using a chimerical strategy. Noticeably, it has shown that cells that are incubated with labeled galanin showed almost no intracellular labeling whereas those incubated with the biotinylated galparan analog were heavily labeled in both the cytoplasm and the nucleus. Therefore a novel peptide was named transportan. Cellular uptake is not blocked by unlabeled transportan and Uptake kinetics exhibit a rapid saturation of the cells. Additionally at concentrations <20 mM, transportan shows no apparent toxicity whereas transportan lowers *GTPase* activity in Bowes cell membranes at IC50 of 50 mM. For preventing this problem, an analog of transportan known as transportan 2, consisting of galanin-1–12 and lysine coupled to the inactive mastoparan analog Mas17 was synthesized. This synthesized product preserved the diffusion property but with a somewhat slower uptake.

3.4.2.2 Protein-Derived Cell-Penetrating Peptides

Cell-penetrating peptides are short and more hydrophilic peptides that get access to the intracellular environment. First objective of CPPs is the cellular internalization, which often involves the crossing of a biological membrane (plasma or vesicular), hence challenging the view of the non-permeability of these structures to large hydrophilic molecules. Secondly, CPPs can force the internalization of hydrophilic cargoes into cells. This step is a rate-limiting step in the development of many therapeutic substances. More interestingly, the two mostly used CPPs, TAT and Penetratin peptides, are derived from natural proteins HIV Tat and Antennapedia homeoprotein, respectively. Through the peptide research it was investigated that penetrating internalization is the most studied penetration mechanism. Internalization at lower temperature suggests that a mechanism that does not require specific chiral receptors or classical endocytosis exists. It's already been proven by internalization of the penetrating D-amino acid and in verso analogs. Reports confirmed that the amphipathicity is sufficient for membrane interaction, which is not necessary sufficient because substitution of the two tryptophan residues by two phenylalanines removes translocation. In conclusion, it was evident that removing one or three amino acids with prolines do not modify the internalization powerfully proposed that a helical structure is not obligatory.

3.4.2.3 Chimeric and/or Synthetic Cell-Penetrating Peptides

Discovery and potential of cell-penetrating peptides was initially demonstrated in 1991. In 1991, researchers demonstrated that Drosophila Antennapedia homeodomain could be internalized by neuronal cells. A 16-amino acid peptide, penetrating, was subsequently derived from this protein. Since then, the number of known natural and synthetic peptides with cell-penetrating capabilities has continued to grow. Peptides which are able to penetrate the cell membrane are known as cell-penetrating peptides (CPPs). CPPs can be broadly classified as protein-derived, chimeric (derived from two or more genes which are coded for separate proteins), or synthetic. CPPs share common features such as positively charged amino acids, hydrophobicity, and amphipathicity. Scheller et al. performed stepwise modifications of an amphipathic a-helix model peptide, previously shown to penetrate cells. In this study, this peptide was altered with respect to its helical parameters and molecular size and charge. Surprisingly, it was found that membrane association was not correlated with uptake efficiency. The results from this structure–function study implicate helical amphipathicity and a length of at least four complete helical turns as the only essential structural requirements for the internalization of amphiphilic a-helical peptides, and thus suggest a mechanism of uptake that differs from that of penetrate. The discovery of CPPs' ability to traverse the cell membrane opens up a new avenue for drug delivery. Attaching therapeutically significant biomolecules to CPPs provides a means to transport them across the cell membrane. A major breakthrough in the field was the delivery of peptide-nucleic acids (PNAs) using the chimeric peptide transportan. A variety of cargo molecules have been attached to CPPs for cellular delivery. These include plasmid, DNA, oligonucleotides, siRNA, PNAs, proteins, peptides, low-molecular-weight drugs, liposomes, nanoparticles, antibodies, enzymes, antibiotics, and enzyme substrates.

3.4.2.4   Model Amphipathic Peptides

In the early 1990s the phenomenon of the cell penetrating ability of peptides was approached in two independent ways. These studies provided the fundamentals for the rather accidental discovery of the cell penetrating ability of a further peptide group, later termed as Model amphipathic peptides by Lindgren et al. MAP group compromised simple synthetic amphipathic peptides able to enter cells in a similar manner, as was now known as special sequences selected from HIV-TAT protein. The third helix of the Antennapedia homoeo domain and fibroblast growth factor. When investigating the proposed direct contact between several peptides and G proteins, as suggested for the poly-cationic peptides mastoparan and substance P, Ohelke et al. designed an 18-mer amphipathic model peptide. This peptide crosses the plasma membranes of mast cells and endothelial cells by both energy-dependent and energy-independent mechanisms, and can act as an efficient transporter for different peptide cargoes. The model peptide shows perforations of the plasma membrane beginning at 4 mM; however, several analogs showing less toxicity and higher uptake have been synthesized.

3.4.2.5   Herpes Simplex Virus Type 1 (HSV-1) Protein VP22

The herpes simplex virus type 1 (HSV-1) VP22, is one of the most abundant HSV-1 tegument proteins with an average stoichiometry of 2400 copies per virion and conserved among *alphaherpesvirinae*. Many functions are attributed to VP22, including nuclear localization, chromatin binding, microtubule binding, induction of microtubule reorganization, intercellular transport, interaction with cellular proteins, such as template activating factor I (TAF-I) and nonmuscle myosin II A (NMIIA), and viral proteins including tegument protein. VP22 is a major structural component of HSV-1 possessing a remarkable property of transport between cells. After introduction of VP22 gene into the cells, the protein is synthesized and located predominantly in the cytoplasm in filamentous pattern colocalized with microtubules, while in the surrounding cells, it concentrates in the nucleus and binds chromatin. VP22 is involved with different functions such as intercellular transport, binding to and bundling of microfilaments, inducing cytoskeleton collapse, nuclear translocation during mitosis, and binding to chromatin and nuclear membrane.

3.4.2.6 Trans-Activating Transcriptional Activator (TAT)

CPPs are peptide vectors that can easily cross through the plasma membrane barrier without disturbing the integrity of the cell. The purpose of its delivery is to deliver various cargoes inside cell. Variety of cargoes can be transported intracellularly by CPPs including a extensive range of hydrophilic molecules and even nanosized entities, encompasses polymer-based systems, solid nanoparticles and liposomes. Several studies have been reported focused on CPPs such as penetrating, VP22, transportan, transactivating transcriptional activator peptide (TATp) and synthetic oligoarginines. These all types are having high inherent potential as intracellular delivery vectors. Nevertheless, the TATp remains the most accepted CPP used for a variety of reasons. TAT is a transcription-activating factor with 86 amino acids and contains a cysteine rich region. This region is significant for metal-linked dimerization in vitro whereas highly basic region involved in nuclear and nucleolar localization and RNA binding. Activity of Tat is dependent on its protein is encoded by two exons. The first exon which codes for the N-terminal 72 amino acids is sufficient for full TAT activity. These TAT proteins when supplemented to cell culture, was internalized by cells. Such internalization is followed by localization to the nuclei where it transactivated the viral promoter. Internalization process is dependent on the concentration and time-dependent in different types of cells. Various receptors (integrins, a 90-kDa cell surface protein or ash integrin) were reported for the binding of TAT to the cell surface. As far as its TAT peptide structure features are concerned, includes residues 37–72 and constitute a region with residues 38–49 that can adopt an α-helical structure with amphipathic features. Moreover, it also contains a cluster of basic amino acids 49–57. These amino acid sequences are minimal protein transduction domain (PTD), which does not overlap with the supposed amphipathic helical structure. Owing to the charge repulsion this cluster of basic amino acids is unstructured and contains a nuclear localization signal (NLS). One of the most essential element which is crucial for TAT uptake is α-helix forming basic domain, TATp (48–60), any deletion or alteration within the 49–57 sequence led to a reduced cellular internalization. Therefore protein transduction sequence for TAT encompasses residues 47–57.

3.4.2.7   Homeodomain of Antennapedia (Antp)

These are the group of transcription factors that efficiently bind DNA through a specific sequence of 60 amino acids. These sequences are called as homeodomain. Structurally it contains three α-helices. Differentiation between these two helices is confirmed by h-turn, which is called a recognition helix since it is concerned in the interaction of the homeodomain with their cognate binding sites. To investigate the exact mechanism of homeoproteins on neural development, it was reported that homeodomain of Antennapedia, a Drosophila homeoprotein, was internalized by mammalian nerve cells and accumulated in their nuclei modifying the morphology of the neurons. Last helix, third helix of the antennapedia homeodomain is contain penetrating, is the 16-m peptide and involved in the translocation process.

3.5   CPPs PENETRATION: IN VITRO AND IN VIVO

Translocation of CPPs across membranes has been investigated by means of cultured cells, artificial lipid vesicles, tissues and in vivo. For this purpose various types of cells have been used. According to earlier reports CPPs can be effectively internalized in primary cells (from rat spinal cord and, rat brain, human umbilical vein endothelium, calf aorta, porcine and in osteoclast culture) however most often, cell lines have been used. For such an investigation no special cell cultivating measures are required for internalization studies with cell lines. In this procedure cells are grown to 70–80% confluence in flasks, cover slips or dishes. Different layers of cells can be incubated with a solution of CPP as such. Or else they can also be detached from the surface and a suspension of cells prepared earlier to incubation with CPP in order to ensure homogeneously available cell surface. Nevertheless reports confirmed that the detachment of cells may have an impact on the kinetics and effectiveness of internalization. Numbers of broken cells obtained from the scraping and trypsin treatment could affect cell surface proteins therefore internalization of CPP is always examined directly by confocal microscopy or other imaging techniques. In this step the cells are usually fixed by formaldehyde or paraformaldehyde-milder fixation agents than acetone and methanol. Earlier criticism on the affect of mild fixation on the internalization of some CPPs has also been reported. Whereas internalization of penetrating, Tat and transportan was observed in both fixed and live cells from several cell lines. Few internalization-based experiments have been in vivo in whole organisms and even fewer ex vivo in isolated tissue. Isolated blood vessels were used for ex vivo tissue experiments. Earlier findings suggested that the endothelial tissue from isolated mouse aorta rings efficiently took up the caveolin-1 scaffolding domain peptide attached to penetrating. This finding resulted in inhibition of acetylcholine-induced vasodilatation and nitric oxide production. It has been observed that most in vivo experiments were carried out with penetrating and Tat. In addition to cells obtained from blood vessel, penetrating was effectively employed in vivo to enter peritoneal exudate cells and to reach brain and spinal cord cells, in some cases also by passing the blood–brain barrier.

3.5.1 MECHANISM OF INTERNALIZATION OF CPP

Internalization mechanics of CPP has not been clearly determined yet. In addition to the some common characteristics of these peptides, their structural diversity, mainly their highly cationic nature, has fuelled the concept that the transduction mechanism is not the same for CPPs of different types. Various reports evidenced CPP internalization in cell lines and their translocation by endocytotic pathways. This internalization have been carried out under conditions that should prevent active transport of CPPs. Translocation by endocytotic pathways was effectively carried out at low temperatures and in the presence of many different inhibitors of endocytosis. It has been proven that the role of endocytosis in internalization of CPP is not negligible; hence CPPs endocytosis could be an exclusive or alternative mechanism of internalization. Recent studies proven that internalization of penetrating and protegrin-1 derived SyrB peptides into the live cells is interrelated to endocytotic processes and that Tat derived CPPs do not enter live cells at low temperature and are not internalized into liposomes. Current findings explored that Tat derived CPPs enter cells mainly by lipid raft mediated macropinocytosis. This is encouraged by cell-surface binding of Tat derived CPPs. Penetratin and Tat follow inverted micelle mechanism. In this mechanism positively charged peptides interact with negatively charged phospholipids to convert part of the membrane into an inverted micelle structure. This inverted micelle structure can release on either the intracellular or the extracellular side of the membrane. In contrast internalization of some CPPs emerges to be very much affected by the membrane composition.

3.6   CELLULAR UPTAKE OF CPPs

Several reports have been focused on the mechanism of CPP uptake across the plasma membrane a exclusive pathway for translocation remains elusive. Various studies demonstrated that different properties of peptides (molecule length and charge delocalization, as well as the properties of the associated cargo, such as size and charge) can have an important influence on the mechanism of peptide uptake. Using multiple cell lines and various incubation procedures it's now possible to select optimal CPP for a given application. It has been evident that CPP may use multiple modes of cellular entry that can depend on the context of the experimental conditions. The different modes of mechanics broadly classified into two groups: energy-independent direct translocation and energy-dependent endocytosis across the membrane bilayer. Because energy-dependent endocytosis can result in endosomal sequestration and decreased bioavailability, a complete knowledge of the effect of peptide and cargo physiochemical properties on the mechanism of uptake is essential for the rational design of potential delivery vehicles. Current study suggested the cellular uptake mechanism of three CPPs (antennapedia/penetrating peptide, nona-arginine, and Tat), focusing on endocytic pathway is not sufficient, and therefore clathrin-mediated endocytosis, macropinocytosis and caveloae/lipid raft mediated endocytosis all occurred. Additionally CPP sequence and concentration utilized also plays a little role in uptake mechanism. Noticeably the two peptides Tat and R9, shared common pathways of both endocytic as well as direct mechanisms of uptake. It has been found that at low peptide concentrations, only inhibitors of clathrin-mediated endocytosis did not have an effect on internalization. This suggests that cellular import was mediated by both macropinocytosis and caveloae/lipid raft mediated endocytosis. Reports justified that higher peptide concentrations for penetrating were not sufficient to induce direct uptake. Incomplete blockage of endocytosis was also essential to direct access to this pathway. Correspondingly, for the Tat and R9 peptides, inhibition of macropinocytosis and caveloae/lipid raft-mediated endocytosis increased direct uptake. However it has been reported that perturbation of endocytotic mechanisms encourage peptide accumulation on the membrane for cellular import via direct uptake. This work differentiates uptake mechanisms in a systematic and controlled fashion. Moreover, it clearly highlights the roles that cargo, cell line, cell density and numerous other factors play in CPP uptake mechanisms. Another CPP translocation mechanism that has been studied by single molecule spectroscopy to monitor cellular uptake allowed the comparisons to be made between molecular transporters that used endocytosis versus those that used direct uptake for membrane translocation. Through this investigation, it was concluded that even at low peptide concentrations (as low as 1 nM), multiple import mechanisms are implicated. We have already mentioned above that direct uptake is the other dominant mechanism of translocation harnessed by CPPs. This direct uptake mechanism is not just passive diffusion across the membrane-rather it is driven by plasma membrane potential. In order to examine this mechanism, studies aimed at understanding the interaction of CPPs with the cellular membrane are required. There are several other ways to understand the internalization mechanism of CPP. To understand CPP uptake it is clear that further work is essential to understand the variables affecting modes of internalization. One of the most attractive approaches would be to use model yeast organisms (*Saccharomyces cerevisiae*) for this purpose. CPPs have been efficiently internalized into *S. Cerevisiae* and this organism is genetically tractable and versatile. Moreover, *S. cerevisiae* has been mainly useful in understanding the mechanisms involved in endocytosis. Hence this model organism could serve as a potential tool for understanding the endocytic mechanisms involved in CPP import. Additionally the foundation of yeast gene deletion arrays has made the view of determining novel factors required to CPP import a possibility by allowing for testing of CPP uptake in multiple models of endocytosis dysfunction in high throughput.

3.6.1 DELIVERY OF PROTEINS

Potential of TAT peptide (residues 1–72 or 37–72) can be defined by their efficiency in delivering heterologous proteins, such as horseradish *peroxidase, RNase* A, h-galactosidase, and domain III of Pseudomonas exotoxin A (PE), into the cytoplasm of different cell types in vitro. In vivo evaluation of TAT-h galactosidase chimeras resulted in delivery to several tissues, with high levels in liver, heart, and spleen, low-to-moderate levels in skeletal muscle and lung, and little or no activity in kidney and brain. This therapeutic peptide showed their potential as a therapeutic and prophylactic vaccine. Delivery of exogenous proteins is limited due to their access the MHC class I processing pathway, hence it is difficult to design a protein-based vaccine that induces class I-restricted cytotoxic T-lymphocyte (CTL) response. Nevertheless after their conjugation with the antigenic protein to TATp (49–57), such as TAT-ovalbumin conjugate, the conjugate was processed by antigen presenting cells. This conjugation leads to the effective killing of the target cells by antigen-specific CTLs.

3.6.2 DELIVERY OF DNA

Peptide mediated gene is promising method delivery for treating human disease by introducing genetic material into targeted cells. Moreover, by this method defective genetic substance can be replaced or it can also provide the genetic coding for functions the cell had been incapable of performing. Major difficulty in gene therapy is the delivery of genetic material across the plasma membrane and then across the nuclear membrane with least toxicity. Genetic transformations through viral vectors are efficient although limited due to safety concerns. Non-viral vectors can be an alternative approach to viral vectors but have problems of endosomal escape and nuclear translocation. Such difficulties can be overcome by using synthetic peptide based gene delivery systems. In peptide based delivery system, DNA is present in condense form which facilitate the release of plasmid from endosomes by destabilizing the lipid bilayers at low pH. There are various examples of such research proving the successful genetic delivery through peptides, for example, synthetic amphipathic peptides such as, KALA, (LARL) and the Hel-peptide facilitated gene delivery in different cell lines in vitro, though these agents are limited by their toxicity and serum instability problems in vivo. Therefore there is an urgent requirement of a novel agent that can facilitate the delivery of genetic material in vivo. Based upon the various research pursuing for protein delivery by PTDs, the potential of such peptides for the delivery of large DNA molecules was discussed.

3.6.3 DELIVERY OF ANTIBODIES

Intracellular delivery of antibodies is required due to their effective tumoricidal effect. However their delivery is limited since immunoglobulins by themselves cannot cross the plasma membrane. Several techniques have been employed for the intracellular delivery of antibodies, such as electroporation or microinjection. But these techniques have clear limitations since these methods result in disruption of cellular membranes and decreased cell viability. These conventional methods for the delivery of antibodies cannot be used in a clinical situation. Hence peptide mediated delivery especially CPPs were considered for the delivery of antibodies inside cells. It has been reported that peptide analogs (37–62-sequence region of the TAT protein) when conjugated to poorly internalizable anti-tumor antibody Fab fragments, enhanced the in vitro cell surface retention and internalization of these fragments to the level of the whole antibodies. In one another report it was explored that TATp (37–72) was used to neutralize tetanus toxin (TET) inside cells. Because of the very slow degradation of TET in nerve cells, anti-TET antibodies are required for the neutralization of TET. According to previous investigation thioether and disulfide conjugates between anti-tetanus F(abV)2 fragments and TATp (37–72) could be taken up by cells, disulfide conjugate being the one that could neutralize tetanus toxin inside the cells. One more approach is to deliver antibodies includes the use of TAT-fused protein. TAT-fused protein consist of two functional domains, the TAT PTD and the B domain of staphylococcal protein A (SpA), which effectively binds to the Fc fragment of IgG. TAT-SpA fusion protein when conjugated to IgG showed the intracellular delivery of antibody in a time and dose-dependent manner.

3.6.4   DELIVERY OF IMAGING AGENTS

Owing to their inability to cross the lipid bilayer of the cellular membranes, intracellular delivery of peptide-based imaging agents is limited. For effective delivery of imaging agents there is a need for a membrane permeating peptide, which can deliver imaging agents intracellularly. Recently complexation between oxotechnetium (v) and oxorhenium (v) with TATp had proven the successful translocation across cell membranes into the intracellular compartments and accumulated inside cells to high concentrations. One more approach attachment of paramagnetic labels to 13-mer TATp and the determination of intracellular concentrations by magnetic resonance imaging, known to be successful for translocation of imaging agents across the plasma membrane. One more application to enable the direct comparison of quantitative radiometric and qualitative fluorescence data both in vitro and in vivo was investigated by synthesis of novel dual-labeled d-TAT-peptides comprising TAT-basic domain, fluorescein, and 99 m Tc-tricarbonyl. Conjugation of superparamagnetic iron oxide (SPIO) particles conjugated with TATp and fluorescein isothiocyanate facilitates its uptake by T cells, B cells, and macrophages, followed by migration of the conjugate primarily to the cytoplasm, which could be tracked readily by MRI.

3.7   APPLICATIONS OF CELL-PENETRATING PEPTIDES

3.7.1   INTRACELLULAR DELIVERY

As mentioned above the potential to deliver large hydrophilic molecules (proteins, peptides, nucleic acids and large particles), into cells is a significant challenge because of the bioavailability restriction imposed by the cell membrane. For non-polar and smaller in size (less than 600 Da), plasma membrane of the cell forms an effective barrier that limits the intracellular uptake. Potential of CPPs in translocation across plasma membrane of cells has opened up fascinating perspectives for the development of cell delivery. Moreover, exploration of CPPs allows the layout of constructs that reach the interior of cells and interact with intracellular proteins. It was evident that the direct conjugation of the AntpHD-derived sequence, or its derivatives, with diverse short motifs could provide new peptidic entities that selectively interfere with diverse cellular mechanisms. It was also investigated that these constructs could specifically inhibit ligand-dependent transduction pathways in various cell lines. Moreover, various reports have targeted on the use of penetrating to encourage the delivery of fragments of protein-inhibiting cyclin-dependent kinase (Cdk), which is involved in the regulation of the cell cycle. For promoting the intracellular delivery of Cdk-inhibiting peptide Tat peptide was utilized which allows the arrest of cell proliferation.

3.7.2 DELIVERY OF OLIGONUCLEOTIDES

CPPs have been used effectively employed to deliver oligonucleotides. Currently the investigation that these oligonucleotide–peptide conjugates could specifically block the translation of a gene into a functional protein (antisense strategy) is of particular interest, for example, it has been reported that transportan and penetrating were able to transport a 21-mer peptide nucleic acid (PNA), which was unable to cross the plasma membrane in its original from, into melanoma cells. Delivery was achieved by blocking the expression of the galanin type I receptor by interacting with the mRNA encoding for this protein. in addition it was also reported that the intrathecal administration of the peptide–oligonucleotide construct in rats resulted in a decrease in galanin binding in the dorsal horn.

3.7.3 DELIVERY OF LARGE PARTICLES AND DNA

Various efforts have been employed to use CPPs for the intracellular delivery of DNA, and even particulates. One report suggested the series of complexes synthesized products containing 1–8 Tat moieties plasmid DNA and its transport in different cell lines. Complexes that contain eight molecules of Tat-derived peptide chain per DNA molecule showed transfection capabilities. In one more report it was reported that Tat is able to interact with plasmid DNA electrostatically, result in the formation of polyelectrolytic complexes at various negative:positive charge ratios of plasmid DNA and Tat peptide. It was also investigated that complexation between DNA–Tat can be used for delivery of plasmid DNA into mammalian cells, however, a low level of transfection was obtained after intravenous injection into mice: this is perhaps because of inactivation of DNA–Tat complexes in the bloodstream. Various other reported application, for example, CPPs conjugation with large cargoes {shell-cross-linked (SCL) nanoparticles, or to magnetic nanoparticles} have shown better cell transduction. These complexes were delivered in a wide variety of cells without modification of viability, differentiation or proliferation. Utilization of CPPs -nanoparticles or liposomes complexes for various therapeutic and clinical applications were also reported in various researches.

3.7.4   CARGO

Major function of CPPs is the possibility of conjugating biologically active cargo and translocating it into cells. There are different ways of attaching CPPs with cargo. They can be covalently linked when the cargo is a peptide or protein. This conjugated system is most often synthesized or expressed in tandem as fusion protein. On the other hand a suitable amino acid side-chain or bifunctional spacer molecule can be used. For the attachment between cargo and transportan the thiol group of cysteine, amino group of lysine, or bifunctional cross-linker SMCC was used. Owing to the reductive environment of the thiol group of cysteine in the cell, cleavage of disulfide bridge between CPP and cargo occurs, resulting in the release of cargo. One another approach of attachment of cargo to CPP can be achieved by non-covalent bonds, employing for instance the interaction of avidin to a biotin-CPP construct. This method has been used very seldom. Large numbers of cargo molecules have been successfully delivered into cells via CPPs, including peptides, fragments of DNA small molecules and RNA, proteins, phages, magnetic nanoparticles and liposomes.

3.7.5 QUANTUM DOTS: DELIVERING ACROSS THE BLOOD–BRAIN BARRIER

A major hurdle for labeling brain tissue is crossing the blood–brain barrier, which presents a series of tight junctions between endothelial cells. To enter this barrier Tat was recently exploit to deliver quantum dots into rat brain tissue. For this purpose a microcatheter was utilized to manage Tat-conjugated quantum dots intra-arterially at a proximal cervical carotid artery in rats. Tat has the potential to rapidly delivered the quantum dots to the brain. At such a high loading that gross fluorescent visualization of the rat brain was possible with a low power hand-held UV lamp. Interestingly this was achieved without manipulation of the blood–brain barrier and the quantum dots were efficiently migrates beyond the endothelial cell line of injection to reach brain parenchyma.

3.7.6 CPPs MEDIATED INTRACELLULAR DELIVERY OF ANTIBODIES

The large biomolecules, such as antibodies, are problematic cargo for delivery vectors. Past research has explored the remarkable translocation potential of Tat was used to deliver antibodies into cells. Though recently this type of approach was used to deliver monoclonal antibodies, extensively used for radio immunotherapy and radio immunodetection, to intracellular targets.

3.7.7 VISUALIZATION OF REAL-TIME VIRAL INFECTION OF LIVING CELLS WITH CPPs

Various imaging application are already discussed briefly above. Recently imaging-related application of CPPs to evaluate the visualization of real-time viral infection of living cells is demonstrated in various studies. It has been reported that molecular beacons are capable to detect viral RNA in infected cells. As this approach is extremely sensitive (detection of a single viral particle was possible), it was only successful in fixed cells permeabilized with Triton to allow uptake of the molecular beacons. For delivering beacons in a less invasive manner, a subsequent study harnessed Tat to deliver the constructs into uncompromised cells. This improvement allowed for real time detection of viral replication and infection. Hence by utilizing CPPs, a sensitive, rapid and real-time system for monitoring viral infection was developed.

3.7.8   CPPs TO DELIVER BIOSENSORS

CPPs potentially used to deliver light-emitting biosensors. Zinc, a cofactor for many enzymes, transcription factors and immune system proteins, essential for the functionality of many enzymes, it is also toxic in certain circumstances, such as during seizures and ischemic insult. Thus methods that are employed for monitoring of zinc levels, distribution of zinc in the body, or incorporation of zinc into proteins are of great interest. Nevertheless various agents such as proteins, glutathione, and histidine bind most cellular zinc, only a very minor fraction remains free and available for detection. Hence various reports till date have been restricted to determine free zinc levels in cell lines that are enriched in zinc, rather than in ordinary resting cell lines. For gaining better knowledge zinc levels in traditional cells, a ratiometric fluorescent zinc biosensor was recently developed. This sensor was human carbonic *anhydrase* which was used as a sensor transducer and its fusion to Tat allowed the construct to be efficiently internalized without the need for cell membrane manipulation. In this study signal was quantitatively used to detect zinc levels as low as 5–10 pM in the nucleus and cytoplasm of cells. This highly sensitive and novel biosensor can now be applied to the study of zinc levels in other commonly used cell lines and can also be used to study the role of zinc in cell biology.

3.7.9   CELLULAR TROJAN HORSES

CPPs in bioactive cargo used to deliver small molecules, proteins and nucleic acids that are capable of modulating cellular function and producing therapeutic effects. Nevertheless, these cargo molecules may exhibit attenuated activities in vivo because cell impermeability hampers the activity they display in vitro. Perhaps this is due to their large size or anionic character. Thus CPPs have been exploited to translocate these agents across the cell membrane.

3.7.10 SMALL MOLECULE DELIVERY

Potential of CPPs lies in is their capacity to transport a extensive variety of compounds and macromolecules into the cytosol in an active form. Various small molecules chemotherapeutics, such as Taxol, cyclosporine A, and methotrexate have shown better activity when conjugated to a CPP. Noticeably, in a purified biochemical system CPP-drug conjugate may show less activity, highly efficient cellular uptake can overcome this shortfall. It has been studied that while a CPP-methotrexate conjugate showed a 20-fold loss in effectiveness than the drug alone, it is a highly efficient cytotoxin of a methotrexate-resistant cell line. It has been investigated that CPP conjugates can successfully enhance the intracellular concentrations of bioactive small molecules and that this can counteract decreases in drug activity that result from conjugation.

3.7.11 CPPs FOR DELIVERY OF PEPTIDES, PEPTIDES AND PROTEINS

Bio active macromolecules usually have unique physicochemical properties that limit intracellular accumulation. Exploiting CPPs as molecular delivery agent's allow these molecules to be delivered intracellularly and at sufficient levels to exhibit potential biological property. Prominent example is the utilization of CPPs for the delivery of peptides and proteins to modulate intracellular processes. Various researches have been explored for this purpose:

- A further example with relevant advance in the CPP field involved the VP22-directed delivery of the GATA4 transcription factor to combat myocardial injury. VP22 is a herpes simplex virus protein with cell-penetrating properties. It was found that co-culture of fibroblasts expressing GATA4-VP22 with mesenchymal stem cells (MSC) activated expression of GATA4-inducible genes in MSCs. Moreover, the intracellular delivery of GATA4-VP22 was shown to have a beneficial effect after myocardial infarction in Lewis rats.
- A recent study has shown that CPP-directed delivery can be enhanced by exposing cells to low voltage electrical pulses. This method of increasing cargo uptake will be highly beneficial, especially for applications where the amount of CPP used cannot be increased for

fear of excess toxicity. Importantly, the authors showed that these low voltage electrical pulses do not cause cellular toxicity and do not induce apoptosis. This technique can also be used to selectively deliver cargo to a particular region of interest, simply by controlling the site of pulse administration.

- Another application of CPP-mediated modulation of the immune response is seen in recent work to overcome bacterial sepsis. In this study CPPs were used for the delivery of anti-apoptotic proteins (Bcl-xL and its BH4 domain) as a treatment for sepsis. Bcl-xL and its BH4 domain were conjugated to Tat and these conjugates were administered to mice suffering from Escherichia coli-induced lymphocyte apoptosis. It was observed that in vivo administration of the conjugate decreased sepsis-induced lymphocyte apoptosis.

- CPP-mediated alteration of transcription factor activity was recently described involving a rationally designed peptide inhibitor of the transcription factor. To accomplish intracellular delivery, the inhibitory peptide was conjugated to the HIV-Tat-derived PTD4 CPP. Internalization of the conjugate exerted a potential anti allergic affect which could be a promising novel treatment for allergic rhinitis and asthma.

- Further work to maximize the therapeutic potential of this peptide may be achieved through optimization of its cell-penetrating ability.

- In addition for some applications, a very high intracellular level of cargo may be required, a level unattainable even by CPPs when cells are in a normal physiological environment.

- In addition to CPPs delivering therapeutically relevant cargo, some CPPs themselves have potentially exhibit biological activity. Prion diseases are fatal neurodegenerative disorders that include bovine spongiform encephalopathy and Creutzfeldt-Jakob disease in humans. In these diseases, neurodegeneration is caused by a misfolded prion protein thought to be infectious and able to convert normal protein isoforms, misfolded prion protein, into the misfolded version. Recently, a peptide composed of the first 28 amino acids of the normal isoform misfolded prion protein was discovered to exhibit cell-penetrating properties.

3.7.12   DELIVERY OF NUCLEIC ACIDS AND SIRNA

The most challenging task in delivery system is the delivery of anionic biomolecules, such as nucleic acids. Nucleic acids delivered solely at very low concentration. Vector-assisted delivery of these molecules often results in endosomal entrapment or degradation by nucleases. Previous findings have explored two strategies have been taken to avoid these road-blocks: engineering endosomal escape, and harnessing neutral nucleic acid analogs, such as peptide nucleic acid (PNA). Former derivates are nucleobase with a peptide backbone. This type of structure results in an uncharged molecule resistant to cellular degradation. Though, in spite of this alteration, PNA maintain the complementary base pairing functionality of natural nucleic acids. Additionally One other such neutral mimic (phosphorodiamidate morpholino oligomers) have been successfully used for the attenuation of nonsense mutations through exon removal. CPPs have currently been exploited for phosphorodiamidate morpholino oligomers delivery into mouse models of Duchenne muscular dystrophy. CPPs have also supported the delivery of small interfering RNA used in the modulation of gene expression. Earlier reports have described the use of Tat for effective delivery of siRNA to cells for gene silencing. Current report suggested that novel CPP, a "peptide for ocular delivery," that is skilled of delivering large and small molecule cargo into ocular tissues. Nevertheless, this work demonstrates the potential utility of CPPs to enhance nucleic acid delivery to ocular tissue.

3.7.13 CPPs: HARNESSED TO IMPROVE ABSORPTION

CPPs not only assist in intracellular delivery, though can also be exploited to improve absorption across skin and intestinal barriers. It's known that both of these barriers are difficult to translocate, so development in this area could be of great therapeutic value. One of the examples of CPP mediated enhanced absorption encompasses CPP-cyclosporine A conjugates where the peptide effectively directed dermal absorption with a simple topical application. In comparison to unconjugated cyclosporine A, absorption of conjugated cyclosporine A, was much more potential than with unconjugated cyclosporine A. Moreover, an in vivo decrease in inflammation was also noted with the CPP-cyclosporine A conjugate but not with unconjugated drug. Such a research may be of significant therapeutic value as drugs can now be administered topically and directly to the skin region of interest. Another barrier called as intestinal barrier is alarming challenge to delivery, since insulin dosing for diabetes treatment has typically been achieved via subcutaneous injections. Recently a novel approach for the efficient intestinal adsorption of this hormone was studied. It was demonstrated that that co-administration of various CPPs with insulin enhanced intestinal uptake, compared to the absence of absorption noted with insulin alone. In addition it was discovered that the co-administration increased plasma insulin levels and reduced blood glucose levels, suggesting that the insulin remained functional.

3.7.14   CPPs IN CANCER THERAPY

CPPs has now become the most efficient tool in the search for more successful cancer therapies. Now days CPPs have been employed to deliver chemotherapeutic drugs to cells as well as to deliver pro-apoptotic proteins. This section will highlight harnessing CPPs with anticancer agents to achieve cancer-cell selectivity. In contrast with the conventional CPPs, like Tat, are unable to deliver chemotherapeutic agents to cells, currently synthesized CPPs efficiently deliver chemotherapeutic agents to cells. This finding allows delivery of drug in tumor and healthy cells, an undesirable situation for anti-cancer treatment. To circumvent this inherent lack of specificity, a Tat-derived CPP was conjugated an anti-Her-2/neupeptide mimetic, AHNP, which allows the selective binding with ErbB2, an epidermal growth factor receptor over-expressed in 30% of breast cancers. Now this novel cancer cell-specific transporter was then harnessed for therapeutic purposes.

3.7.15   APPLICATION IN VACCINE DELIVERY

Cell-penetrating peptides exhibit various applications in vaccine delivery. Current researches have seen a tremendous rise in interest in CPP as an efficient means for delivering therapeutic targets into cellular compartments. Its known that cell membrane is resistant to hydrophilic substances nonetheless linking to CPP can allow delivery into cells. Hence the sole translocatory feature of CPP ensures they remain an attractive carrier, with the potential to deliver cargoes in an efficient manner having applications in drug delivery, gene transfer and DNA vaccination. For establishing a successful vaccine delivery of antigen epitopes to antigen-presenting cells, ensuing processing and presentation and induction of an immune response is required. Recent research proven that vaccination with proteins or synthetic peptides including CTL epitopes have proven restricted due to the failure for exogenous antigens to be presented efficiently to T cells. Conjugation of antigens with CPP overcomes such difficulties by allowing cellular internalization, processing and presentation of exogenous antigen for the induction of potent immune responses.

3.8 CELL-TARGETING PEPTIDES (CTPs)

In addition to CPPs, various peptides composed of a few amino acids that are cell specific and are internalized by endocytosis are emerging in the literature. Among these peptides new classes of peptides called as CTPs and have been already utilized in different therapeutic assays. Prominent feature of CTPs is their high specificity and strong affinity for a given targeted cell line upon interactions with a receptor that is exclusively over-expressed by these cells. Among all CTPs the most extensively studied is possibly the RGD peptide. The RGD peptide is the first tumor-homing peptide discovered.

3.8.1 IDENTIFICATION OF CTPs

Various monoclonal antibodies that efficiently target cell surface receptors, such as the anti-CD20 antibody Rituximab, have been approved for cancer treatment. Although various challenging factors such as large size of the antibody (160 kDa), the high cost of its production and characterization as well as its relatively non-specific binding to the reticulo-endothelial system limits their utilization when any cytotoxic drugs or radionuclides are coupled to the antibody. In contrast binding to the RES can be considered as benefit when the antibody is exposed as it can stimulate an immunological response which can easily kill tumor cells. It was also believe that macrophage binding with Fc fragment of these antibodies will also favor the immune response. Hence, owing to their easier preparation and their good affinity or specificity, short peptides or peptide-mimetics are smart alternative targeting agents for either cancer imaging or therapy. Different types of peptides that target exclusively one known cell line have been recognized using different techniques aimed at defining the shortest peptide with the highest specificity and affinity.

3.8.1.1   Structure–Activity Relationship of Ligands

The most preferable strategy for synthesizing a specific binding peptide for a known receptor is to create either from either structural data base or from the structure–activity relationship study of the molecular interactions between a circulating protein and its cellular receptor. Nevertheless structural data for each and every peptide is not affirmative and structure–activity relationship studies not often advance to a linear well-defined sequence which could be directly used as a receptor-specific binding peptide. This is mainly because the tertiary structure scaffolding frequently hides the ligand-receptor interactions. But still some small peptide sequences have been defined by these conventional methods. Various peptides were structurally defined (by X-ray crystallography) and explored such as erbB2 receptor-binding hepta-peptide for erbB2 receptor and vascular endothelial growth factor for Neuropilin-1.

3.8.1.2   Phage Display

There two techniques available for the targeting peptides against cancer cells or tumor blood vessel endothelial cells. These techniques are based on the combinatorial approach. The first technique known as phage display technique based on the combinatorial generation of short peptide sequences introduced in the extracellular protein of a filamentous phage. After interacting with a particular extracellular receptor of a known cell type, the phage is amplified following cell infection. Various dilutions of the transfected suspension are then made, and one single phage type can be isolated after several rounds of selection. Later on, the most "active" combinatorial sequence is recognized by sequencing. This technique allows the identification of peptide sequences ranging from 8 to 12 amino acids. After complete characterization, the peptide itself is determined for its potential to bind with high specificity and strong affinity to the specific receptor of the targeted cell. For synthesizing CTP, the sequences described by phage display are attractive and their production is not difficult. In reality, currently, chemistry of peptides is easy and comparatively economic for peptides up to 15 amino acids, and they can be prepared with rather good yields. However there are some longer peptides, the synthesis of such large peptides is more difficult and requires more expensive production techniques. Lengthy of the peptide could also be a major concern when acknowledging the structural characteristics of the peptide-ligand interactions. Apart from few cases peptides of less than 15 amino acids are poorly structured, hence their affinity or their specificity might be somehow deceased when extracted from the displaying protein-phage. This can be improved by inserting cysteine residues at both ends of the identified sequence by genetic engineering. This is only possible when a structural constraint through the formation of a disulfide bridge occurs, leading to a cyclic peptide within the phage protein. This cyclization usually improves the attraction of the peptide towards the target receptor in comparison to the linear form. One of the major limitation of the phage display technique is that because of its biological origin, only natural L-amino acids can be inserted in the peptide sequence, hence ligands with higher affinity can be missed out. Nevertheless after the identification of native sequence, it is also possible to carry out a structure–activity relationship study on analogs anchorage non-natural synthons.

3.8.1.3 Chemical Strategies

Recently the chemical generation of libraries, which include both D- or non natural amino acids, has developed. This development offers the possibility of discovering new ligands with either higher affinity or better specificity for a given cell receptor. Second technique mentioned in above section is a synthetic technique which is based on the combinatorial synthesis of one compound on one single solid bead, hence producing random peptides, each with its own distinctive amino acid sequence on the same bead. there this strategy is defined as the "One Bead One Compound" strategy. This strategy has led to the discovery of a large number of specific ligands. After the complete assembling of library, different cell lines can be developed in the presence of the beads. Development of particular cell type on one bead clearly signifies the interaction of a membrane receptor with the peptide assembled on that particular bead. Peptide sequence attached on the bead is then evaluated by sequencing. One of main advantage of this technique is that non-natural amino acids can be inserted within the sequence and the main benefit of such non-natural ligands is their high stability against proteases.

3.8.2 IMPROVING CTPs

There are various strategies to improve the affinity and/or avidity of CTPs for their respective receptor. Generally cyclization and multimerization of the peptide have been employed to promote the interaction with the target receptor. For illustrating this RGD tri-peptide is known to be best-studied example that presents a specific interaction with the integrin receptors.

3.8.2.1 The RGD Peptide

According to the previous reports αvβ3 integrin receptor plays an important role in angiogenesis of solid tumors, invasion, cell migration and also metastatic activity and its expression depends on the type and stage of the tumor. Thus for current researchers αvβ3 integrin became a target of choice for cancer treatment. During 1980s, sequence known as the Arg-Gly-Asp (RGD), which is present in several circulating proteins, was shown to be responsible for the binding to the αvβ3 receptor either on its own or included in a penta-peptide. After the discovery of these sequences and its role much work was focused on the definition of the requirements needed to produce a ligand with very strong affinity/ selectivity towards αvβ3. This resulted in the development of a molecule, cilengitide, which is now in phase II study. Current researches are still focusing to explore more specific ligands to reduce the currently used pharmacologic doses. As it's already been discussed that little modification within natural sequence of integrin could deeply improve their activity profiles, for example, in the c-RGDfV peptide, where f stands for α D-amino acid, each amino acid was replaced by its α-N-methylated form. Based on the structure–activity refinement of the N-methylation position [the specific binding (IC50) varied from 10^{-5} M for the c(RGD(NMe)fV) to 10^{-8} M for the c(RGDf(NMe)V)] which represents a gain of more than 1000 folds. It was also demonstrated that replacement of the aspartate residue by glutamate within the RGD peptide could improve the activity.

3.8.2.2 Cyclization

It's widely accepted that peptide ligand and its receptor interaction shows higher affinity when the peptide is under a controlled conformation rather than simply linear. Additionally after cyclization resistance against proteases is largely increased. This aspect can be demonstrated by Arg-Gly-Asp (RGD) tri-peptide. Main functionality of RGD peptide is to recognize its target receptor for initially acting as a targeting peptide. Later on this peptide is introduced in the turn localized at the edge of a β-sheet within two of the natural ligands of the integrin receptor, fibronectin and vitronectin. Such twist involves 3 to 4 residues, thus the RGD peptide was frequently prepared in a cyclized form in order to mimic its natural environment within the protein. Initial cyclization was achieved by bridging two cysteine side chains placed at the two extremities of the peptide. During this time cyclization using chemical closure was also introduced. Effective cyclization can be achieved by cyclo-peptides with 5–6 amino acid residues. This arrangement represents the best compromise between flexibility and reactivity to obtain cyclization with good yields. Various other types of ring closure have been also experimented via various chemical strategies, and peptido-mimetics have been inserted at the place of different native residues. This often results to molecules that actually far away from the native tri-peptide, in which the charge locations of the ionic side chains have to be placed and oriented accordingly.

3.8.2.3 Multimerization

Dimeric ligand affinity for its respective receptor is generally greater than that of the subsequent monomeric molecule. Most of the research highlighted the enhanced avidity is the mechanical result of a greater local concentration of ligands because of the direct linkage between two moieties. As ligand binding is an active event, in case if one molecule becomes separate from the receptor, it is statistically more likely to observe the instantaneous re-attachment of the second half of the dimer to the same receptor molecule. In contrast a dimeric ligand could also proficiently bind to two adjacent receptors, thus triggering the overall potential of the actual affinity of the two receptors. Noticeably early multimerization of targeting sequences has been achieved by repeating two or more adjoining recognition sequences and this characteristic fortify the hypothesis of a higher local concentration, since it shows doubtful that two adjacent receptors could bind to two sequences so closely linked along the same primary sequence. Conclusively the use of multimeric RGD structures develops the obvious affinity, or more accurately the avidity of the ligand for the integrin receptor evaluated against the monomeric ligand. Most of the studies restricted the investigation to 8 RGD moieties at most, and generally tetra- di- and monomers were used. The potential was 4N2N1 RGD peptide, but it has been also demonstrated a 2N4N1RGD peptide efficiency. These disagreements propose that the data depend also on other structural and compositional parameters within the macromolecular complex. It has been discovered that four targeting moieties appear to enough to enhance the binding forces over the monomer. During this time this also restricts the complexity of the molecule and the potential solubility troubles, which could be overcome when administered them in vivo. In certain cases the number of targeting peptides could be higher, for example, liposomes or nanoparticles intended at delivering drugs into targeted tissues can be coated with up to several hundred of targeting units. Certainly, in this case, the overall surface of the drug transporter is significant to explore degree of functionalization

KEYWORDS

- **Cancer**
- **Cargo**
- **Cell delivery**
- **Cell-penetrating peptide**

- **Cell-targeting peptide**
- **Drug**
- **Protein**

REFERENCES

1. Lindgren, M., Hällbrink, M., Prochiantz A., Langel, Ü. Cell-penetrating peptides. TiPS 2000, 21.
2. Gupta, B., Levchenko, TS., Torchilin VP. Intracellular delivery of large molecules and small particles by cell-penetrating proteins and peptides. Advanced Drug Delivery Reviews 2005, 57, 637–651.
3. Fonseca, SB., Pereira, M.P., Kelley S.O. Recent advances in the use of cell-penetrating peptides for medical and biological applications. Advanced Drug Delivery Reviews 2009, 61, 953–964.
4. Vivès, E., Schmidt, J., Pèlegrin A. Cell-penetrating and cell-targeting peptides in drug delivery. Biochimica et Biophysica Acta 2008, 1786, 126–13.
5. Vivès, E., Schmidt, J., Pèlegrin A. Cell-penetrating and cell-targeting peptides in drug delivery. Biochimica et Biophysica Acta 2008, 1786, 126–138.
6. Temsamani, J., Vidal, P. The use of cell-penetrating peptides for drug delivery. DDT Vol. 9, No. 23 December 2004.
7. Milletti, F. Cell-penetrating peptides: classes, origin, and current landscape Drug Discovery Today 2012, 17.
8. Koren, E., Torchilin V.P. Cell-penetrating peptides: breaking through to the other side. Trends in Molecular Medicine 2012, 18, 7.
9. Grdisa M. The Delivery of Biologically Active (Therapeutic) Peptides and Proteins into Cells. Current Medicinal Chemistry 2011, 18.
10. Bolhassani A. Potential efficacy of cell-penetrating peptides for nucleic acid and drug delivery in cancer. Biochimica et Biophysica Acta 2011, 1816, 232–246.

INDEX

Printed and bound by CPI Group (UK) Ltd, Croydon, CR0 4YY

23/10/2024

01777696-0010